M. Bodanszky

Principles of Peptide Synthesis

Principles of Peptide Synthesis

by

Prof. Dr. Miklos Bodanszky
Cleveland, Ohio (USA)

AKADEMIE-VERLAG · BERLIN
1984

Die Originalausgabe erscheint im Springer-Verlag
Berlin · Heidelberg · New York · Tokyo

Vertrieb ausschließlich für die DDR und die sozialistischen Länder

Erschienen im Akademie-Verlag, DDR-1086 Berlin, Leipziger Straße 3—4
Lizenznummer: 202 · 100/557/83
Printed in the German Democratic Republic
Gesamtherstellung: VEB Druckerei „Thomas Müntzer“, 5820 Bad Langensalza
Umschlaggestaltung: Karl Salzbrunn
LSV 1275
Bestellnummer: 763 316 3 (6807)
14800

To the memory of my parents

Preface

A look at the shelves of a major library awakens doubts in the author of this small volume about the importance of writing a new introduction to peptide synthesis. This rather narrow area of bio-organic chemistry already has received considerable attention. A whole series of books deals with the synthesis of peptides. Some of these are textbooks written to support lecture courses on peptide synthesis. Others try to help the beginner, otherwise well versed in organic chemistry, to embark on some experimental project that requires the construction of peptide chains. Not less useful are the monographs which were compiled to aid the adept practitioner and to provide him with references to the growing literature of a very active field. Is there any need for a new book on peptide synthesis? Should we add a new volume to an already impressive and certainly useful series? The answer is not obvious.

The author has already participated in two similar endeavors. The first edition[1] of "Peptide Synthesis", with M. A. Ondetti as coauthor, was meant to serve as an introduction for the beginner. It was rather well received by researchers who joined the field of peptide chemistry and were looking for initiation. While working on the second edition[2] with Drs. Klausner and Ondetti, we became painfully aware of the impossibility of the task. In the ten years elapsed between the two editions, the body of knowledge in peptide synthesis has grown to such dimensions that it could not be condensed into a small book. As a compromise, many potentially important new developments were presented only as entries in one of the numerous tables. Also, the obvious lack of personal experience with several relatively recent procedures excluded the critical evaluation that was the strength of the first edition. A mere rendering of methods of protection, activation and coupling cannot help the uninitiated who has to chose between alternatives without the benefit of personal experience. The beginner was better served when the authors, based on their own experience, treated the material more subjectively. The explosive growth of the literature made, however, a brief and yet essentially complete and also critical account of peptide synthesis nigh impossible. The heroic efforts of E. Wünsch and his associates, resulting in two impressive volumes,[3] were indeed necessary to approach a much desired full presentation. Such an admirable completeness, however, could be achieved only at the expense of simplicity: the reader has to master the handling of the extensive material. Also, the ongoing enrichment of the field with novel procedures and reports of new experience gained with old methods must continuously reduce the completeness, the up-to-dateness and thereby one of the most important merits of any definitive work.

There is probably no entirely satisfactory solution for this dilemma. A new book that reflects the rapid growth of knowledge in peptide synthesis cannot be more than a snapshot, outdated within a few years. On the other hand, just because of the rich harvest of new procedures, an introductory monograph becomes more and more necessary. It seemed, therefore, not unreasonable to write, as an experiment, instead of a third edition of "Peptide Synthesis" rather a small volume on the general *principles* that govern the development of methodology. Such general lines or fundamental thoughts — if they can be discerned — could help in the understanding and perhaps even in the critical evaluation of alternative procedures. This emphasis on principles should also absolve the author from an obligation toward attempts for exhaustive literature references or a complete account of all methods known in peptide synthesis. Perhaps by concentrating on the ideas behind the procedures, he can produce an intellectually more satisfactory and therefore more readable book.

Somewhat disappointingly, peptide synthesis resulted in only a few compounds which contribute to the armament of modern medicine. Not too many peptides are produced on an industrial scale. At the same time, the number of known, biologically-active peptides increases almost exponentially and many important research tools can be obtained through synthesis in the laboratory. This requires the participation of numerous well informed, well trained, competent peptide chemists. It is unlikely that the revolution brought to this area by the production of peptides via recombinant DNA would seriously alter this picture in the near future. Thus, a sizeable number of new colleagues might join the field of peptide synthesis. The author hopes that they will find this volume, "Principles of Peptide Synthesis", helpful during their initiation.

1. Bodanszky, M., Ondetti, M. A.: Peptide Synthesis, New York: Wiley-Interscience 1966.
2. Bodanszky, M., Klausner, Y. S., Ondetti, M. A.: Peptide Synthesis (2nd Edition), New York: Wiley-Interscience 1976.
3. Wünsch, E., ed., Houben-Weyl, Methoden der Organischen Chemie, Vol. 15, Parts I and II, Stuttgart: Thieme 1974.

Cleveland/Ohio, August 1983 Miklos Bodanszky

Acknowledgements

The author expresses his gratitude to his wife for her help in the search of the literature. He thanks Dr. John C. Tolle for reading and correcting the manuscript, to Ms. Elsie Hart and Ms. Roberta Sladeck for preparing the typescript and drawing the figures. Part of this book was written in 1981 when the author, as a recipient of the A. v. Humboldt Award, was the guest of the Deutsches Wollforschungs-institut in Aachen, Germany. His work was both encouraged and helpful by the Director of the Institute, Professor Dr. Helmut Zahn.

Table of Contents

I. Introduction

The recognition by Hofmeister [1] and by Emil Fischer [2] that the structure of proteins is best represented by chains of amino-acids linked to each other through amide bonds, was preceded by the syntheses of the first simple peptide derivatives by Curtius [3] and later by Fischer [4]. The challenge that led to these endeavors can be discerned throughout the history of organic chemistry: reproduction or perhaps re-creation of the work of nature. Subsequently, the aims of peptide synthesis became more pragmatic. Preparation of small, well-defined peptides turned out to be indispensable for the study of the specificity of proteolytic enzymes. The synthetic peptides as substrates were models of complex proteins. In turn, the difficulties experienced in the synthesis of even such simple model compounds stimulated a sustained effort toward improvements in the methodology of peptide synthesis. With the discovery of biologically active peptides, the objectives of synthesis underwent a dramatic change. The isolation of oxytocin in pure form in Vincent du Vigneaud's laboratory [5], the determination of its structure [6, 7] and, last but not least, its total synthesis [8] gave an unprecedented impetus to the development of synthetic procedures. The elucidation of the structure of insulin by Sanger and his associates [9] projected new and still higher aims for peptide synthesis: it should serve medicine in the study of peptide hormones and other peptides which play a role in the regulation of life processes. The classical role of organic chemistry to provide, through unequivocal synthesis, independent proof for the correctness of structures determined by degradation, remained one of the principal motivations in the development of peptide synthesis. Yet, it was also expected that the newly introduced procedures will be applicable for the practical production of peptides such as human insulin or important proteins, e.g. human growth hormone, crucially needed tools of medicine but scarcely available from natural sources. One has to admit, however, that over and above such rational reasons, synthesis of complex peptides was and still is attempted, like the climbing of mountains, because of the challenge of the endeavor.

Formation of the peptide bond requires *activation* of the carboxyl group of an amino acid:

$$\text{R—COOH} \longrightarrow \text{R—CO—X}$$

where X is an electron-withdrawing atom or group. Reactive intermediates of carboxylic acids, acid chlorides or acid anhydrides were known and used in the preparation of amides well before the advent of peptide synthesis, but protection (Y) of the amino group and of the carboxyl group which were not meant to be part of the amide

$$\text{Y—NH—CHR—CO—X} + \text{H}_2\text{N—CHR'—COOY'} \longrightarrow \text{Y—NH—CHR—CO—NH—CHR'—COOY'} \quad (+\,\text{HX})$$

remained problematic for a long time. The first peptide derivatives secured by synthesis, benzoylglycyl-glycine [3] and ethoxycarbonylglycyl-glycine ethyl ester [4] carried blocking groups which could not be removed without the destruction of the newly formed peptide bond. It became obvious that peptide synthesis requires *easily removable protecting groups.* A major breakthrough toward the solution of this problem was the discovery of the benzyloxycarbonyl (or carbobenzoxy or Z) group by Bergmann and Zervas [10] in 1932. This was a remarkably lasting contribution. The new protecting group could be removed by catalytic hydrogenation, at room temperature and ordinary pressure, a process that leaves the peptide bond and the various side chain functions unaffected and generates only relatively harmless by-products, toluene and carbon dioxide:

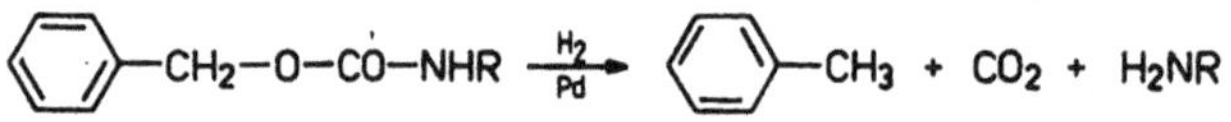

Deprotection can be carried out without special apparatus or special skills, usually in quantitative yield. It is a most elegant method. The Z group is removable also by several alternative procedures. Among them reduction with sodium in liquid ammonia [11] and acidolysis [12] are particularly noteworthy. Together with these remarkable features, the ability of the Z group to protect during synthesis the chiral integrity of the amino acid to which it has been attached, explain why this by now classical method of protection remains a cornerstone of peptide synthesis. In several respects the benzyloxycarbonyl group is still unsurpassed.

The acidolytic removal of the Z group stimulated further research toward acid sensitive protecting groups. This led to the development of a long series of blocking groups which are cleaved under mild conditions. To mention a few, the tert. butyloxycarbonyl (Boc) group [13–15], the *o*-nitrophenylsulfenyl (Nps) group [16, 17] and the biphenylisopropyloxycarbonyl (Bpoc) group [18] come to mind as of major importance. More recently, deprotection with nucleophiles appears to be an attractive alternative to acidolysis in

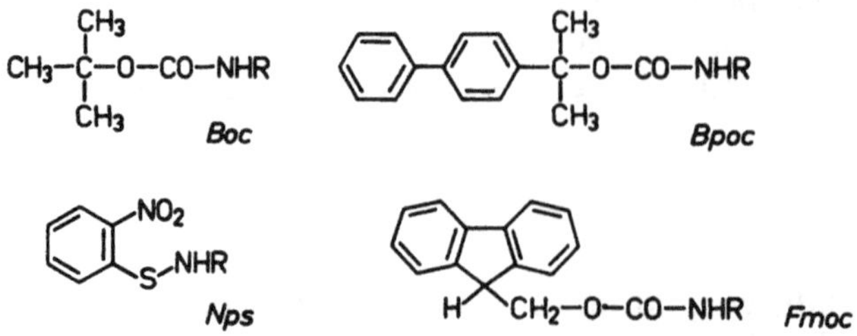

the case of the Nps group [19, 20] and the cleavage of the 9-fluorenylmethyloxycarbonyl (Fmoc) group by weak bases [21] seem to gain significance in the synthesis of complex peptides.

Parallel to the discovery of new methods of protection, new procedures for the activation of the carboxyl group and novel methods of coupling became equally numerous. The most classical approach, the azide method of Curtius [22], remains a valuable tool in peptide synthesis, while the once important acid chloride method of Fischer [23] is used now only exceptionally. Coupling through anhydrides, however, became popular. Mixed or unsymmetrical anhydrides, due to the variability of the component used for the activation of the protected amino acid, were proposed by many investigators and the literature is rich in procedures which are, essentially, different versions of a general approach to peptide bond formation. Excellent results were achieved with anhydrides composed of a protected amino acid and an alkyl carbonic acid [24, 25]. An optimum in terms of yields and purity of the products may have been reached with isobutyl or sec. butyl carbonic anhydrides [26]. The formerly less favored symmetrical anhydrides [27] seem to enjoy a certain revival [28].

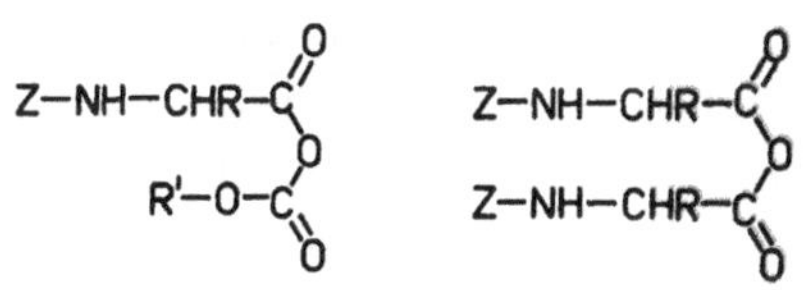

Activation of the carboxyl group can also be achieved by the conversion of protected amino acids to their reactive (or "activated" or "active") esters:

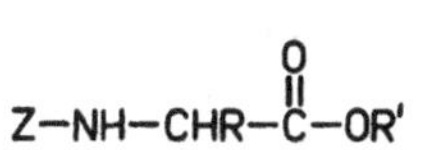

where R′ is an electron-withdrawing group. From the numerous active esters in the literature, nitrophenyl esters [29], 2,4,5-trichlorophenyl esters [30] and N-hydroxysuccinimide esters [31] gained practical application.

An attractive approach to the formation of the peptide bond is the use of "coupling reagents". These are compounds which can be added to the mixture in which both a partially protected amino acid or peptide with a free carboxyl group and a second partially protected amino acid or peptide with a free amino group are present. The most successful coupling reagent, dicyclohexylcarbodiimide (DCC or DCCI), was introduced by Sheehan and Hess [32]. In spite of numerous attempts to replace this powerful reagent with

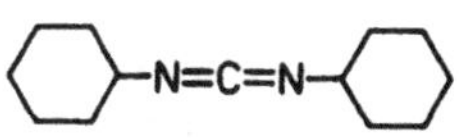

DCC

$CH_3–CH_2–N=C=N–CH_2–CH_2–N(CH_3)_2 \cdot HCl$

water soluble carbodiimide

$(CH_3)_2CH–N=C=N–CH(CH_3)_2$

diisopropylcarbodiimide

more efficient or less drastic materials, DCC, together with its water soluble variants [33], are leading the field with diisopropylcarbodiimide [34] emerging as a possible competitor.

The mixed anhydride producing compound 1-ethoxycarbonyl-2-ethoxy-1,2-dihydroquinoline [35] (EEDQ) and its improved version, 1-isobutoxy-carbonyl-2-isobutoxy-1,2-dihydroquinoline [36] (IIDQ), are very promising coupling reagents. They are readily prepared, easily stored and cause little racemization or other side reactions.

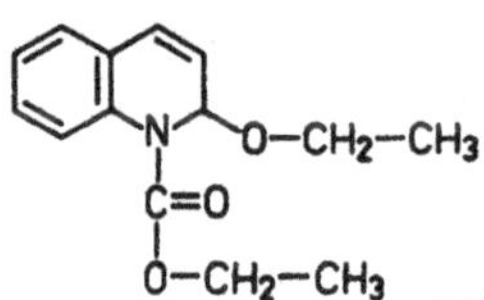

EEDQ

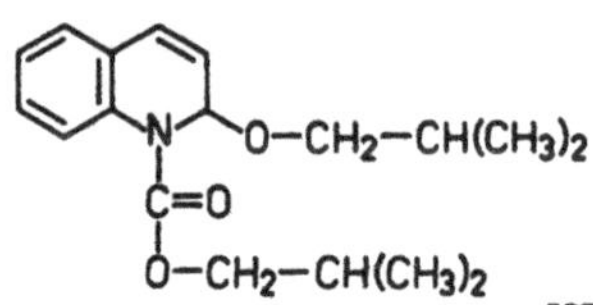

IIDQ

If the methods of protection, activation and coupling, and deprotection were based on unequivocal chemistry, synthesis of peptides could be accomplished with perfection and the resulting products would be single entities, homogeneous materials. In reality, however, the desired reactions are accompanied by competing, undesired side reactions. Hence, the products of synthesis are usually contaminated with byproducts. This necessitates a study of side reactions, an unrelenting effort toward their understanding, elimination or, at least, suppression. The individuality of amino acids and peptides renders this effort rather arduous. In this area generalizations can be quite misleading and side reactions may have to be considered anew for almost each combination of amino acid residues. Some side reactions were well studied: alkylation of the thioether sulfur atom in methionine or the indole nucleus in tryptophan, electrophilic substitution of the aromatic ring in tyrosine. Ring closure involving the side chain carboxyl groups in aspartyl or glutamyl residues or the guanidino group in arginine received considerable attention. Unintentional acylation of the hydroxyl group in the side chains of serine, threonine and tyrosine was encountered time and again and requires special remedies. These are only a few examples of the problems surrounding a seemingly simple process, formation of an amide bond. Over and above such disturbing side reactions, a more general risk, the loss of chiral purity, is present in practically all steps of a synthesis. Therefore, studies concerning the mechanisms of racemization, its detection and prevention, must be discussed in detail in any monograph on peptide synthesis. We shall deal with this problem in a chapter (VI) on diverse side reactions. The consequences of side reactions, including racemization, are byproducts which have to be detected and eliminated if a homogeneous material is the aim of the synthesis. Hence, analytical procedures and methods of purification belong to the armament of the peptide chemist, but a substantial discussion of analytical methods is impractical within the confines of a small volume and we cannot devote a separate chapter to this area.

Difficulties in the execution of peptide synthesis and the substantial time required for the preparation of a longer peptide chain prompted the development of techniques which allow facile chain-building by simple, repetitive operations and which, therefore, are amenable to mechanization and perhaps automation. Particularly successful is Merrifield's technique of solid-phase peptide synthesis (SPPS) [37]. Other approaches to facilitation of synthesis, e.g. the picolinic ester "handle" method [38, 39] or the "in situ" technique [40, 41], might gain significance in the future. Some special problems related to such techniques of facilitation, for instance the nature of the polymeric support in solid phase peptide synthesis, are beyond the scope of this volume. The chemistry, however, in the formation of the peptide bond, the role of protecting groups, the complicating side reactions, briefly, the principles, are the same whether a peptide is assembled in solution or on a polymeric support. Therefore, we will treat these problems independently from the technique in which they may appear.

The techniques developed for the facilitation of peptide synthesis can greatly reduce the drudgery in execution. The latter can indeed be overwhelming when chains of fifty or more residues have to be constructed. The efforts necessary for the synthesis of ribonuclease A [42–44] suggest that the limits of peptide or protein synthesis might have been approached, if not reached. Yet, some proteins, such as human growth hormone, urokinase, or interferon, are badly needed for important medical purposes. Also, the smaller but complex molecule of human insulin could be synthesized, recently in a well designed, elegant manner [45], but without the promise that the methods which made such an impressive achievement possible could be improved to the point where they can be applied for the production of the much needed hormone on an industrial scale. The question must be raised whether or not peptide synthesis can be reduced to a mere routine and, hence, is an automation of the process really feasible? One cannot answer such questions in an unequivocal manner at this time. An automated procedure might turn out to be satisfactory if it is complemented with built-in analytical controls and programmed instructions for the correction of errors. It is unlikely, however, that the educated judgement of a well-trained and experienced peptide chemist could be fully replaced by artificial intelligence. The extent of side reactions varies from case to case and depends not only on the individual amino acids involved, but also on the sequence of the building components in a chain. Variations in the properties of peptides, caused sometimes by long-range intramolecular interactions along complicated, folded chains, superimposed on the influence of solvents, create problems which are probably still beyond the reach of electronic computation.

The limits of the total synthesis of peptides in the organic chemical laboratory have already been transgressed by the transformation of natural substances through *partial syntheses*. The conversion of pork or whale insulin to human insulin by enzymatic means [46] or by a combination of enzymatic and synthetic methods [47] are examples of this probably very practical approach. An even more revolutionary event in the history of peptide synthesis is the adaptation of the nucleic acid code and the known details of protein biosynthesis for the commercial production of peptides and proteins. The incorporation of new information into the genetic make-up of microorganisms through recombinant DNA opens new vistas for medicine. The potential of genetic engineering or its significance for peptide and protein chemistry are question which certainly transcend the contents of a book dealing with the chemical principles of peptide synthesis. An answer, however, to the question whether or not peptide synthesis by the means of organic chemistry remains a viable avenue or will soon be displaced by biosynthesis in living cells, must be sought before we conclude this introduction. If nucleic acid technology renders the methods of organic chemists obsolete, then there is no justification for one more book on peptide synthesis, an already doomed or at best ephemeral subject. Yet, it is our firm belief that no single approach is sufficient for the preparation of a plethora of biologically-active peptides and their analogues. For large molecules which are probably beyond the reach of practical organic synthesis, biosynthesis should be a welcome alternative. For relatively small peptides such as oxytocin, vasopressins, corticotropin or calcitonin, to mention just a few but important examples, synthesis is already the source of the substantial amounts needed in medicine and it will not be easy to replace well-established procedures of organic synthesis by the creation of new kinds of microorganisms and by fermentation. Separation of small peptides from carrier proteins, isolation of the target compounds from fermentation media in the presence of proteolytic enzymes, are problems which have to be solved one by one for each individual compound. Also, analogues of biologically-active peptides, much needed for studies of structure-activity relationships or as longer acting variants of their parent molecules, sometimes as their antagonists, often contain unusual building stones, among these amino acids of the D-configuration. Without posttranscriptional changes the recombinant DNA approach is probably not applicable for the preparation of such analogues. Even the formation of terminal carboxamide groups, a frequently found feature in peptide hormones, presents a yet unsolved problem of protein biosynthesis. Finally, both proof of structure by synthesis and the challenge posed by complex molecules remain strong motivation for the peptide chemist. All in all, there should be little doubt about the future of peptide synthesis by methods of organic chemistry. In order to carry out specific biological objectives, nature used the principle of peptide chains, the combination of amino acids in a well defined sequence, quite regularly. The number of known, biologically-active peptides is ever increasing; it seems to grow, at this time, almost exponentially. Thus, there are ample opportunities for the construction of the same or similar molecules in the laboratory and also sufficient reasons to do that. This probably justifies a search for the *principles* involved in the synthesis of peptides. It is our hope that the readers of this volume will agree with this view.

References of Chapter I

1. Hofmeister, F.: Ergeb. Physiol. Biol. Chem. Exp. Pharmacol. *1*, 759 (1902)
2. Fischer, E.: Ber. dtsch. Chem. Ges. *39*, 530 (1906)
3. Curtius, T.: J. Prakt. Chem. *24*, 239 (1881)
4. Fischer, E.: Ber. dtsch. Chem. Ges. *35*, 1095 (1902)
5. Pierce, J. G., Gordon, S., du Vigneaud, V.: J. Biol. Chem. *199*, 929 (1952)
6. du Vigneaud, V., Ressler, C., Trippett, S.: J. Biol. Chem. *205*, 949 (1953)
7. Tuppy, H.: Biochim. Biophys. Acta *11*, 449 (1953)
8. du Vigneaud, V., Ressler, C., Swan, J. M., Roberts, C. W., Katsoyannis, P. G., Gordon, S.: J. Amer. Chem. Soc. *75*, 4879 (1953)
9. Sanger, F.: Nature *171*, 1025 (1953)
10. Bergmann, M., Zervas, L.: Ber. dtsch. Chem. Ges. *65*, 1192 (1932)
11. Sifferd, R. H., du Vigneaud, V.: J. Biol. Chem. *108*, 753 (1935)
12. Ben-Ishai, D., Berger, A.: J. Org. Chem. *17*, 1564 (1952)
13. Carpino, L. A.: J. Amer. Chem. Soc. *79*, 98; 4427 (1957)
14. Anderson, G. W., McGregor, A. C.: J. Amer. Chem. Soc. *79*, 6180 (1957)
15. McKay, F. C., Albertson, N. F.: J. Amer. Chem. Soc. *79*, 4686 (1957)
16. Goerdeler, J., Holst, A.: Angew. Chem. *71*, 775 (1959)
17. Zervas, L., Borovas, D., Gazis, E.: J. Amer. Chem. *85*, 3660 (1963)
18. Sieber, P., Iselin, B.: Helv. Chim. Acta *51*, 614; 622 (1968)
19. Juillerat, M., Bargetzi, J. P.: Helv. Chim. Acta *59*, 855 (1976)
20. Tun-Kyi, A.: Helv. Chim. Acta *61*, 1086 (1978)
21. Carpino, L. A., Han, G. Y.: J. Amer. Chem. Soc. *92*, 5748 (1970); J. Org. Chem. *37*, 3404 (1972)
22. Curtius, T.: Ber. dtsch. Chem. Ges. *35*, 3226 (1902)
23. Fischer, E.: Ber. dtsch. Chem. Ges. *36*, 2094 (1903)
24. Wieland, T., Bernhard, H.: Liebigs Ann. Chem. *572*, 190 (1951)
25. Boissonnas, R. A.: Helv. Chim. Acta *34*, 874 (1951)
26. Vaughan, J. R., Jr.: J. Amer. Chem. Soc. *73*, 3547 (1951); Vaughan, J. R., Jr., Osato, R. L.: J. Amer. Chem. Soc. *74*, 676 (1952)
27. Wieland, T., Schäfer, W., Bokelmann, E.: Liebigs Ann. Chem. *573*, 99 (1951)
28. Wieland, T., Flor, F., Birr, C.: Liebigs Ann. Chem. *1973*, 1595
29. Bodanszky, M.: Nature *175*, 685 (1955)
30. Pless, J., Boissonnas, R. A.: Helv. Chim. Acta *46*, 1609 (1963)
31. Anderson, G. W., Zimmerman, J. E., Callahan, F.: J. Amer. Chem. Soc. *85*, 3039 (1963); *86*, 1839 (1964)
32. Sheehan, J. C., Hess, G. P.: J. Amer. Chem. Soc. *77*, 1067 (1955)
33. Sheehan, J. C., Hlavka, J. J.: J. Org. Chem. *21*, 439 (1956)
34. Sheehan, J. C.: Ann. N.Y. Acad. Sci. *88*, 665 (1960)
35. Belleau, B., Malek, G.: J. Amer. Chem. Soc. *90*, 1651 (1968)
36. Kiso, Y., Kai, Y., Yajima, H.: Chem. Pharm. Bull. *21*, 3507 (1973)
37. Merrifield, R. B.: J. Amer. Chem. Soc. *85*, 2149 (1963)
38. Gamble, R., Garner, R., Young, G. T.: Nature *217*, 247 (1968)
39. Gamble, R., Garner, R., Young, G. T.: J. Chem. Soc. (C), 1911 (1969)
40. Bodanszky, M., Funk, K. W., Fink, M. L.: J. Org. Chem. *38*, 3565 (1973)
41. Bodanszky, M., Kondo, M., Yang-Lin, C., Sigler, G. F.: J. Org. Chem. *39*, 444 (1974)
42. Gutte, B., Merrifield, R. B.: J. Amer. Chem. Soc. *91*, 501 (1969)
43. Hirschman, R., Nutt, R. F., Veber, D. F., Vitali, R. A., Varga, S. L., Jacob, T. A., Holly, F. W., Denkewalter, R. G.: J. Amer. Chem. Soc. *91*, 507 (1969)

44. Yajima, H., Fujii, N.: J. Chem. Soc. Chem. Commun., 115 (1980)
45. Sieber, P., Kamber, B., Hartmann, A., Jöhl, Riniker, B., Rittel, W.: Helv. Chim. Acta *57*, 2617 (1974)
46. Morihara, K., Oka, T., Tsuzuki, H.: Nature *280*, 412 (1979)
47. Inouye, K., Watanabe, K., Morihara, K., Tochino, Y., Kanaya, T., Emura, J., Sakakibara, S.: J. Amer. Chem. Soc. *101*, 751 (1979)

II. Activation and Coupling

Formation of an amide bond between two amino acids is an energy-requiring reaction. Carboxylic acids do react with amines at elevated temperatures and amides can be produced this way. For instance, ammonium acetate can be converted to acetamide by heating. The temperatures, however, at which such transformations occur far exceed the limits considered safe for complex peptides. In fact, peptide synthesis is usually performed at or below room temperature and coupling methods which involve heating of the reaction mixture are regarded as not generally useful. Therefore, in order to form a peptide bond, one of the groups that will produce the desired amide, either the carboxyl or the amino group, must be activated.

A. Activation

Activation of the amino group is a challenging problem for which no practical solution has been found so far. Electron-releasing substituents should enhance the nucleophilicity of the nitrogen atom, but appropriate substitution, e.g. by the tert. butyl group, will also decrease the rate of acylation because of the bulkiness of the substituent. An enhancement of the electron-density around the amine nitrogen by substitution with trialkylsilyl groups should suffer from similar disadvantages and these groups might also be lost prior to the actual acylation reaction.

Some increase in the reactivity of the α-amino group, a degree of *N-activation*, can be achieved by esterification of the carboxyl group of an amino acid with tert. butanol [1]. The effect of the tert. butyl group, however, is considerably reduced by the two carbon atoms which separate the nitrogen atom from the electron-releasing group:

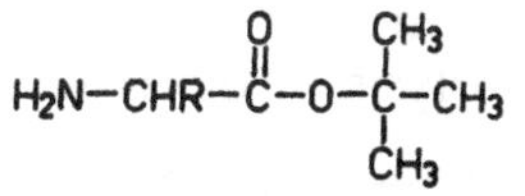

As shown by these examples, activation of the amino group requires further studies before it could be considered a viable approach to peptide bond formation. For the time being, activation of the carboxyl group (*C-activation*) remains the underlying principle of all coupling methods in use:

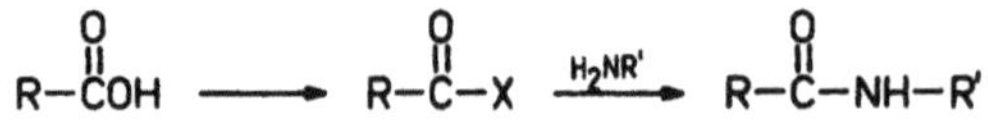

where "X" is an electron-withdrawing atom (e.g., chlorine) or group (such as the azide group) which renders the carbon atom of the carboxyl sufficiently electrophilic to facilitate the nucleophilic attack by the amino group. The tetrahedral intermediate thus formed is stabilized by the elimination of X^-, which is usually a good leaving group:

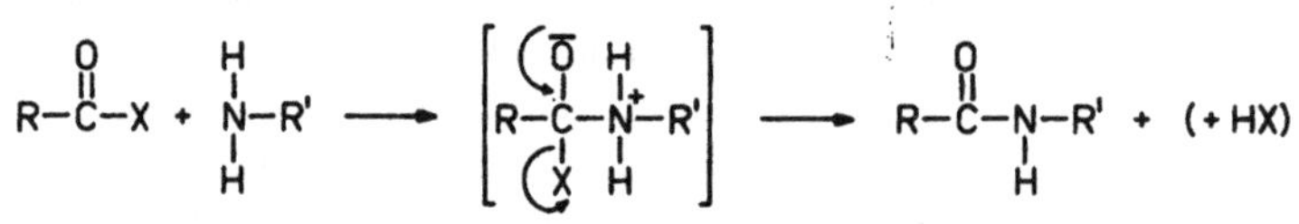

There is, of course, an unlimited choice of electron-withdrawing X-groups and therefore it is relatively easy to find new methods of activation and coupling. The large number of activating groups already proposed proves this point and the literature abounds in further additions of new coupling methods to the armament of peptide chemists. One could question why investigators should experiment with various X-groups, when even the earliest methods of activation, the azide procedure of Curtius [2] or the acid chloride approach of Fischer [3], provide simple and efficient ways to peptide bond formation. To answer this question one has to call the attention to the process of activation, to reactions which convert the carboxyl group to a reactive derivative. For the preparation of acid chlorides, for example, the protected amino acid or peptide is treated with phosphorus pentachloride or thionyl chloride. Such highly reactive materials can affect side chain functions, e.g. they can convert the carboxamide group in asparagine residues to a nitrile:

It is obvious, therefore, that not only the coupling reaction itself has to be carried out under mild conditions, but the process of activation as well. In this respect the acid chloride method is rather unattractive. Even if less drastic reagents are used for the preparation of carboxylic acid chlorides, the reactivity of the chlorides themselves is still too high. This renders them sensitive also to nucleophiles which are less reactive than amines, including water. Unless anhydrous conditions are maintained, acylation of an amine with a carboxylic acid chloride is accompanied by hydrolysis of the latter:

$$R{-}CO{-}Cl + H_2NR' \longrightarrow R{-}CO{-}NH{-}R' \qquad R{-}CO{-}Cl + HOH \longrightarrow R{-}COOH$$

Even more disturbing is the possibility of *intramolecular* attack on the acid chloride grouping by a weak but favorably placed nucleophile within the carboxyl component. This occurs in benzyloxycarbonylamino acid chlorides, which on standing, or faster on heating, eliminate benzyl chloride and give rise to the formation of N-carboxyanhydrides (Leuch's anhydrides):

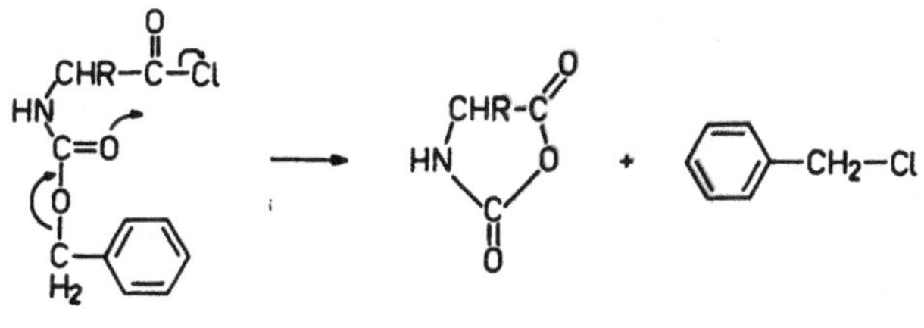

In fact, treatment of benzyloxycarbonylamino acids with thionyl chloride, if carried out above room temperature, is a preparative method for the production of Leuch's anhydrides.

The reactivity of acid chlorides is obviously more than what is needed in peptide synthesis. We see here a clear case of "*over-activation*" as defined by Brenner [4]. On the other hand, a replacement of the hydroxyl group in the carboxyl of protected amino acids or peptides with only moderately activating groups leads to poorly reactive intermediates. One could argue that low reactivity is an attractive feature in an acylating agent since it will be compensated by enhanced selectivity toward the amino group of the amino-component. The practical execution, however, of the synthesis of longer peptide chains requires that the individual coupling reactions should not take more than a few hours, at most a day. With slower reactions, building of a peptide chain of more than just a few amino acids assumes more patience than can be expected from most peptide chemists. The situation worsens when the peptides have reached a certain length: the increase in molecular weight and the often concomitant decrease in solubility force the practitioner to work with solutions of low molar concentration. In such solutions, acylations, as bimolecular reactions in general, proceed with markedly reduced rates. In turn, when coupling rates drop below desirable limits, the risk of side reactions, which are often unimolecular and hence independent of concentration, greatly increases. Some side reactions are negligible and easily overlooked in model experiments because the simple models are relatively small molecules and the coupling reactions are carried out at high concentration of the reactants. The same side reactions might become quite conspicuous and can produce by-products in unacceptable amounts when, instead of small model compounds, peptides of fairly large size have to be linked to each other. Thus, the peptide chemist sails between Scylla and Charybdis [4]: he has to avoid over-activation, but also coupling methods which do not achieve peptide bond formation in reasonably short time. Obviously, the large and ever increasing array of coupling procedures must be subjected to careful critical evaluation and only those that can withstand such scrutiny should be applied in demanding endeavors.

Conversion of the carboxyl-component into an acid anhydride requires the application of powerful reagents such as acid chlorides or alkyl chlorocarbonates:

$$R{-}COO^- + R'{-}CO{-}Cl \longrightarrow R{-}CO{-}O{-}CO{-}R' \ (+Cl^-)$$

$$R{-}COO^- + R'{-}O{-}CO{-}Cl \longrightarrow R{-}CO{-}O{-}CO{-}O{-}R' \ (+Cl^-)$$

Thus, beyond the high reactivity of the anhydrides produced in the activation reaction, the process of activation itself might again give cause for concern. The side chain functions in the carboxyl-component are exposed to acid chlorides or to alkyl chlorocarbonates, certainly not harmless reagents. It is possible to reduce the reactivity of acylating agents by replacing the strong electron-withdrawing groups with more subtly electronegative substituents. For instance, aryl esters are good acylating agents, albeit generally less reactive than acid chlorides or anhydrides. The selectivity of the acylation reaction. to wit, a pronounced tendency to react with the amino-components amino group rather than with side-chain hydroxyl groups or with water, is improved in such esters. The loss of reactivity can be restored, to some extent, by simple modification of aryl esters, like the addition of electron-withdrawing substituents to the aromatic ring. In such active esters, over-activation and thus the ready formation of by-products are considerably reduced and still practical rates can be achieved with them in coupling reactions. It is somewhat unfortunate that, in the preparation of active esters, it is necessary to use highly reactive derivatives of the carbonyl component to acylate the phenolic or alcoholic component, e.g.:

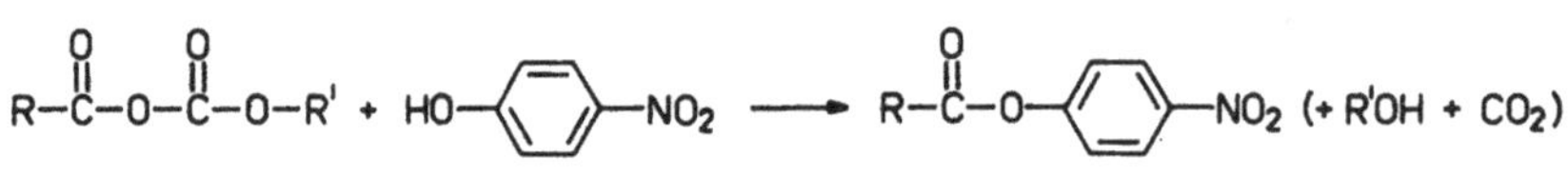

Therefore, in coupling reactions with active esters, the amino-component is sheltered from overactivated derivatives of the carboxyl-component, yet, in the process of activation, in the preparation of active esters, the carboxyl-component still has to be exposed to powerful reagents or to conditions more drastic than desirable in peptide synthesis. Appropriately protected amino acids can tolerate such treatment, but the well-known tendency for racemization of protected peptides manifests itself under such circumstances. In the following schematic representation, we try to indicate the energy levels in the preparation of moderately active intermediates, such as active esters. The overactive derivative of the carboxyl-component is attacked, instead of the amino-component, rather by an alcohol or phenol and the resulting active ester is brought, subsequently, in reaction with the amine:

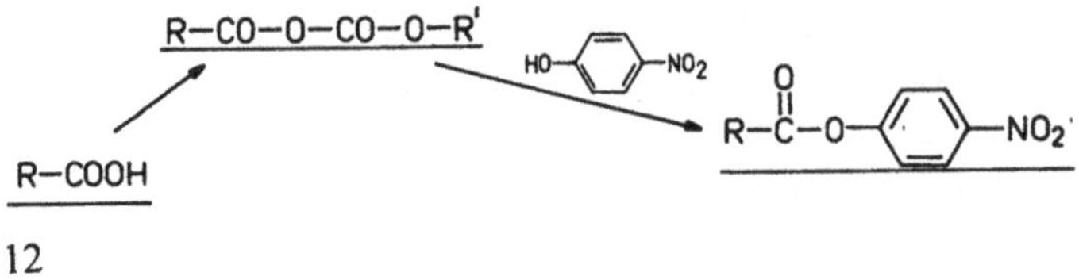

A possible alternative for the preparation of active esters could avoid the over-activation of the carboxyl component. This approach, which could be designated as "*O-activation*", consists of a reaction in which the yet unactivated carboxyl-component is esterified with a reactive derivative of the phenol or alcohol that plays the role of the activating group (or *hydroxyl-component*) in the ester:

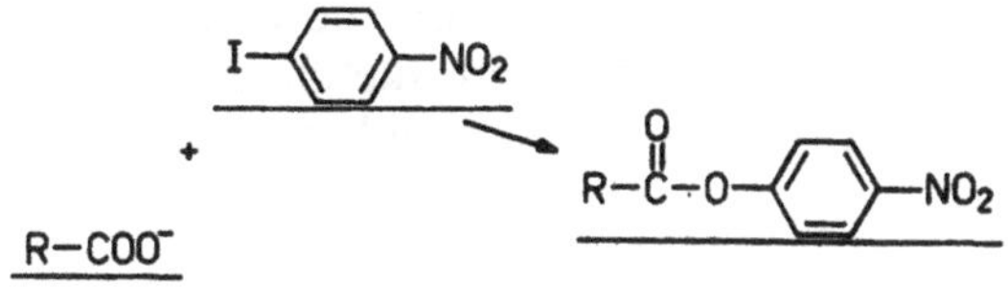

In such a sequence of events, activation of the carboxyl component does not exceed the level present in the active ester. Because of the insufficient reactivity of aryl halogenides, this approach yet awaits practical realization, perhaps through the discovery of suitable catalysts for the esterification reaction. Some indications for the feasibility of O-activation can be found in the literature, e.g. in a method proposed by Taschner and his associates [5] for the preparation of acyl derivatives of substituted hydroxylamines:

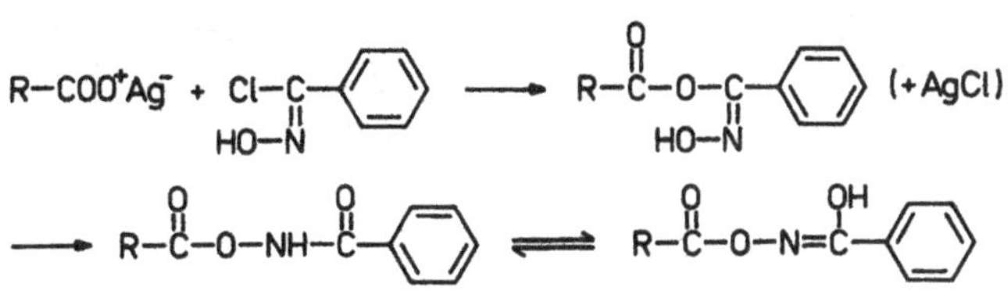

Also, the formation of peptides during dinitrophenylation of amino acids [6] suggests the presence of 2,4-dinitrophenyl esters generated by the reactive form of a phenol:

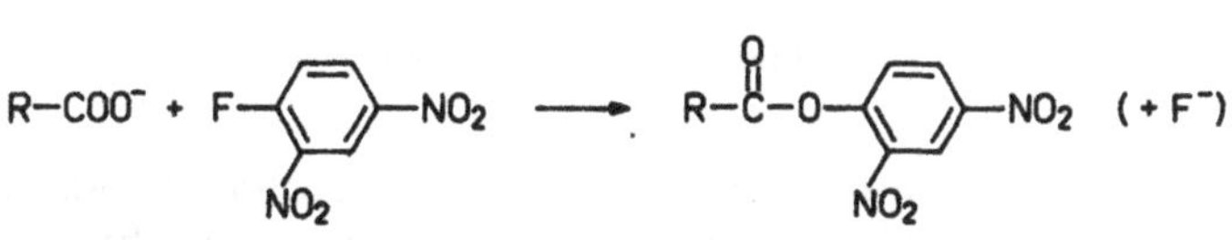

Finally, in the pioneering studies of Schwyzer and his associates [7], activated methyl esters were prepared by the reaction of salts of protected amino acids with alkyl halogenides. The esters of choice, the cyanomethyl esters, were secured by the metathesis of carboxylates with chloroacetonitrile, a reactive derivative of cyanomethanol:

$$R{-}COO^- + ClCH_2{-}CN \longrightarrow R{-}\overset{O}{\overset{\|}{C}}{-}O{-}CH_2{-}CN \quad (+Cl^-)$$

Replacement of $ClCH_2CN$ with the more reactive bromoacetonitrile [8] can further reduce the exposure of the carboxyl-component and thus suppress the racemization of sensitive amino acids. The moderate reactivity of cyanomethyl esters prevented, so far, their application in major syntheses of complex natural products, but these esters could become more significant if efficient catalysts can be found for the acceleration of coupling.

The expression "O-activation" proposed here suggests activation of the alcohol or phenol component of an active ester to be formed. The use of this term could be extended for reactive derivatives of thioalcohols, thiophenols and of substituted hydroxylamines [9] as well. While no truly important application of O-activation could be quoted, the here sketched examples suggest that it is a principle which deserves more thought and more research.

B. Coupling

The crucial step in peptide synthesis, the formation of the peptide bond, seems to be, by necessity, a bimolecular reaction between the carboxyl-component and the amino-component:

$$R{-}CO{-}X + H_2N{-}R' \longrightarrow R{-}CO{-}NH{-}R' + HX$$

Yet, ingenious schemes were designed by Brenner and his coworkers [10–12] and by Ugi [13, 14] in which the peptide bond results from *intramolecular rearrangements.* In the first amino acid *insertion* method of Brenner [9–11], transformation of O-aminoacyl-salicylamides takes place under the influence of weak bases:

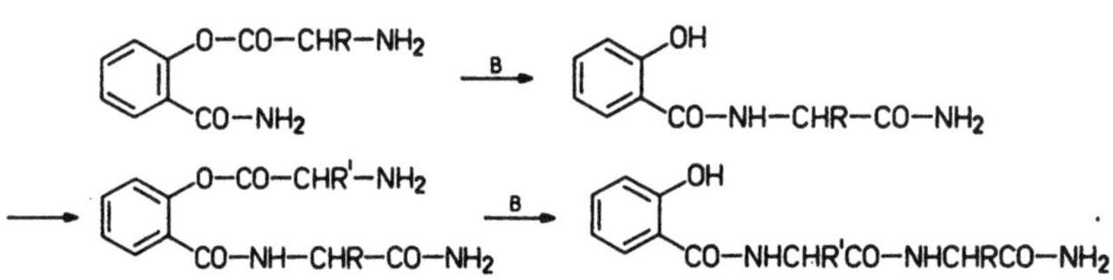

In the subsequently published [12] alternative, Brenner and Hofer proposed the acid-catalyzed rearrangement of N-aminoacyl-N′-acylaminoacyl hydrazines for a similar insertion of amino acid residues:

$$Z{-}NH{-}CHR{-}CO{-}NH{-}NH{-}CO{-}CHR'NH_2 \xrightarrow{H^+} Z{-}NH{-}CHR{-}CO{-}NH{-}CHR'{-}CO{-}NH{-}NH_2$$

The full potential of the insertion principle has not yet been explored, although it is a thought-provoking and challenging concept.

The four-component condensation (4CC) method of Ugi [13, 14] is complex both in mechanism and execution. Nevertheless, it has already received much attention and was tried in laboratories actively engaged in major syntheses [15]. The reactive intermediate in this most interesting approach is the condensation product of an amine, an isonitrile, an aldehyde and a carboxylic acid:

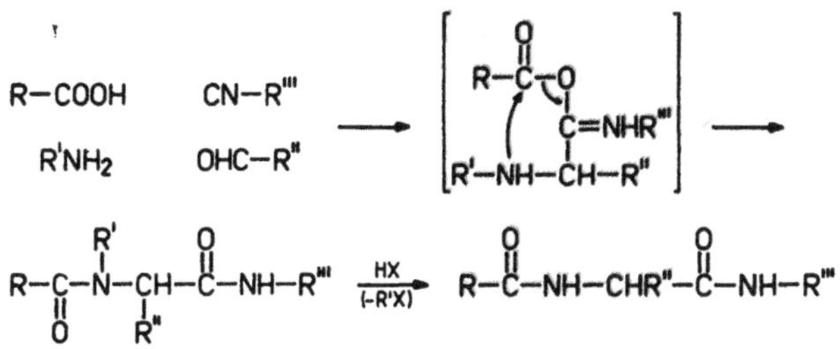

The newly formed amino acid residue (with R″ in its side chain) is, of course, achiral unless special measures are taken to provide it with an optically active and chirally pure center. This can be done by the application of chiral primary amines which have strong asymmetric inducing power. Alkylamines with ferrocene as substituent on their α-carbon atom fulfill this requirement [16]. Alternatively, the segments[1] may be so selected that glycine is the newly formed residue. This circumvents the problem of induced chirality. An excellent review on the four-component synthesis was written by Ugi [14].

The significance of intramolecular formation of peptide bonds should not be underestimated. With reactants of low molecular weights, intermolecular reactions proceed at practical rates. When, however, large segments have to be linked to each other, their higher molecular weights and often also their poor solubility prevent the execution of the coupling reaction at concentrations which allow reasonably fast condensation unless the carboxyl-component is highly activated. With moderately activated carboxyl-components, coupling rates are satisfactory only if the concentration of the reactants is at least 0.1 molar. Nonetheless, intramolecular formation of the peptide bond is not the only way of escape from the dilemma caused by a lower molar concentration of the reaction components. Application of one of these components in excess permits the execution of coupling as a pseudo-unimolecular reaction. This "*principle of excess*" [18] found application in syntheses where chains were built by incorporation of single amino acid residues [19]. The same principle is readily recognizable in the practice of solid-phase peptide synthesis [20]. Only with an excess on acylating agent can the usually more valuable amino-component be completely utilized. Equimolar amounts of the reactants cannot achieve that. Yet, any unreacted amino-component represents a painful loss and also requires its careful removal from the crude product, otherwise it will be acylated in the next coupling reaction and will be the source of a "deletion sequence". Separation of the unreacted amino-component is not necessarily a simple procedure. It may require extensive purification steps, e.g. chromatography. On the other hand, elimination

[1] The conventionally used term "fragment" will be replaced in our discussions by the more precise expression "segment" recommended by Pettit [17]

of excess acylating agent, particularly if it is an active ester of a protected amino acid [19], or a reactive derivative of a small peptide [21], can be carried out in most cases simply by washing the product with judiciously selected solvents. Thus, the use of acylating agents in considerable excess is a powerful device when the synthesis of pure peptides is essential.

C. Coupling Methods

1. The Azide Procedure

It is interesting to note that the earliest method applied for the formation of the peptide bond, the azide process of Curtius [2], survived the scrutiny of numerous investigators, resisted the challenge of many alternative procedures and is still widely used in peptide synthesis (cf. e.g. ref. [21]). One of the reasons for this unusual permanence is the resistance of azide-activated peptide derivatives to racemization. For a long time it was generally believed that protected peptides, activated in the form of their acid azides, can be coupled to amino components without any racemization. For acylation with derivatives of single amino acids, there were other known solutions for racemization-free coupling, but the azide method was regarded as the sole approach which allows retention of chiral purity during the coupling of peptides. More recent experience [22] shows that, while peptide azides are usually not readily racemized, they can lose optical purity, particularly in the presence of excess base. Nevertheless, the azide process offers some unusual advantages. Among these, the possibility to convert a carboxyl-protecting group to an activating group is particularly attractive. The carboxyl group can be protected in the form of an alkyl ester, e.g. methyl ester, and the latter changed to an acid hydrazide simply by exposure to a solution of hydrazine in an organic solvent such as methanol. The hydrazide, in turn, is transformed to the reactive azide by treatment with nitrous acid or an alkyl nitrite:

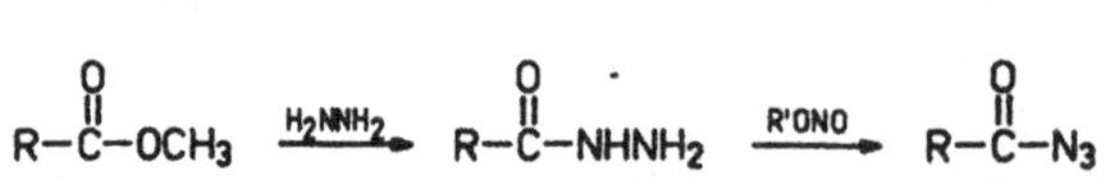

The fact that alkyl esters are useful protecting groups is combined here with the absence of a need for the actual removal of the blocking group from a carboxyl prior to its activation. Conversion of esters to hydrazides is a simple process, although the conditions of hydrazinolysis vary from peptide to peptide. Acid hydrazides are mostly tractable materials, often insoluble in organic solvents from which they separate as they form. Many of them are crystalline and thereby assist in the characterization of intermediates. The conversion of acid hydrazides to acid azides in an extremely facile reaction which proceeds rapidly even in dilute solutions and requires essentially no excess on nitrites. It is possible, therefore, to use the azide process for the preparation of cyclic peptides. Nitrous acid, applied in calculated amount, reacts preferentially with the hydrazide group and leaves the free amino group at the N-terminus practically intact. On addition of base, cyclization takes place:

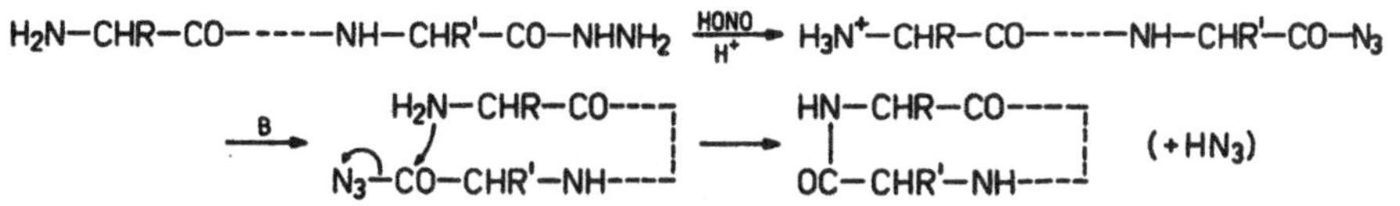

Preparation of hydrazides as intermediates for azide coupling is by no means restricted to hydrazinolysis of esters. This can be a welcome change from the previous routine, because in some cases the conditions of hydrazinolysis are too drastic. For instance, in peptides which have valine or isoleucine at their C-terminal residue, the reactivity of the ester group is greatly reduced by the combination of electron release by the branched alkyl side chain and steric hindrance caused by branching at the β-carbon atom. Such peptide esters are resistant to hydrazinolysis and complete conversion to hydrazides may require many hours of heating to the boiling point of the mixture (e.g., a peptide ester in a 10% solution of hydrazine in ethanol). Less hindered amino acid or peptide esters will afford the desired hydrazide within hours at room temperature, but difficulties can arise from the insolubility of protected amino acid or peptide esters in solvents such as alcohol or dioxane. In dimethylformamide, the most commonly used solvent in peptide synthesis, hydrazinolysis is far from unequivocal. The solvent participates in the reaction and formic acid hydrazide can be isolated from the mixture.

A further complication in the hydrazinolysis of peptide esters is the unexpected removal of tert. butyl groups from esters of side chain carboxyls. Also, when a free carboxyl group is present in the side chain of acidic residues, these are neutralized to hydrazinium salts and thus complicate the conversion of the hydrazide to the azide. The total amount of "hydrazine" in the intermediate can be determined by oxidation and determination of the volume of N_2 liberated in the process [23] or by oxidimetric titration [24]. Of course, it is possible to remove hydrazine bound only by ionic forces by extraction with acid-containing solvents.

As an alternative to ester hydrazinolysis, the use of *protected hydrazides* gained major significance. In this approach chain building starts with an acylhydrazine such as benzyloxycarbonylhydrazine [25] which, in turn, is acylated on its remaining free NH_2 group with a protected amino acid, the C-terminal residue. Obviously the protecting group on this amino acid must be selectively removable by methods which leave the benzyloxycarbonyl (Z) protection intact. For example, the tert. butyloxycarbonyl (Boc) group can be chosen for this purpose since it is cleaved by relatively weak acids which do not affect the Z group. The chain is then lengthened by the incorporation of the next Boc-amino acid and so on until the segment is completed. At this point the Z group is eliminated from the hydrazide grouping and the latter converted to the reactive

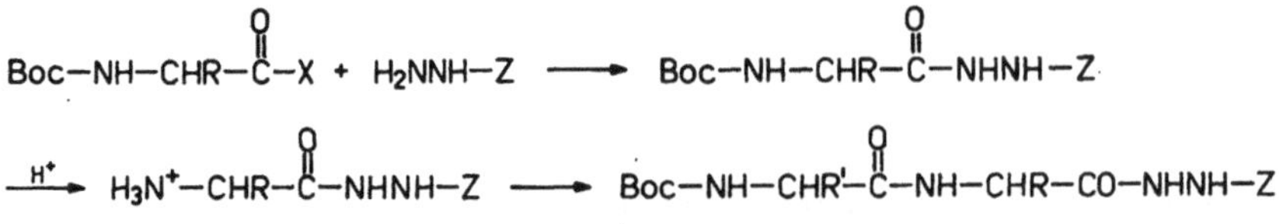

azide by treatment with nitrous acid or alkyl nitrites. Of course, other combinations of protecting groups can also be adopted for the same purpose. Thus, the protected hydrazide principle could be implemented with the Boc group on the hydrazine moiety and Z on the attached amino acids. Formyl, trifluoroacetyl and trityl hydrazine were similarly applied.

An obvious advantage of the protected hydrazine strategy lies in the mild conditions under which the semipermanent protecting group can be removed from the hydrazide. This is in contrast with the "cooking" of some alkyl esters with a solution of hydrazine. Yet, there might be some price to be paid for this elegance in execution. The preparation of the C-terminal portion of a segment, a protected amino acid protected hydrazide, is more laborious than the preparation of a methyl ester. Furthermore, a semipermanent protecting group has to be left in place until the completion of the synthesis of the segment and removed only prior to the conversion of the hydrazide to azide. This leads to serious restrictions in the choice of protecting groups for the blocking of side chain functions. Last, but not least, one must feel some concern about possible Brenner-rearrangements [12] or insertions since N-aminoacyl-N'-acylaminoacyl hydrazines are intermediates in the process. Still, all these problems did not diminish the significance of the protected hydrazide approach. It has been applied in numerous syntheses and led to success in the preparation of complex molecules.

An interesting question has to be raised at this point: is protection of the second amino group in mono-acyl hydrazines indeed necessary? In "unprotected" acylamino acid hydrazides, the superb nucleophilic character of the hydrazine

Z—NH—CHR—CO—NH—NH_2

molecule is greatly reduced and the free NH_2 group is only moderately reactive toward acylating agents. No reaction should be expected with activated derivatives of protected amino acids if activation is kept under certain level [26]. Accordingly it is possible [27] to build peptide chains starting with an amino acid hydrazide and acylating it with the *p*-nitrophenyl ester of the next protected amino acid:

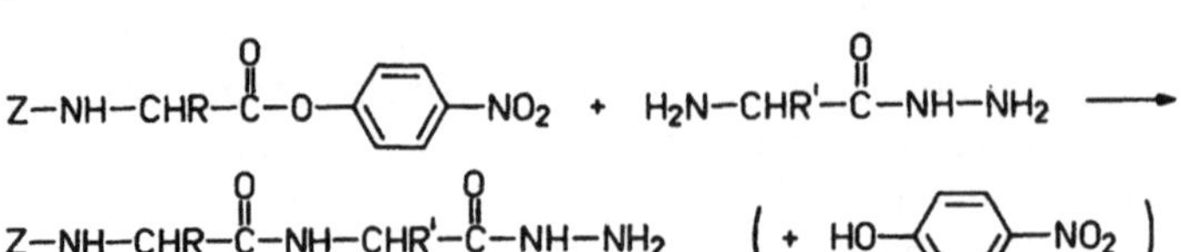

This seems to be a rather attractive approach for the building of segments which subsequently can be coupled to a second segment via the azide procedure, but there is probably a delicate balance here between failure and success. Undesired acylation of the free NH_2 group of the hydrazide still might occur [28] and whether or not this happens depends on the nature of the residues, the activating groups and the conditions of the acylation reaction.

Independently from how the peptide hydrazide was obtained, it has to be converted to the reactive azide. The original method of Curtius [2] consists of treatment of a solution of the hydrazide in aqueous acid with a solution of sodium nitrite in water. This is a simple and practical process if the hydrazide is soluble in the medium (that can contain also acetic acid to improve solubility) and if the resulting azide can be readily separated from the reaction mixture by filtration or extraction with an organic solvent. Yet, to keep the possible Curtius rearrangement (cf. below) at a minimum, these operations must be carried out at low temperature and expeditiously. The procedure can be considerably simplified by adding the aqueous solution of sodium nitrite to a solution of the hydrazide in dimethylformamide containing the required amount of hydrochloric acid and by using this solution, after the addition of base, for the acylation of the amino-component also dissolved in dimethylformamide [29, 30]. The essence of this simplification is the omission of isolation of the azide. This can be particularly useful in the coupling of large segments which cannot be extracted with organic solvents or where the separation of the azide by filtration is difficult and, hence, time consuming. Time is an important factor in processes involving not entirely stable intermediates.

An interesting and very useful modification of the Curtius method was introduced by Honzl and Rudinger [31], who in the conversion of hydrazides to azides replaced sodium nitrite by alkyl nitrites:

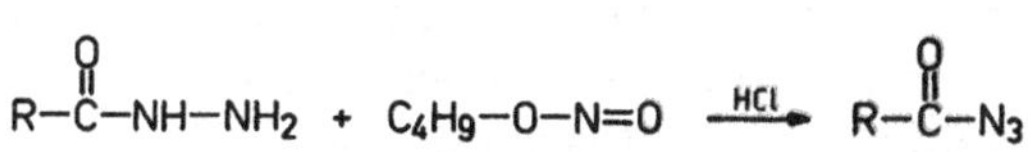

The required acid is applied as a solution of HCl in an organic solvent, usually dioxane. Hence, the presence of water and thereby some side reactions are avoided. A relatively recent, but already popular, method [32] broke even more with the tradition in the preparation of acid azides. Protected petides with a free carboxyl group at their C-termini can be treated with diphenylphosphorazidate and the desired acid azides are obtained in smooth

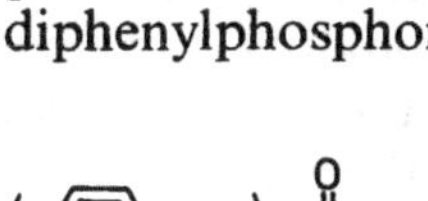

reaction and in good yield.

So far in our discussions, we have avoided the problem about the structure of acid azides. Instead of the classical cyclic formula or the more contemporary bipolar ion representation:

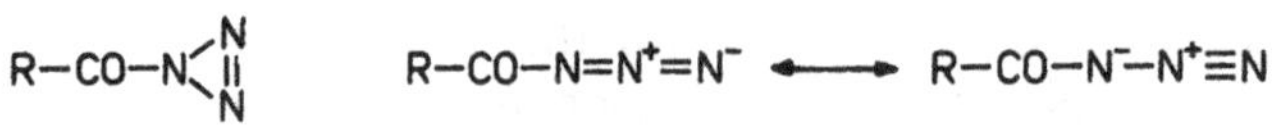

we used the noncommittal symbol R—CO—N_3. Without going into details about the chemistry of acid azides, one of their reactions, the Curtius rearrângement [33], has to be pointed out, since it has direct bearing on peptide synthesis:

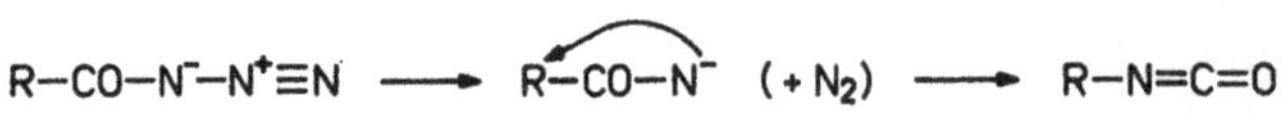

This is an unusually noxious side reaction because the resulting isocyanates are quite reactive. With water they produce amines, but this is not the main reason for concern. More disturbing is the reaction of isocyanates with amines, such as the amino-component in the coupling reaction. Addition of the NH_2 group to the N=C double bond leads to urea derivatives which resemble the desired peptides:

$$\text{R—CO—N}_3 + \text{H}_2\text{N—R}' \longrightarrow \text{R—CO—NH—R}'$$

$$\text{R—N=C=O} + \text{H}_2\text{N—R}' \longrightarrow \text{R—NH—CO—NH—R}'$$

When larger segments are coupled, the similarity between the target compound and the urea-derivative formed as the result of Curtius rearrangement can be so close that their separation amounts to a major problem. Detection of the presence of such urea derivatives is relatively simple. In the quantitative amino acid analysis of the product, the C-terminal amino acid of the original carboxyl-component does not appear since it is decomposed in the process.

It is obvious that Curtius rearrangement must be suppressed in azide couplings. This can be done by carrying out both the preparation of the azide and the coupling reaction at low temperature. Also, high concentrations of the reactants favor the rate of peptide bond formation and thus diminish the extent of the Curtius rearrangement, which is independent of concentration. Yet, for those who use the azide method, it is good news [34] that the urea derivatives, the by-products resulting from the rearrangement, are sensitive to acidolysis and are destroyed during the removal of acid labile protecting groups.

An additional difficulty in the azide method is the slow formation of the peptide bond. Coupling via azides can require considerable time, even several days, particularly if the reaction is carried out at low temperature (e.g. 4 °C) to avoid Curtius rearrangement. Such difficulties notwithstanding, the azide method remains a classical contribution. In contrast to the acid chloride method of Emil Fischer [3] which, by now, has only historical significance, the procedure introduced by Theodor Curtius at the turn of the century is still one of the mainstays of peptide chemists.[2]

---

[2] Side reactions which accompany the azide procedure were pointed out by Schnabel [35]. The role of the azide method in the coupling of large segments was treated in detail by Meienhofer [36]. For a review of the azide method cf. also ref. [37].

2. Anhydrides

The earliest formation of a peptide derivative, in the laboratory of Curtius [38], was due to an unexpected side reaction. In the acylation of the silver salt of glycine with benzoyl chloride, in addition to the target compound benzoylglycine (hippuric acid), the blocked dipeptide benzoylglycylglycine was also present among the products of the reaction. Obviously, a *mixed anhydride* formed from hippuric acid and benzoyl chloride and this anhydride acylated a still unreacted part of glycine to give the dipeptide derivative:

$$C_6H_5COCl + H_2NCH_2COOAg \longrightarrow C_6H_5CO{-}NHCH_2COOH \ (+AgCl)$$

$$C_6H_5CO{-}NHCH_2COOH + C_6H_5COCl \longrightarrow C_6H_5CO{-}NHCH_2CO{-}O{-}CO{-}C_6H_5$$

$$C_6H_5CO{-}NHCH_2CO{-}O{-}CO{-}C_6H_5 + H_2NCH_2COOH \longrightarrow$$

$$C_6H_5CO{-}NHCH_2CO{-}NHCH_2COOH + C_6H_5CO{-}NHCH_2COOH + C_6H_5COOH$$

Strangely enough, the anhydride idea remained unexploited for several decades. First, in 1947, we can find a report [39] on the application of mixed (or "unsymmetrical") anhydrides for the synthesis of peptides. The first anhydrides were composed from protected amino acids and esters of phosphoric acid (cf. refs. [40–43]), and the experiments were stimulated by biochemical analogies. A more systematic reinvestigation of applicability of mixed anhydrides in peptide synthesis was initiated by Wieland and his associates [44] who soon recognized the direction to be followed in the development of potent acylating agents producing a minimum of by-products.

In mixed anhydrides of protected amino acids with benzoic acid (cf. the above example), the difference between the two carbonyl groups with respect to electrophilic character is not too pronounced. Accordingly, the attack of the nucleophile (the amino-component, glycine in the above example) will occur about equally on both electrophilic centers and two acylation products are obtained in nearly equal amounts. For a more unequivocal course of the acylation reaction, it is necessary that a considerable difference exist between the electron densities on the two sides of the anhydride grouping and that the carbonyl carbon of the protected amino acid or peptide be the stronger electrophile. Since the protected amino acid or peptide cannot be modified, the second acid, used for activation, must be selected with this difference in mind. When benzoate was replaced by acetate in mixed anhydrides, the electron-release by the methyl group had a beneficial effect on the ratio of the desired product to the second acylation product, an acetyl derivative:

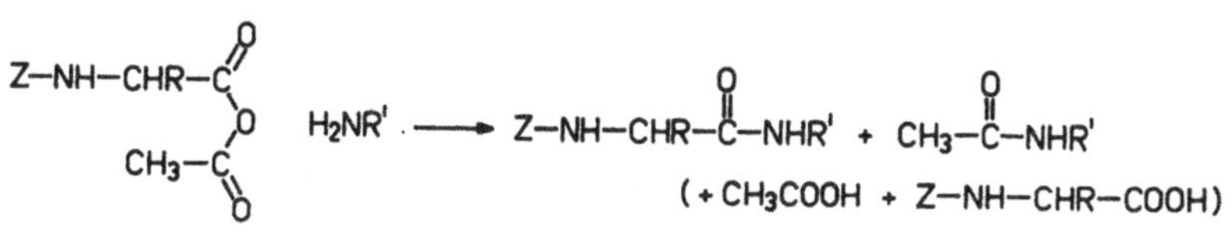

The relative amount of the second acylation product can be further reduced by enhancing the electron-releasing effects in the activating acid. Thus, isovaleric acid mixed anhydrides [45] and trimethylacetic (pivalic) acid mixed anhydrides [46] should be much superior to those in which acetic acid is the partner

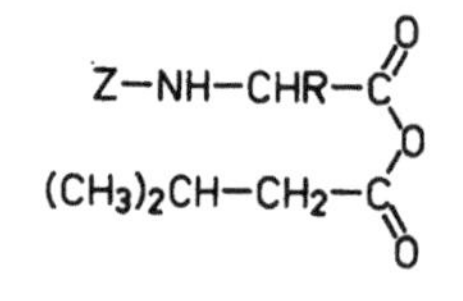

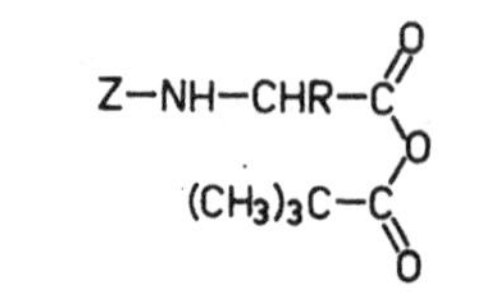

of the protected amino acid. The pronounced electron-release by the branched aliphatic chain in isovaleric acid and an additional steric effect in pivalic acid inhibit the attack of the nucleophile on the carbonyl of the activating group. Therefore, essentially only the desired acylation, with Z—NH—CHR—CO—, will take place. Yet, care must be taken, particularly with pivalic acid mixed anhydrides, that in the activation reaction the reagent, pivalyl chloride, be completely consumed, otherwise the nucleophile will attack the unreacted part of the acid chloride to yield the second acylation product:

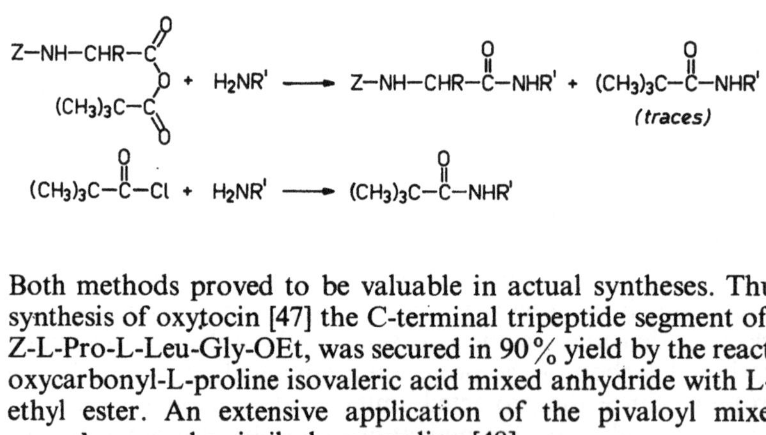

Both methods proved to be valuable in actual syntheses. Thus, in the first synthesis of oxytocin [47] the C-terminal tripeptide segment of the molecule, Z-L-Pro-L-Leu-Gly-OEt, was secured in 90% yield by the reaction of benzyloxycarbonyl-L-proline isovaleric acid mixed anhydride with L-leucyl-glycine ethyl ester. An extensive application of the pivaloyl mixed anhydrides procedure can be similarly rewarding [48].

Replacement of carboxylic acids with half esters of carbonic acid [49–51] is an important point in the development of the mixed anhydride methods. In the presence of an acid binding agent, such as triethylamine, a rapid reaction between protected amino acid (or peptide) and ethyl chlorocarbonate affords a mixed anhydride in which the reactivity of the carbonic acid carbonyl is diminished by

$$\text{Z—NH—CHR—COO}^- + \text{Cl—}\overset{\text{O}}{\overset{\|}{\text{C}}}\text{—OC}_2\text{H}_5 \longrightarrow \text{Z—NH—CHR—}\overset{\text{O}}{\overset{\|}{\text{C}}}\text{—O—}\overset{\text{O}}{\overset{\|}{\text{C}}}\text{—OC}_2\text{H}_5 \quad (+\ \text{Cl}^-)$$

the unshared pairs of electrons on the neighboring oxygen atom. Hence, only little second acylation product (a urethane) can be expected:

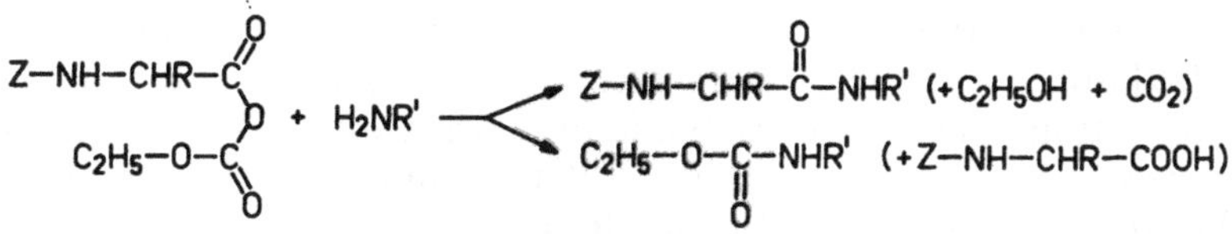

It is a particular advantage of the method that the by-products formed in the decomposition of the leaving group, ethyl carbonate, are alcohol and carbon dioxide, which do not interfere with the isolation of the desired product, a protected peptide. Participation of the "wrong" carbonyl group in the acylation reaction was further reduced by the modification proposed by Vaughan [50] who introduced isobutyl chlorocarbonate, one of the most widely used activating reagents in peptide synthesis:

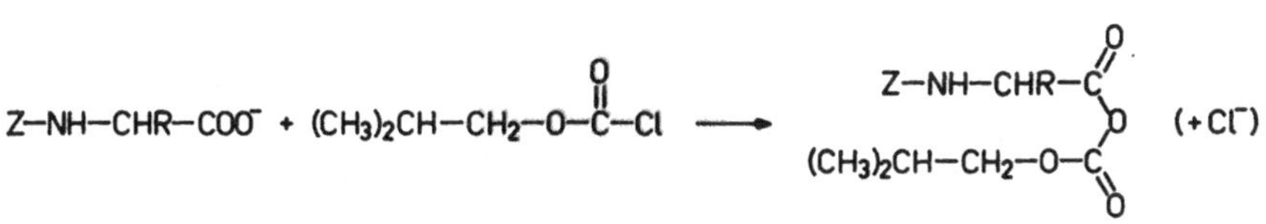

In this method, the undesired attack on the carbonic acid carbonyl usually yields less than 1% urethane. Yet, when the carboxyl belongs to a hindered amino acid such as valine or isoleucine, both the electron-release caused by branching at the β-carbon atom and the steric hindrance which follows from the same branching reduce the reactivity of this electrophilic center, while at the same time they leave the carbonic acid carbonyl unaffected. In such cases, the amount of urethane can reach 6 to 8% in the mixture [52]. Since the isobutyloxycarbonyl derivative of the amino-component is permamently blocked, it remains unchanged in the subsequent steps of a synthesis and can be removed, relatively easily, at a later stage or at the conclusion of the chain building process. Therefore, mixed anhydrides with isobutyl chlorocarbonate as activating agent were applied for systematic chain lengthening in several laboratories [53–55].

Anhydrides as activated derivatives of protected amino acids or protected peptides are relatively easily constructed. Almost any acid can play the role of the electron-withdrawing (X) group in the reactive intermediate R—CO—X. Hence, the literature is rich in procedures based on various mixed anhydrides, but only a few of these have some special advantage. For instance, anhydrides involving diesters of phosphorous acid [56–58] produce only one acylation product because the second acylation product is reconverted to the mixed anhydride:

$$\text{Z-NH-CHR-CO-O-P(OR')}_2 + \text{H}_2\text{NR''} \longrightarrow \text{Z-NH-CHR-CO-NHR''} + \text{(R'O)}_2\text{P-NHR''}$$

$$\text{(R'O)}_2\text{P-NHR''} + \text{Z-NH-CHR-COOH} \longrightarrow \text{Z-NH-CHR-CO-O-P(OR')}_2 + \text{H}_2\text{NR''}$$

Only few mixed anhydrides entered the general praxis of peptide synthesis. Also, most of them do not represent a new principle. Therefore, instead of a more detailed discussion, we try to illustrate this area of activation with a selection of mixed anhydrides in Table 1. An early review on mixed anhydrides by Albertson [59] and a somewhat more recent ony by Tarbell [60] provide additional information. The mixed carbonic anhydride method was treated in depth by Meienhofer [61]. An essentially full account of mixed anhydrides can be found in the monograph by Wünsch [62].

A special category of mixed anhydrides is represented by the *N-carboxy or Leuchs' anhydrides* [92]. In these compounds activation and protection are combined in a single —CO—O— grouping:

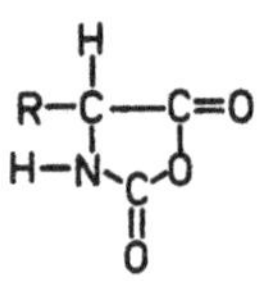

Leuchs' anhydrides can be prepared by fairly simple procedures such as treatment of the suspension of an amino acid in a non-polar solvent with phosgene or thermal elimination of benzyl chloride from benzyloxycarbonyl-amino acid chlorides:

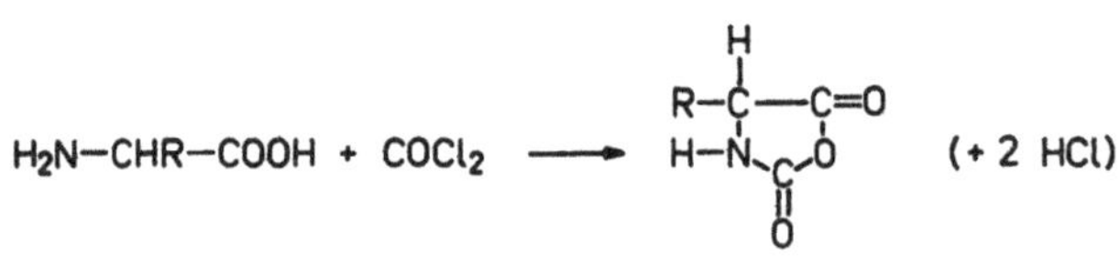

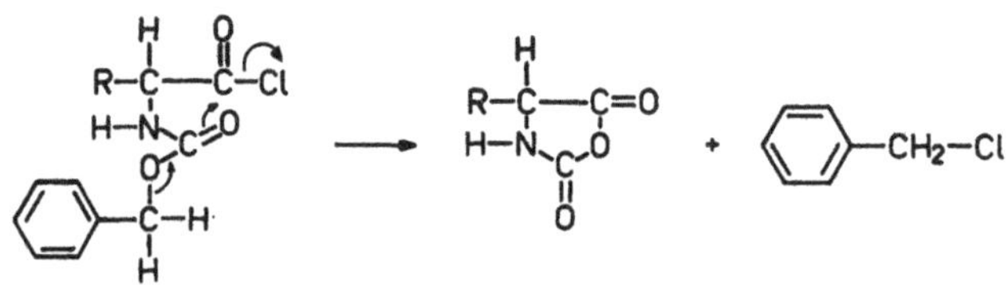

Attacks by nucleophiles occur primarily on the carbonyl of the amino acid and acylation is immediately followed by decarboxylation of the thus formed carbamoic acid derivative. The regenerated amino group is ready for acylation by a second N-carboxyanhydride:

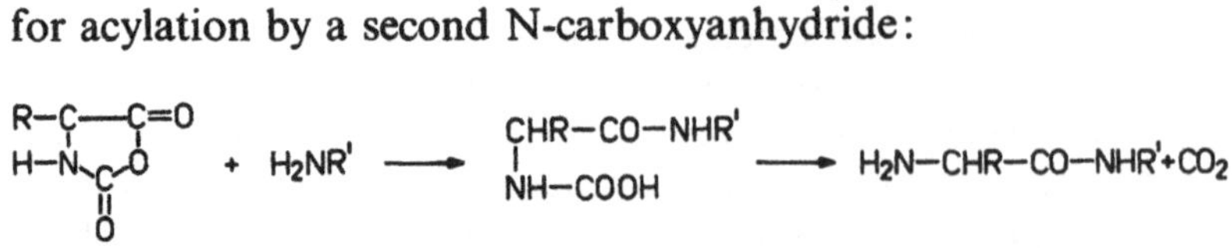

The only by-product, carbon dioxide, escapes from the reaction mixture. This is an elegantly simple scheme and it is understandable, therefore, that the possibility of chain-building with Leuchs' anhydrides attracted the interest of investigators time and again [93–97]. The apparent elegance of the method, however, is counterbalanced by several shortcomings, such as the sensitivity

Table 1. Mixed Anhydrides

Activating Reagent	Mixed Anhydride	Refs.	Activating Reagent	Mixed Anhydride	Refs.
a Derivatives of carboxylic and carbonic acids					
$(CH_3)_2CH-CH_2-C(=O)-Cl$	$R-C(=O)-O-C(=O)-CH_2-CH(CH_3)_2$	45	$CH_3-CH_2-CH(CH_3)-O-C(=O)-Cl$	$R-C(=O)-O-C(=O)-O-CH(CH_3)-CH_2-CH_3$	50
$(CH_3)_3C-C(=O)-Cl$	$R-C(=O)-O-C(=O)-C(CH_3)_3$	46	$(CH_3)_2CH-CH_2-O-C(=O)-Cl$	$R-C(=O)-O-C(=O)-O-CH_2-CH(CH_3)_2$	50
$CH_3-CH_2-O-C(=O)-Cl$	$R-C(=O)-O-C(=O)-O-CH_2-CH_3$	49,51			
b Derivatives of phosphorous and arsenous acids					
PCl_3	$R-C(=O)-O-PCl_2$	63	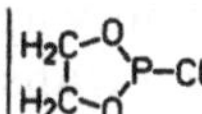	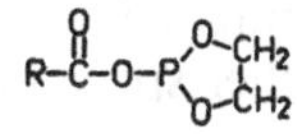	56-58
$C_2H_5-O-PCl_2$	$R-C(=O)-O-P(Cl)(OC_2H_5)$	64	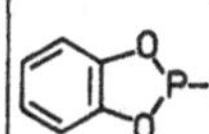	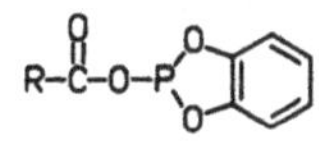	67,68
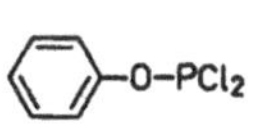	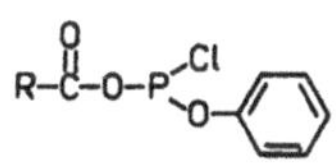	65	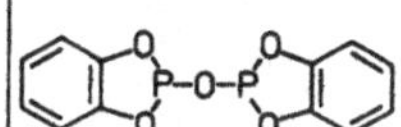	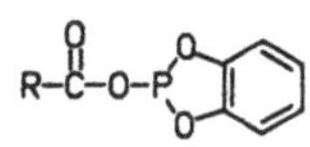	67,68
$(C_2H_5O-)_2P-Cl$	$R-C(=O)-O-P(OC_2H_5)_2$	56-58,66	$(C_2H_5O-)_2As-Cl$	$R-C(=O)-O-As(OC_2H_5)_2$	69
c Derivatives of phosphoric acid					
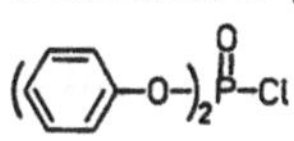	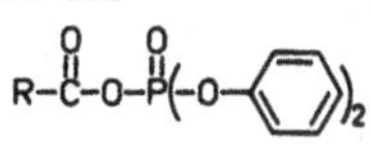	39 42	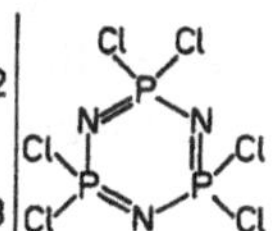	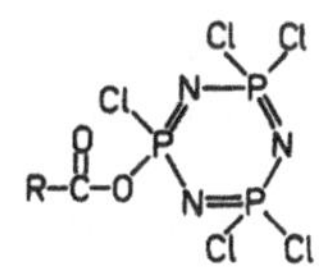	72
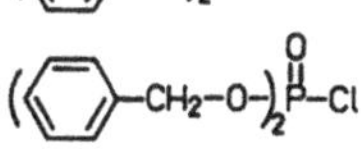	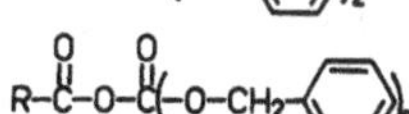	43			
$POCl_3$	$R-C(=O)-O-P(=O)Cl_2$	70,71			
d Acylphosphonium salts					
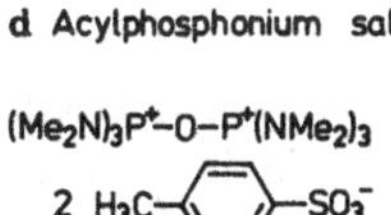	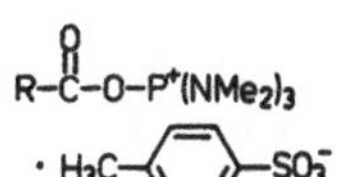	73-75	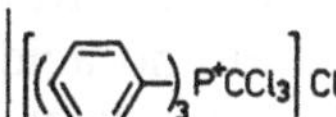	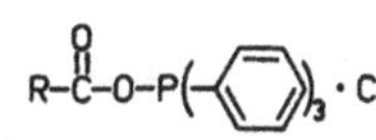	76-78
			$[(Me_2N)_3PCl]^+ CCl_3^-$ [a]	$R-C(=O)-O-P^+(NMe_2)_3 \cdot Cl^-$	76,79-83
e Pyridinium and imidazolium salts					
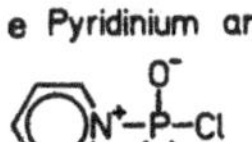	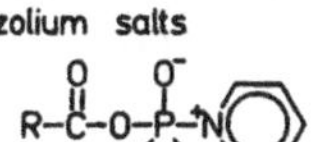	84,85	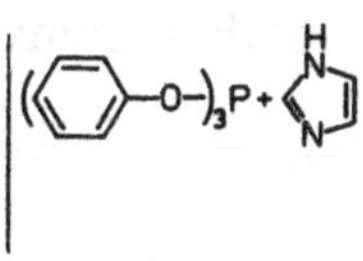	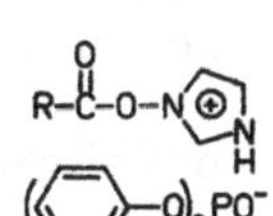	84,85
f Sulfuric acid derivatives					
$SO_3 + H-C(=O)-N(CH_3)_2$ (+NaOH)	$R-C(=O)-O-S(=O)_2-ONa$	86-88	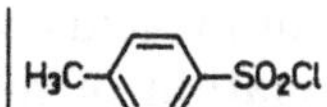	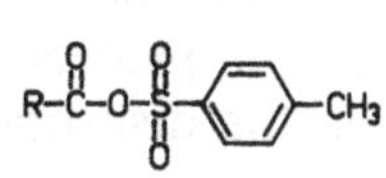	89
g Thiol acids					
$R-C(=O)-X + H_2S$	$R-C(=O)-S-H$	90,91			

of the reactive intermediates to water. To prevent polymerization, Leuchs' anhydrides have to be stored under careful exclusion of moisture. More serious limitations are caused by the products formed in the attack on the "wrong" carbonyl. A particularly noxious side reaction, double incorporation of the same amino aid residue occurs if decarboxylation takes place prematurely, that is during the acylation reaction itself [96]. The amino group is exposed then too early and will attack any unreacted N-carboxyanhydride. These side reactions can be kept at a minimum, but only under rigorously maintained special conditions including a narrow pH range (if the reaction is carried out in aqueous media), very short reaction times and extremely rapid stirring [97]. Such difficulties explain why the use of Leuchs' anhydrides remains limited. They are frequently applied in the preparation of polyamino acids, a rather special area which transcends the objectives of this book.

The cyclic anhydrides of aspartic acid and glutamic acid

have only limited usefulness [99–101].

The most obvious problem connected with mixed anhydrides is the formation of an, undesired, second acylation product. An equally obvious solution for this problem is the application of *symmetrical anhydrides* which, on reaction with an amino component, generate a single amide:

$$\text{R–CO–O–CO–R} + \text{H}_2\text{NR}' \longrightarrow \text{R–CO–NHR}' + \text{R–COOH}$$

Yet, the regeneration and possible loss of one mole of protected amino acid or protected peptide, which necessarily occurs in this reaction, seemed to be unacceptable for a long time. The drudgery and expense connected with the preparation of a protected amino acid and its conversion to a symmetrical anhydride were too great to allow such a sacrifice. Therefore, early studies in which such anhydrides were prepared and applied for peptide bond formation found no echo in the laboratories of practitioners. Thus, symmetrical anhydrides [102, 103] were generated from protected amino acids with the aid of ynamines such as methylethynyldiethylamine [104]

$$2\ \text{R–COOH} + \text{CH}_3\text{–C≡C–N(C}_2\text{H}_5)_2 \longrightarrow \text{R–CO–O–CO–R} + \text{CH}_3\text{–CH}_2\text{–CO–N(C}_2\text{H}_5)_2$$

and were used in the stepwise elongation strategy of chain building. A less expensive method for the preparation of symmetrical anhydrides is the treatment of protected amino acids with phosgene [105]. The reaction is based on the disproportionation of mixed anhydrides:

$$2\ \text{R–COOH} + \text{COCl}_2 \longrightarrow \text{R–}\overset{\text{O}}{\overset{\|}{\text{C}}}\text{–O–}\overset{\text{O}}{\overset{\|}{\text{C}}}\text{–O–}\overset{\text{O}}{\overset{\|}{\text{C}}}\text{–R} \longrightarrow \text{R–}\overset{\text{O}}{\overset{\|}{\text{C}}}\text{–O–}\overset{\text{O}}{\overset{\|}{\text{C}}}\text{–R} + \text{CO}_2$$

A major revival of the symmetrical anhydride idea could be seen, however, when the known formation of symmetrical anhydrides in the reaction of protected amino acids with dicyclohexylcarbodiimide, a reaction postulated [106] by Khorana in 1955, was utilized. The application of two moles of carboxylic acid and one mole of carbodiimide favors the production of symmetrical anhydrides. They, rather than the O-acyl-isourea intermediate, become the dominant species in the reaction mixture, which is then used, without isolation of the anhydride, for acylation:

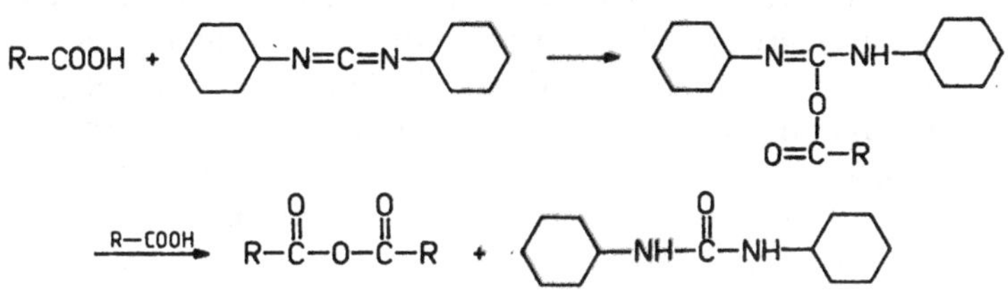

These developments coincided with a period when Boc-amino acids were already commercially available and became less and less expensive. Hence, they were often used in considerable excess and the loss of a part of the acylating agent caused no more concern. The successful application of symmetrical anhydrides and particularly their widely accepted use in solid-phase peptide synthesis led to further improvements in their preparation. For instance, the reaction between protected amino acids and water-soluble carbodiimides [107, 108] gives much better results [109] than the conventionally used dicyclohexylcarbodiimide. Symmetrical anhydrides of Boc-amino acids could be secured, in crystalline form,[3] by extraction of

$$2\ \text{Boc–NH–CHR–COOH} + \text{CH}_3\text{–CH}_2\text{–N=C=N–CH}_2\text{–CH}_2\text{–N(CH}_3)_2 \longrightarrow$$

$$\text{Boc–NH–CHR–CO–O–CO–CHR–NH–Boc} + \text{CH}_3\text{–CH}_2\text{–NH–CO–NH–CH}_2\text{–CH}_2\text{–N(CH}_3)_2$$

the reaction mixture with water. It is not obvious whether isolation of symmetrical anhydrides is of major significance, but there is no doubt about their value in chain elongation. The second acylation product problem is completely circumvented with them and the reactivity of these potent intermediates allows facile coupling at a rate which exceeds the rates achievable with active esters. The latter, discussed next in this chapter, were once thought to be eminently suitable for stepwise chain-lengthening [19]. It seems now that symmetrical anhydrides might play an equally important role in stepwise syntheses in the future.

[3] Symmetrical anhydrides of protected amino acids have been reported before (cf. Ref. [110] as crystalline materials.

3. Active Esters

Unequivocal coupling reactions can be achieved only if a single electrophilic center is present in the acylating agent and thus the nucleophile (the amino component) is acylated in a unique way. The inherent presence of two electrophilic groups in mixed anhydrides suggests that they cannot entirely fulfill this requirement of an ideal acylating agent. Mixed anhydrides obtained with derivatives of phosphorous acid might be, as already mentioned, exceptional in this respect. Symmetrical anhydrides of protected amino acids also have two reactive carbonyl groups, but because of the symmetry of the molecule, the two acylation products are identical. There is, however, an additional approach to unequivocal acylation. If the electron-withdrawing substituent used for the activation of the carboxyl group cannot play the role of an acylating agent, then only the amino acid carbonyl can become part of the newly formed peptide bond. Examples of such substituents already occurred in our discussions: acid chlorides and azides of carboxylic acids, although prone to other side reactions, do not form a second acylation product since the leaving groups cannot produce amides. Yet, the same can be said about alcohols if they can be used for activation. In esters, when used for acylation, the leaving group, an alcoholate or after protonation an alcohol, does not combine with the amino component:

$$R{-}CO{-}OR' + H_2NR'' \longrightarrow R{-}CO{-}NHR'' + R'{-}OH$$

Even simple alkyl esters, such as methyl or ethyl esters, can be ammonolyzed or aminolyzed, but generally the amides form slowly and practical rates can be obtained only at elevated temperature or by the use of the amine in large excess. Exceptions can be found, e.g. in the facile ammonolysis of methyl nicotinate in which the reactivity of the ester carbonyl is enhanced by the neighboring aromatic system. In acylation with protected amino acids, no such enhancement is possible. On the other hand, the ester group is not restricted to methyl or ethyl esters and activation of the ester carbonyl can be accomplished by the selection of alcohol-components in which an electron-withdrawing group is present. This was the underlying idea in the experiments of Schwyzer and his associates [7, 111] who investigated the aminolysis of a series of modified methyl esters of hippuric acid, each carrying a different electron-withdrawing substituent on the methyl group. The reactivity of the cyanomethyl esters, determined with benzylamine as the nucleophile, seemed to be sufficient for application in peptide synthesis:

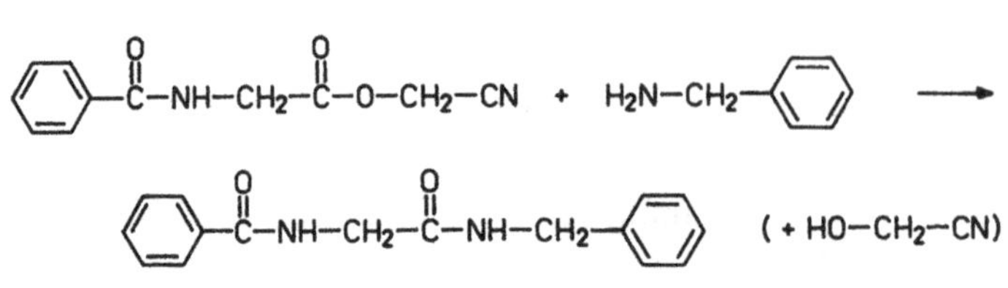

One particular advantage of the cyanomethyl ester method was already discussed in the introduction of this chapter. No overactivation of the carboxyl-component is necessary in this procedure since a reactive derivative of the hydroxyl-component is available: the esters are prepared by metathesis of the triethylammonium salt of a protected amino acid with a reactive halogen derivative of the alcohol (cyanomethanol), to wit, chloroacetonitrile:

$$R{-}COO^- + ClCH_2CN \longrightarrow R{-}\overset{O}{\overset{\|}{C}}{-}O{-}CH_2CN \quad (+Cl^-)$$

Thus the activated carboxyl-components can be secured under favorable conditions, particularly if the even more reactive bromoacetonitrile is used [8]. Nevertheless, cyanomethyl esters were applied only occasionally in actual syntheses of important peptides. The reason for this must lie in their moderate reactivity. Activation of methyl esters by the potent electron-withdrawing CN group is still insufficient and when couplings have to be carried out in dilute solutions the reaction rates become impractically low. Probably an efficient catalyst [112] is needed for a revival of this potentially valuable method.

In two papers dealing with the effect of electron distribution in esters on their ammonolysis, Gordon, Miller and Day [113, 114] pointed to the rate increase caused by electron-withdrawing substituents either in the alcohol or in the acid component of the ester molecule. In a footnote they mention that the ammonolysis rates observed with vinyl and with phenyl esters far exceed those obtained with alkyl esters. In fact, these rates were too high for exact measurements. This footnote prompted this author to reconsider the reactivity of thiophenyl esters proposed by Wieland and his associates [115] for peptide synthesis. Their proposal was based on an analogy: the reactive acetyl group present in the form of a thiol ester in coenzyme A [116]. The comments of Gordon, Miller and Day [113, 114] made it plausible that thiophenyl esters of protected amino acids owe only a part of their reactivity to the fact that they are esters of thiol-acids. To a major extent their ability to form amides under relatively mild conditions must be due to the circumstance that they are *aryl esters*:

A logical step from here was to replace the sulfur atom by oxygen in the Wieland-esters and to increase the reactivity of these moderately active acylating agents by adding electron withdrawing substituents to the aromatic ring. The most accessible negatively substituted phenols, *o*-, *p*- and *m*-nitrophenol were examined [117] as was 2,4-dinitrophenol. Esters of the latter were too reactive and accordingly also too sensitive to hydrolysis by water in the reaction mixture. From three monosubstituted phenols the para derivative was selected

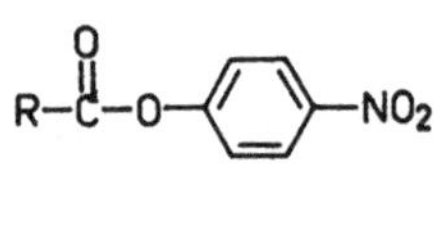

R—CO—ONp

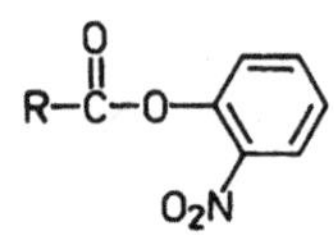

R—CO—ONo

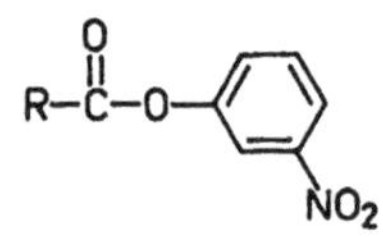

R—CO—ONm

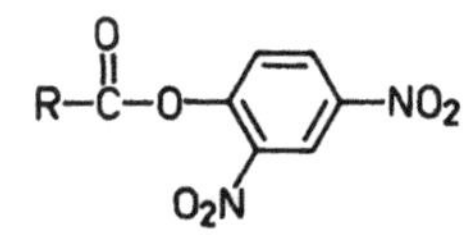

R—CO—ODNp

for the praxis of peptide synthesis mainly because of the readiness of its esters to crystallize. Some advantages of the ortho-nitro analogs were overlooked at that time and were recognized only many years later [118–121] and *p*-nitrophenyl esters of Z-amino acids were used in practical syntheses [122, 123]. More definitive was in this respect the exclusive application of *p*-nitrophenyl esters for the incorporation of the amino acid residues in the molecule of oxytocin [124, 125] in a synthesis which was also used for the demonstration of a new strategy, the stepwise elongation of a peptide chain by the addition of single residues. Active esters seemed particularly well suited for this strategy [19].

About the same time that the first paper on nitrophenyl esters appeared. Farrington, Kenner and Turner [126] proposed the application of *p*-nitrophenyl thio-esters. These are extremely potent acylating agents, although this asset is somewhat counter-balanced by the unpleasant properties of the leaving *p*-nitrothiophenol. In subsequent years, numerous aryl esters with electron-withdrawing substituents in the aromatic ring were recommended. Of these, the pentachlorophenyl esters introduced by Kupryszewski [127, 128] excel with high reactivity, but suffer from the steric effect of the bulky activating groups. Hence, these esters are less potent in crowded environments, such as those encountered in solid-phase peptide synthesis. A logical remedy for this shortcoming, replacement of the five chlorine atoms by fluorine [129–131], produced very potent active esters, with less steric hindrance, which retain their reactivity in the matrix of peptidyl polymers. The full potential of pentachloro- and pentafluorophenyl esters still awaits further evaluation by practicing peptide chemists.

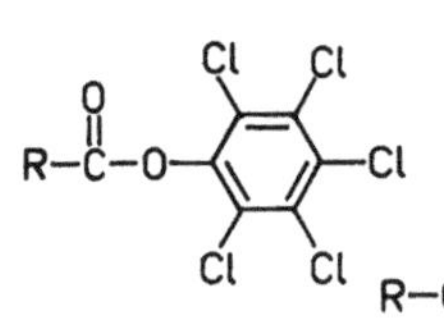

R—CO—OPcp

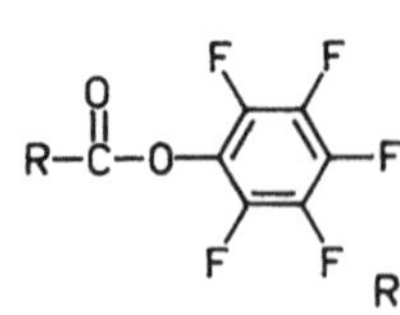

R—CO—OPfp

An extensive study of the relationship between the acidity of phenols and the reactivity of active esters derived from them led Pless and Boissonnas [132] to the selection of 2,4,5-trichlorophenyl esters. These trichlorophenyl esters do not suffer from the steric hindrance observed in pentachlorophenyl esters. Such an interference by bulky substituents becomes pronounced only when *both* ortho positions are occupied by them. This is not the case in the 2,4,5-

trichlorophenyl esters and thus they are among the most frequently applied aryl esters at this time.

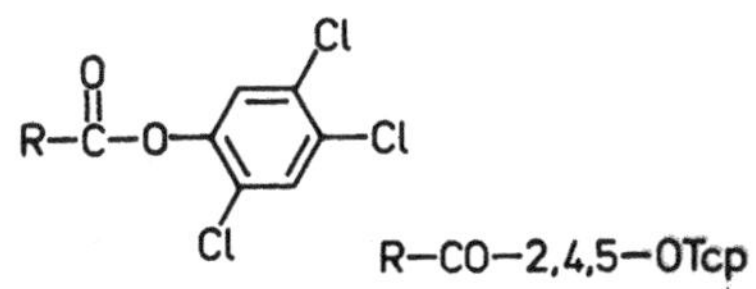

The quite remarkable discovery by König and Geiger [133] that 1-hydroxybenzotriazole can efficiently catalyze the aminolysis of *p*-nitrophenyl- and 2,4,5-trichlorophenyl esters further enhanced the usefulness of these aryl esters. Other catalysts, such as imidazole [134–136] or pyrazole [137], gained less significance in praxis. Intramolecular catalysis by neighboring hydroxyl groups is also quite promising: esters of protected amino acids with (mono) O-benzyl catechol can be activated by the hydrogenolytic removal of the benzyl group [138]. In a second version [139] of the same activation principle, the phenacyl group is applied for the temporary protection of one of the two hydroxyl groups in catechol:

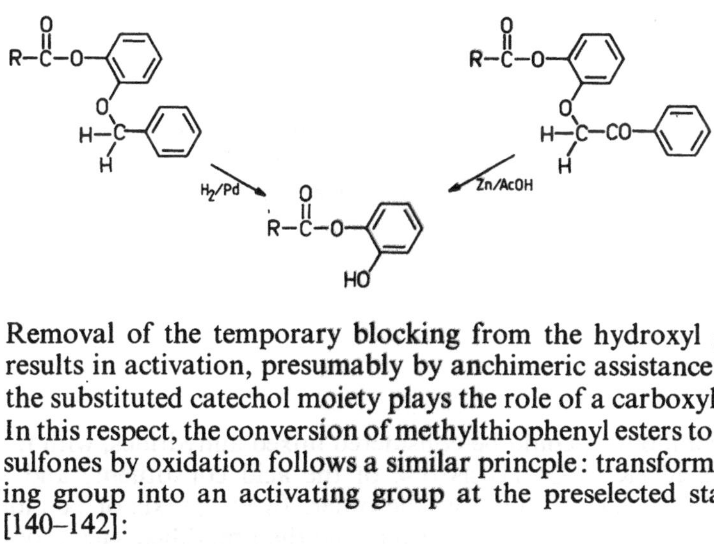

Removal of the temporary blocking from the hydroxyl group of catechol results in activation, presumably by anchimeric assistance. Prior to this step the substituted catechol moiety plays the role of a carboxyl-protecting group. In this respect, the conversion of methylthiophenyl esters to the corresponding sulfones by oxidation follows a similar princple: transformation of a protecting group into an activating group at the preselected stage of a synthesis [140–142]:

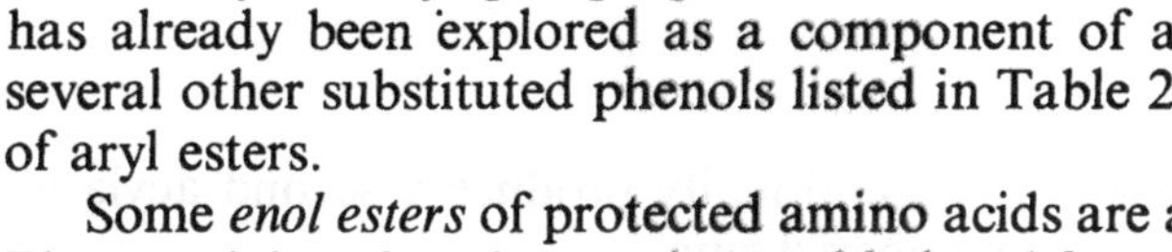

The methylsulfonyl grouping as electron-withdrawing substituent of phenol has already been explored as a component of active esters [143], as were several other substituted phenols listed in Table 2, used for the construction of aryl esters.

Some *enol esters* of protected amino acids are also good acylating agents. The reactivity of enol esters in peptide bond formation is not suprising since the earlier cited studies of Gordon, Miller and Day [113] revealed the high ammonolysis rates of vinyl esters. It is noteworthy, however, that certain substituted vinyl esters can be generated from tertiary alcohols by dehydration, e.g. with trifluoroacetic acid [144, 145]:

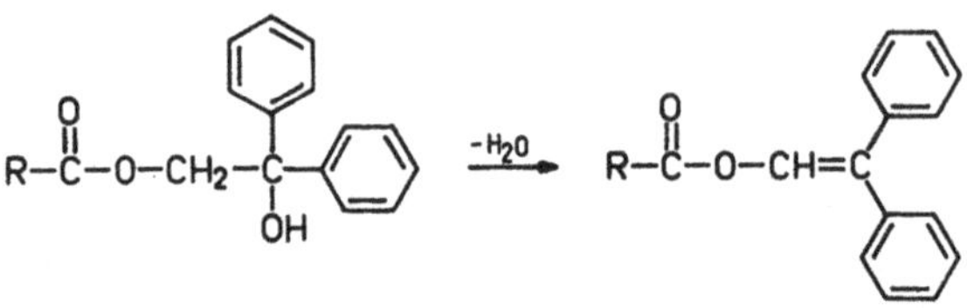

Once again we can discern a realization of the principle of converting a protecting group to an activating group. An analogous process starts with the phenacyl group used before [146] for the protection of the carboxyl function. Addition of HCN and dehydration of the resulting cyanohydrin yields a reactive vinyl ester [147]:

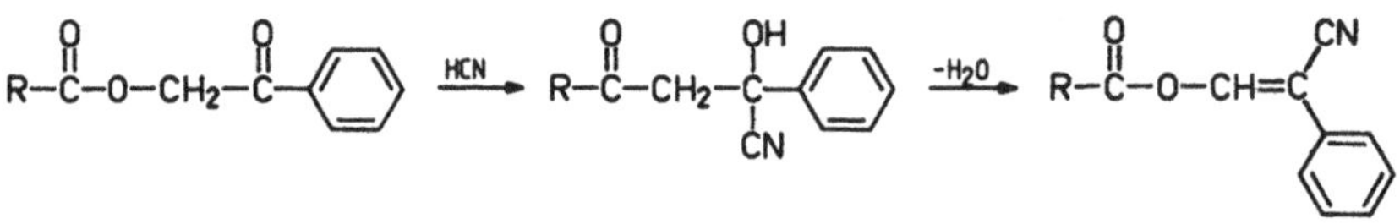

An entirely new class of active esters, *O-acyl* derivatives of substituted *hydroxylamines* gained considerable importance. The first representatives of this class, esters of N-hydroxyphthalimide, were discovered by Nefkens and Tesser [165]:

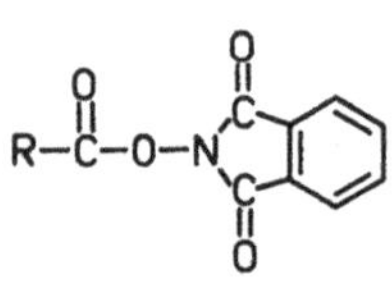

In a sense, these compounds could be considered mixed anhydrides, with the protected amino acid (R—COOH) as one of the acid constituents and a hydroxamic acid (R′—CO—NHOH] as the other. In the praxis of peptide synthesis, however, hydroxyphthalimide esters and their modified successors, the esters of N-hydroxysuccinimide [166], behave like active esters. They do not disproportionate to

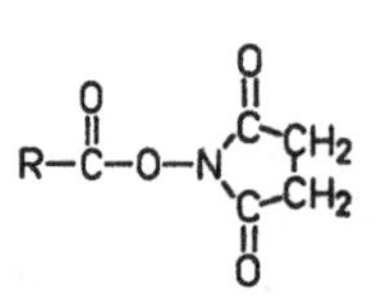

(R—CO—OSu)

symmetrical anhydrides and only exceptionally produce a second acylation product (via opening of the five-membered ring).

Table 2. Active Esters

	Refs.		Refs.		Refs.
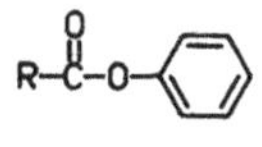	148	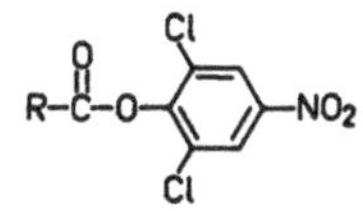	152	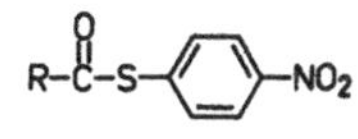	126
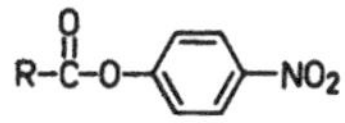	117	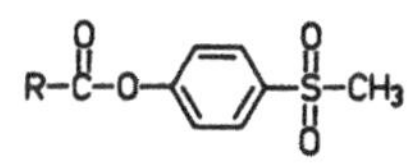	143	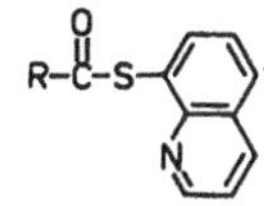	158
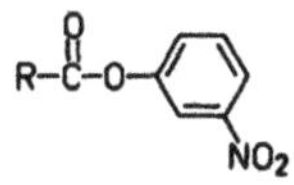	117	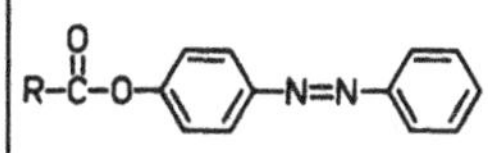	153	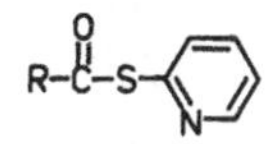	159
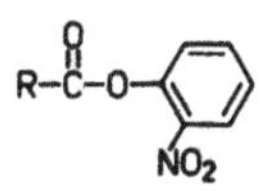	117-121	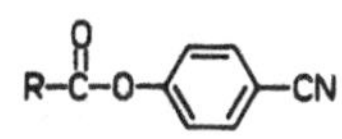	143	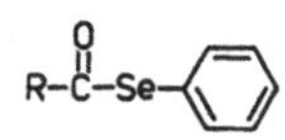	160
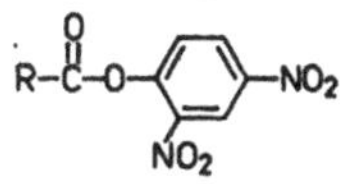	117	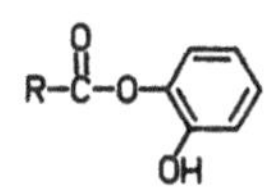	138	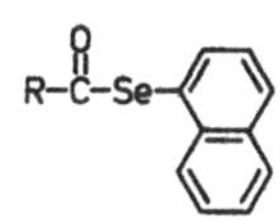	161
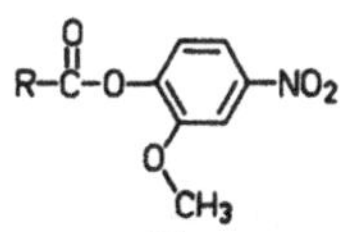	149-151	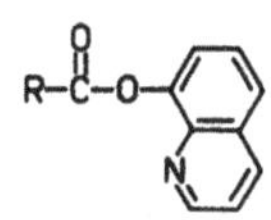	154	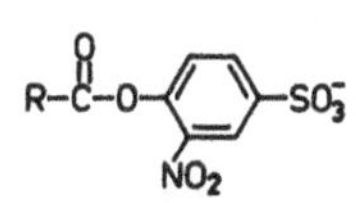	162
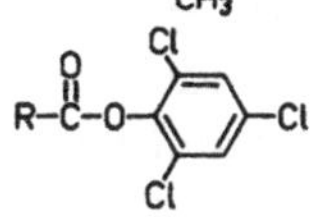	127,128	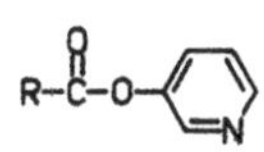	155	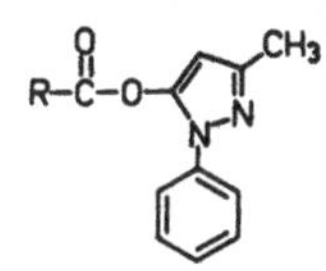	163
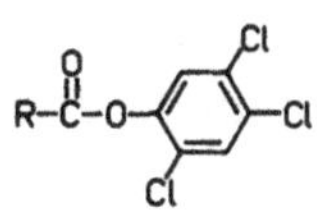	132	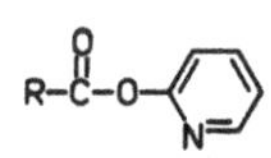	156	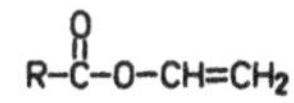	164
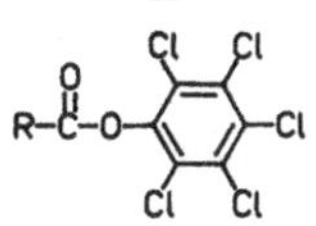	127,128	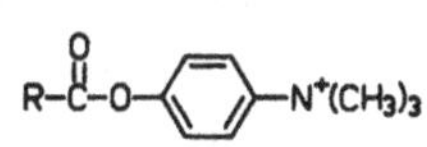	157	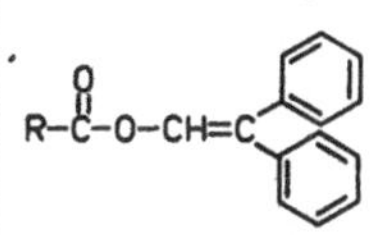	144
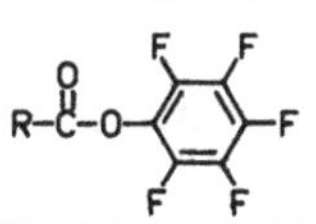	129	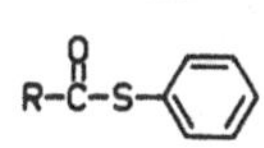	51,71,115	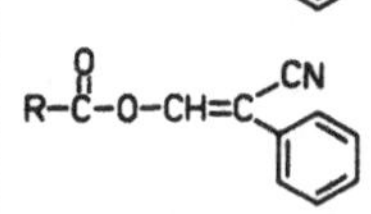	147

An entire series of O-acyl hydroxylamines carry no acyl group on the nitrogen atom and are nonetheless good acylating agents. Thus, the reactivity of hydroxyphthalimide and hydroxysuccinimide esters should not be attributed solely to their anhydride character. There are no hydroxamic acid components in esters of the type [167]

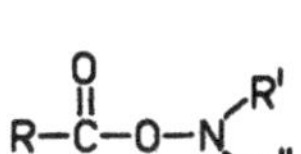

of which the best known representatives are the N-hydroxypiperidine esters [168]. The reactivity of O-acyl hydroxylamines is usually explained with anchimeric assistance provided by the nitrogen atom next to the ester oxygen:

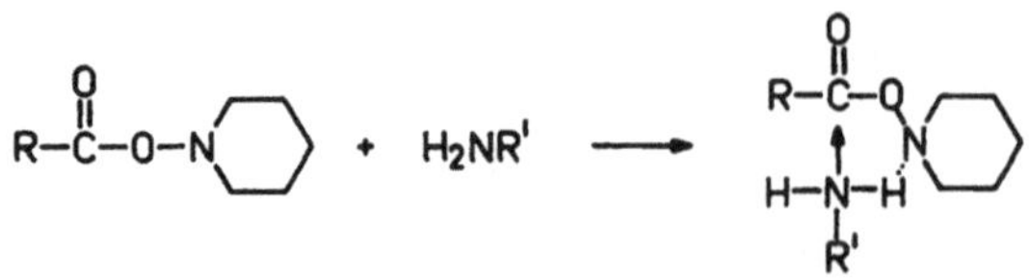

In this connection we should mention also the O-acyl derivatives of 1-hydroxybenzotriazole (HOBt) [169], compounds which are mostly not prepared as such, but which play the role of the acylating agent when HOBt is used as *additive* [169] in couplings with dicyclohexylcarbodiimide [170] or when HOBt is applied as catalyst [133] in acylations with otherwise only moderately reactive aryl esters:

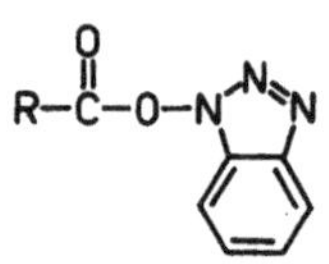

While many O-acyl-hydroxylamine derivatives were proposed for peptide synthesis (Table 3), only the esters of N-hydroxysuccinimide are widely used. They are crystalline, stable compounds which have excellent reactivity in aminolysis reactions. Also, the leaving N-hydroxysuccinimide is readily soluble in water and thus easily separated from the usually water-insoluble product, a protected peptide. In this respect N-hydroxysuccinimide esters look somewhat superior to their predecessors, the esters of N-hydroxyphthalimide which for the removal of the by-product require extraction with an aqueous solution of bicarbonate. A certain ambiguity still may exist in the reactions of N-hydroxysuccinimide esters. The strained five-membered ring is fairly sensitive to nucleophiles which can open it. Such an undesired acylation was recognized by Šavrda [171], who noted the opening of the succinimide ring when the rigid geometry of proline interfered with the attack on the active ester carbonyl:

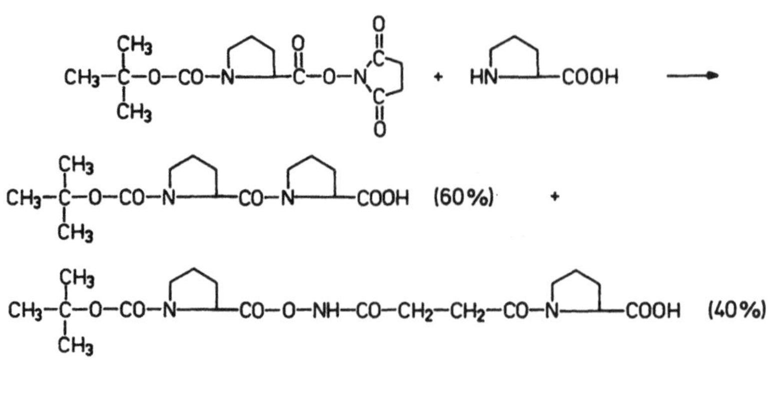

Table 3. Reactive Hydroxylamine Derivatives

	Refs.		Refs.		Refs.
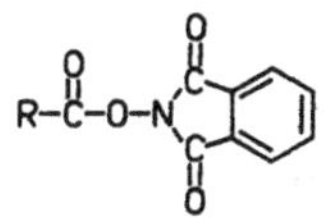	165	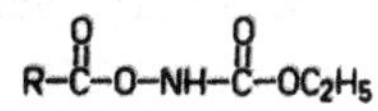	173	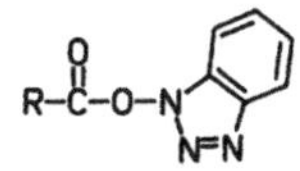	169
		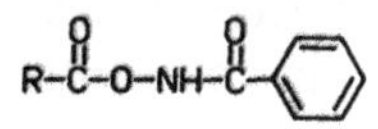	5,174		
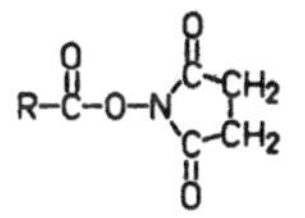	166	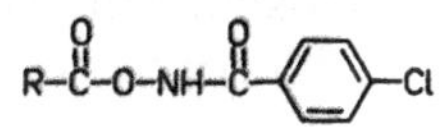	175	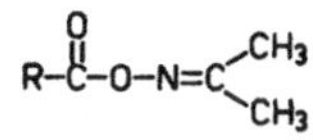	167
		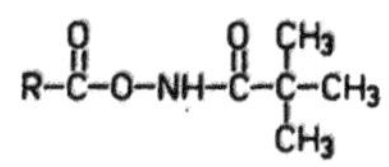	176	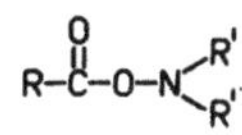	168
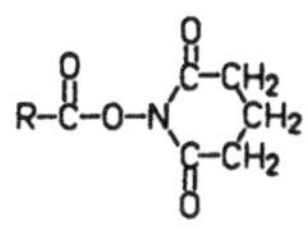	172	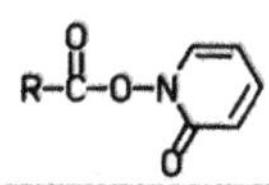	177-179	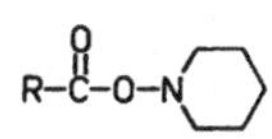	168

So far we have left without mention the methods of preparation of active esters and also the procedures applied in their use as acylating agents. The synthesis of aryl esters and of O-acyl hydroxylamines is not different from the formation of a peptide bond. The activated derivative of a protected amino acid is allowed to react with a substituted phenol or with the hydroxyl group of a hydroxylamine derivative rather than with an amine. For instance, aryl esters can be obtained through mixed anhydrides [115]:

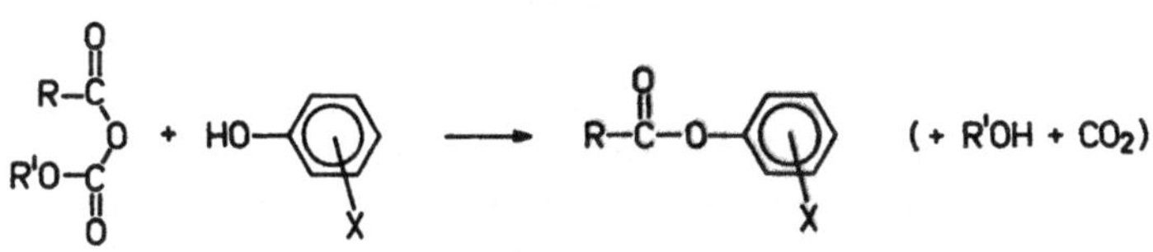

The most commonly used approach, however, is the esterification of a protected amino acid with the help of a condensing agent, particularly dicyclohexylcarbodiimide [125, 180, 181]:

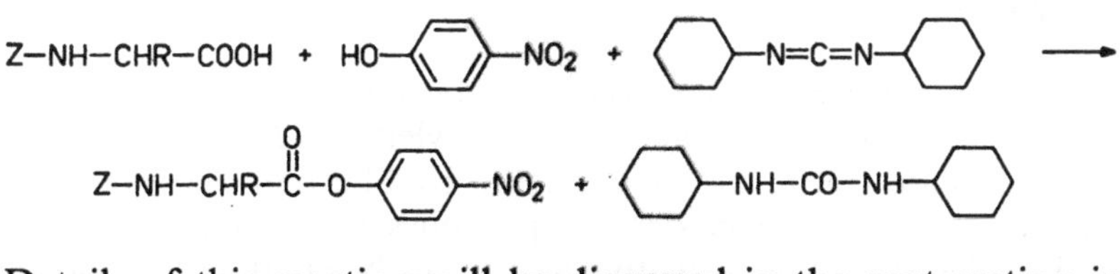

Details of this reaction will be discussed in the next section in connection with coupling reagents. Other condensing agents, e.g. ethoxyacetylene [182], have also been proposed [183] for the preparation of aryl esters. More practical might be, however, esterification with aryl phosphites [126], aryl sulfites [143, 184] or with aryl trifluoroacetates [185]. These reactions are carried out in pyridine. Thus, the possibility that instead of simple base catalyzed

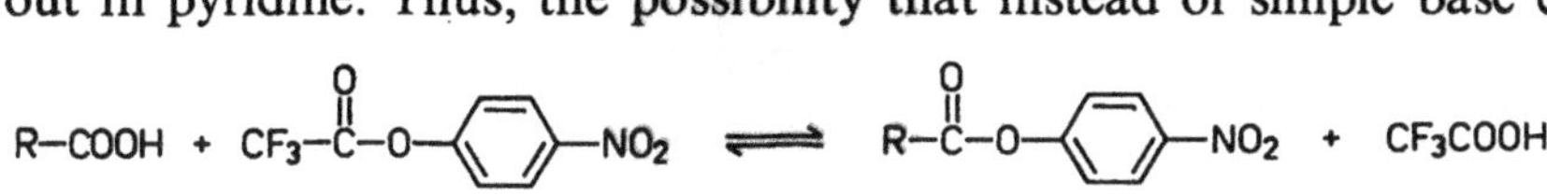

transesterification a mechanism involving mixed anhydrides [143, 184, 186] or one that proceeds through acylpyridinium ions [187] is operative, must also be considered. The various pathways may even compete with each other. Similar principles can be recognized in more recent methods proposed for the preparation of active esters. Hexachlorocyclophosphatriazine [188] as condensing agent gives good results in the synthesis of *o*-nitrophenyl esters, while dichlorotris(dimethylamino)phosphorane [189] could be used for the preparation of *p*-nitrophenyl, pentachlorophenyl and N-hydroxysuccinimide esters:

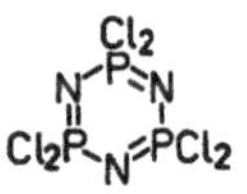

$[((CH_3)_2N)_3PCl]^+Cl^-$

Acylation with active esters is usually a simple procedure. The reactants are dissolved and the reaction mixture is allowed to stand at room temperature until a spot test with ninhydrin or fluorescamine indicates that no more unreacted amine is present. To ensure that this occurs in reasonable time, e.g. within a few hours or overnight at most, the active ester is applied in excess [18]. Practical rates are achieved if the initial concentration of the acylating agent is at least 0.1 molar. The rate of the reaction depends also upon the solvent used in the reaction. For instance, *p*-nitrophenyl esters react quite rapidly in dimethylformamide or dimethylsulfoxide, only moderately well in tetrahydrofurane, dioxane or ethyl acetate, and rather poorly in chloroform or dichloromethane. This is a fortunate coincidence with the needs of the peptide chemist since larger peptides which can cause problems are generally more soluble in dimethylformamide than in less polar solvents such as dichloromethane. Some active esters, e.g. derivatives of 2-hydroxypyridine and pyridine-2-thiol, show an opposite dependence of their acylation rates from various solvents and react only slowly in dimethylformamide. Solvent effects, thus, have considerable importance; they can have a major influence on the practical value of an active ester.

4. Coupling Reagents

In 1955 two compounds were simultaneously proposed as reagents that can effect the formation of peptide bonds: ethoxyacetylene [182] by Arens and dicyclohexylcarbodiimide (DCC or DCCI) [170] by Sheehan and Hess. In both cases, activation of the carboxyl group occurs through its *addition*, to a triple bond in the acetylene derivative and to an N=C double bond in carbodiimides:

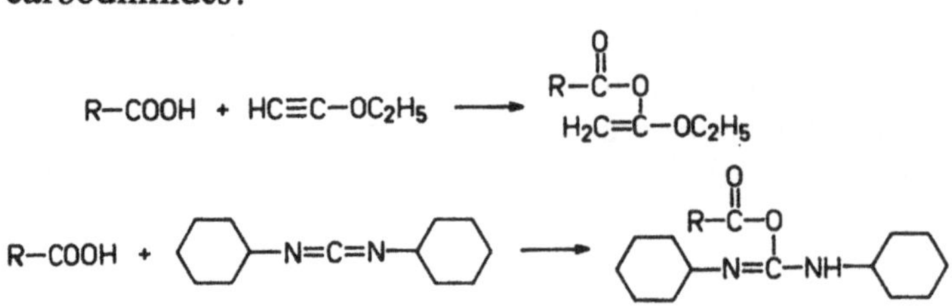

A characteristic feature of both procedures is the application of the carboxyl-activating compound in the presence of the amino-component. Condensing agents which can be added to a mixture of both components are more numerous by now and are called *coupling reagents*. An obvious prerequisite of a coupling reagent is inertness toward primary and secondary amines. This requirement is not completely fulfilled in the case of carbodiimides, which can combine with amines to give guanidine derivatives [190], but under the usual conditions of peptide synthesis, this reaction is too slow to compete with the rapid addition of the carboxyl group. Therefore, carbodiimides can be used as coupling reagents and are usually added to the mixture of the carboxyl and amino-components:

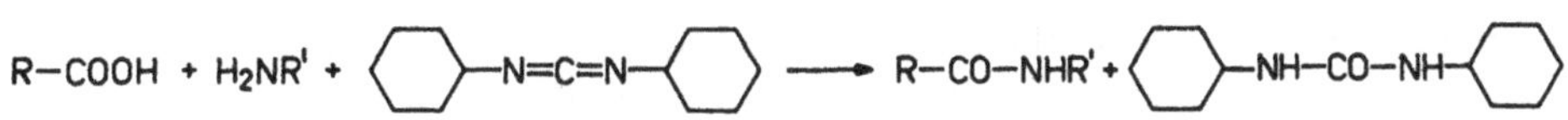

The same is true for alkoxyacetylenes:

$$R{-}COOH + H_2NR' + HC{\equiv}C{-}OC_2H_5 \longrightarrow R{-}CO{-}NHR' + CH_3CO{-}OC_2H_5$$

The by-products of the coupling reactions, N,N′-dicyclohexylurea in DCC-mediated couplings and ethyl acetate in the ethoxyacetylene procedure, are readily removed from the reaction mixtures. A particular advantage of the Arens method is that the reagent and the by-product are quite volatile. Nevertheless, alkoxyacetylene remains an interesting curiosity. The method inspired further research toward better acetylene derivatives, such as ynamines [104, 191, 192], but has not been used so far for the preparation of larger peptides. The reason for this must lie in the moderate reactivity of the reagent or of the enol ester intermediate. Thus, practical rates can be achieved with ethoxyacetylene only if it is used in considerable excess or at elevated temperature. In contrast, dicyclohexylcarbodiimide became and still is a mainstay of peptide chemists.

There are several shortcomings of the DCC method. The N,N′-dicyclohexylurea by-product, while indeed insoluble in most organic solvents (except in alcohols) and thus removable by filtration, is not entirely insoluble, particularly in the presence of other dissolved materials and therefore it frequently contaminates the product of coupling. A remedy for this imperfection could be in the use of water-soluble carbodiimides [107, 108], such as (salts of)

$$CH_3{-}CH_2{-}N{=}C{=}N{-}CH_2{-}CH_2{-}N(CH_3)_2$$

The salts of the urea-derivative formed in couplings with such modified reagents are extracted with water. A more disturbing side reaction is the intramolecular rearrangement of the O-acyl isourea derivative. The attack on the activated carbonyl group by the nearby nucleophile (NH) results in an O → N shift yielding an N-acylurea derivative as by-product. Such ureides are

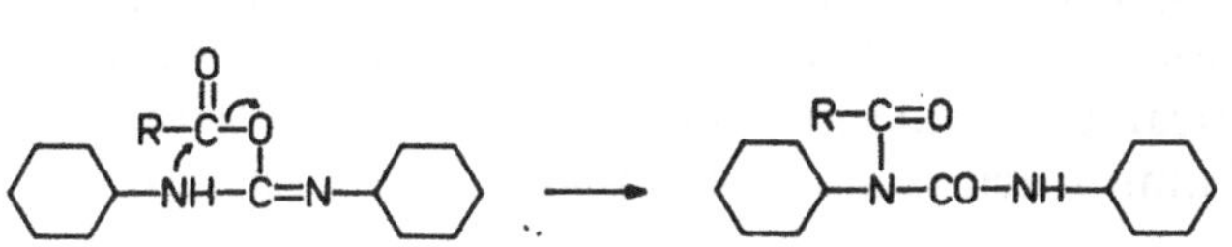

undesirable not only because they represent a loss of valuable carboxyl-component, but also because their separation from the main product of the reaction might be difficult, especially in the coupling of larger peptide segments. Last, but not least, activation by DCC causes racemization of the carboxy-terminal residue. These problems, inherent in the DCC method, did not deter the peptide chemists from its application, but rather stimulated new research toward the elimination of such shortcomings. This attitude is readily explained by the facile execution of couplings with DCC. The commercially available reagent, a solid, is added, at or below room temperature, to a solution of the two components to be linked together by a peptide bond. The reaction, which can be carried out in a large variety of solvents, proceeds rapidly and will yield the desired product without fail. A note of caution has to be added here: dicyclohexylcarbodiimide is a powerful reagent. It is also allergenic and should be handled, therefore, with proper care.

So far we have considered only one of the pathways through which DCC participates in the formation of the peptide bond, namely the nucleophilic attack of the amino-component on the O-acylisourea intermediate:

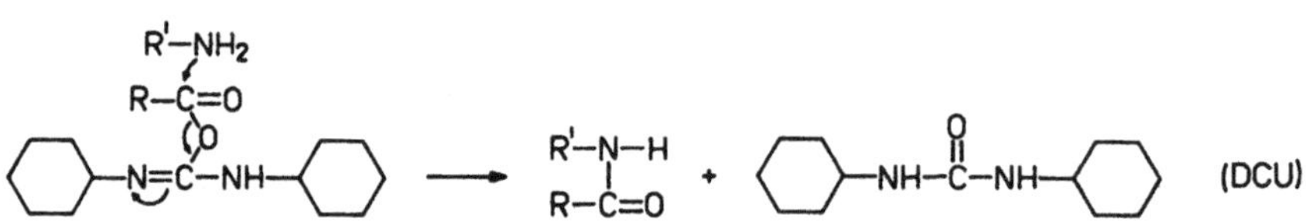

An alternative mechanism, however, is equally important. Reaction of the O-acylisourea with unreacted carboxyl-component yields a symmetrical anhydride, a potent acylating agent [106]:

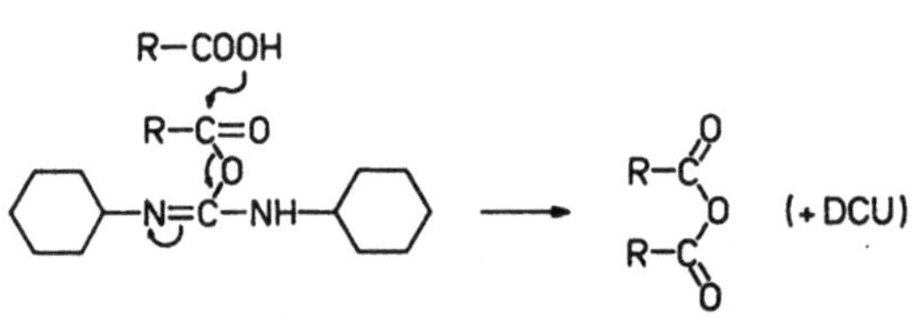

If the carboxyl-component is a protected amino acid, it is possible and often desirable to use two moles of the carboxylic acid for one mole of DCC. Such a ratio will, of course, favor the formation of symmetrical anhydrides and diminish at any given time the concentration of the O-acylisourea derivative. In turn, the extent of O → N acyl migration and, consequently, the production of ureides (cf. above) is suppressed. For an even more perfect execution of the coupling reaction, symmetrical anhydrides can also be isolated [110], but in this case DCC plays the role of an activating agent and not that of a coupling reagent. At this point we should mention once more the extensive analogy between DCC and alkoxyacetylenes. The latter are also quite effective in the generation of symmetrical anhydrides [183].

In the praxis of peptide synthesis, carbodiimides are applied mostly without isolation of the symmetrical anhydride intermediates. Acylation with a 2:1 mixture of the protected amino acid and carbodiimide is often referred to as acylation with symmetrical anhydrides. This designation is not fully warranted, since participation of some O-acylisourea in the acylation is still likely. The assumption that symmetrical anhydrides are the reactive intermediates mainly responsible for the coupling is perhaps more justified in the so-called "preactivation" approach, where the protected amino acid (2 moles) and DCC (1 mole) are allowed to react with each other for a short time before the mixture is brought into contact with the amino-component.

In the activation of the carboxyl group, carbodiimides and alkoxyacetylenes are attractive materials because they are reasonably selective: they are reactive toward carboxylic acids and sufficiently inert to amines. Also, DCC is commercially available and convenient to use. In terms of mechanism, however, neither of these coupling reagents offer a novel pathway in amide bond formation. The reactive intermediates are enol esters or symmetrical anhydrides. According to recent evidence [193, 194], the reaction between protected amino acids and water-soluble carbodiimides generates — at least with certain amino acids — azlactones as well. These are well known, potent acylating agents. Thus, the value

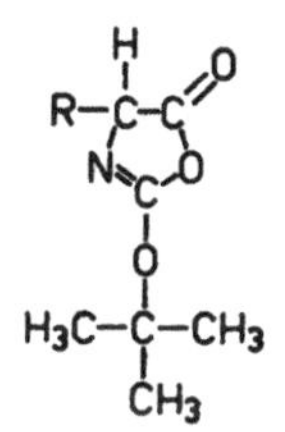

of DCC as a coupling reagent is based not on any novelty in the reaction mechanism, but on its selectivity. Not all activating reagents show such discrimination. For instance, isobutyl chlorocarbonate [50], which is perhaps the most widely used reagent for the activation of the carboxyl group, reacts readily with amines to yield urethanes. The rates of the two reactions are

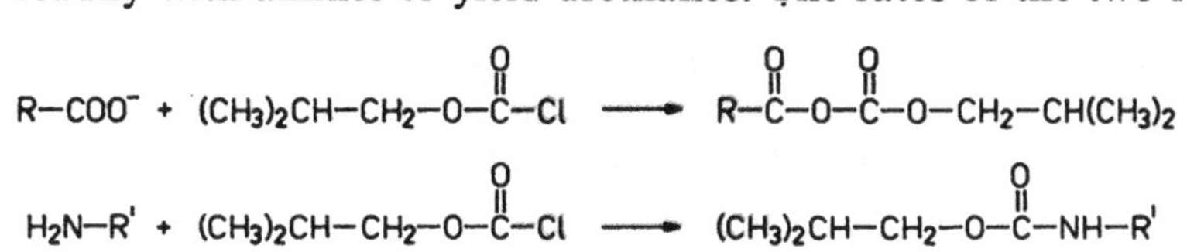

not different enough to allow the utilization of chlorocarbonates in the manner of coupling reagents. Yet, an interesting alternative was discovered by Belleau and Malek [195]. In their method, ethyl chlorocarbonate is brought into reaction with quinoline and ethanol in the presence of a tertiary base as acid binding agent, and the product, 1-ethoxycarbonyl-2-ethoxy-1,2-dihydroquinoline (EEDQ), is then used as coupling reagent:

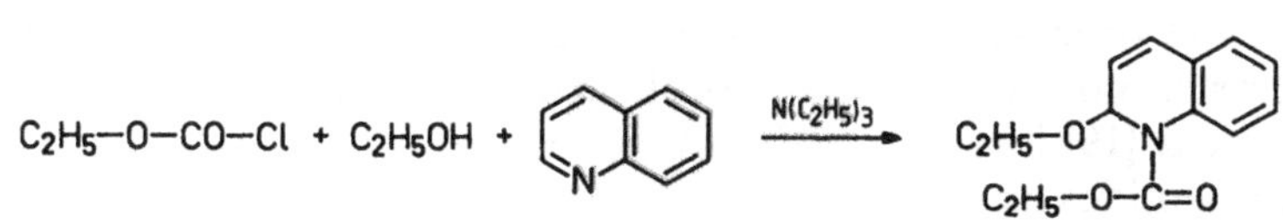

EEDQ

EEDQ is a stable, crystalline material, by now commercially available. It is not entirely inert toward amines [196] but the rate of urethane formation is low when compared with the rate of the reaction between the reagent and carboxylic acids. The latter reaction starts with the displacement of ethanol and leads, via an intramolecular attack, to the formation of a carbonic acid mixed anhydride:

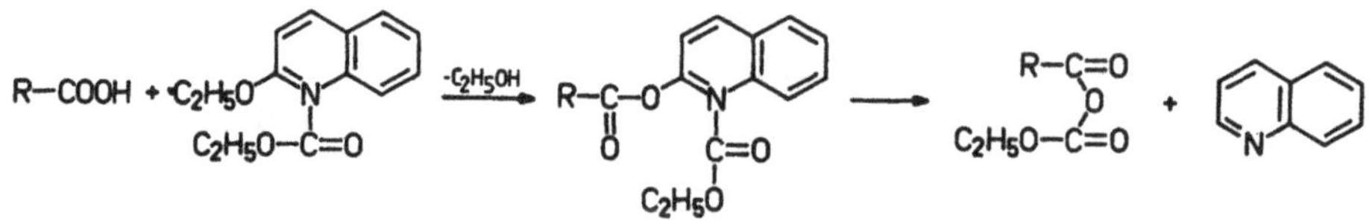

The value of such mixed anhydrides has been well documented before [49–51]. A major advantage of EEDQ over ethyl chlorocarbonate lies in the absence of a tertiary base in the anhydride forming reaction. The regenerated quinoline is of negligible basicity. It is also readily removed from the reaction mixture as are ethanol and carbon dioxide, the by-products in the acylation of the amine. Of course, if the reaction indeed proceeds through a mixed anhydride, then some second acylation product, a urethane, must also be expected. This undesired side reaction can be reduced by a modification of the coupling reagent: 1-isobutyloxycarbonyl-2-isobutyloxy-1,2-dihydroquinoline, IIDQ [197], seems to be quite suitable for segment condensation. An important feature of the EEDQ and IIDQ mediated couplings is the fair conservation of chiral purity during coupling. In this respect DCC and several other coupling reagents are less satisfactory [198]. This might be related to the absence of proton abstractors in EEDQ and IIDQ and in the reactive intermediates. Also, no tertiary base is needed for acid binding in the reaction mixture, while in the preparation of mixed anhydrides through chlorocarbonates, some base must be added to neutralize the HCl liberated in the reaction.

Execution of a coupling with IIDQ is a simple process. The reagent is added to the mixture of the carboxyl- and amino-components at room temperature and the reaction is allowed to proceed until completion:

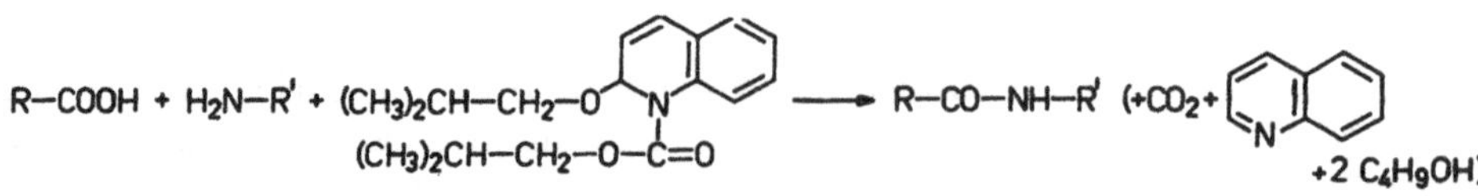

Since activation of the carboxyl-component requires two consecutive steps, of which the first, displacement of the alkoxy group from position 2 of the quinoline moiety, is bimolecular and only the second, the anhydride formation, is intramolecular, the process reaches practical rates at relatively high concontration of the reactants. With properly protected peptides, however, IIDQ can be used in excess [196] and thus the rates become satisfactory for the synthesis of larger peptides as well.

Mixed anhydrides as reactive intermediates must be assumed also in couplings with diphenylketene [199]. This method of activation appears to be quite attractive. Thus, addition of the carboxyl group to the unsaturated system requires no base, except in catalytic amounts. Also, the mixed

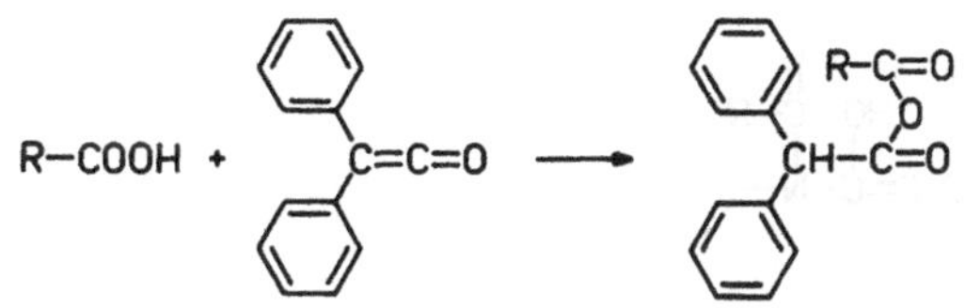

anhydride formed in the addition could possibly react with the amino-component in an unequivocal manner, since the "wrong" carbonyl is shielded by the bulky phenyl groups from the attack by the nucleophile. In reality, however, steric hindrance is outbalanced by the electronic effect of the same two phenyl rings. Therefore, the diphenylacetyl-derivative of the amino-component is a not negligible by-product in coupling with diphenylketene:

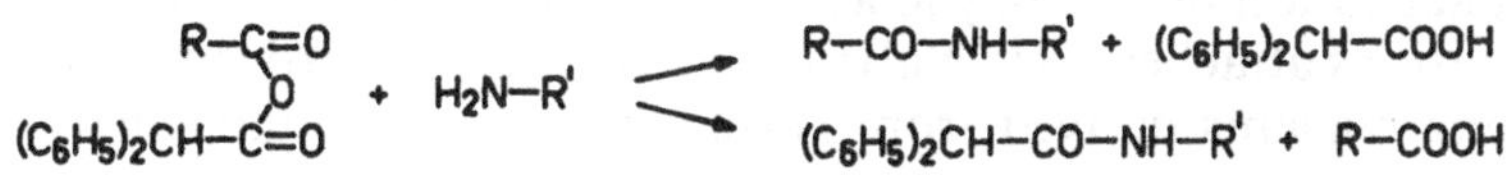

A common feature in the activation with diphenylketene and with DCC is the formation of an acylating agent through the addition of a carboxyl group to a system of vicinal double bonds. The same principle can be recognized in the method proposed by Stevens and Munk [200] involving the coupling reagent diphenylketene *p*-tolimine:

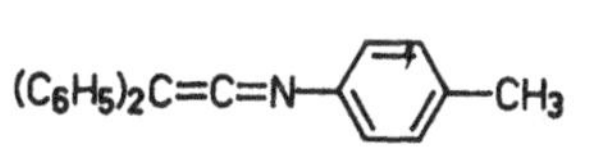

and also in the application of Woodward's reagent K [201, 202]. The latter, an isoxazolium salt, when treated with base, opens up to a ketene-imine:

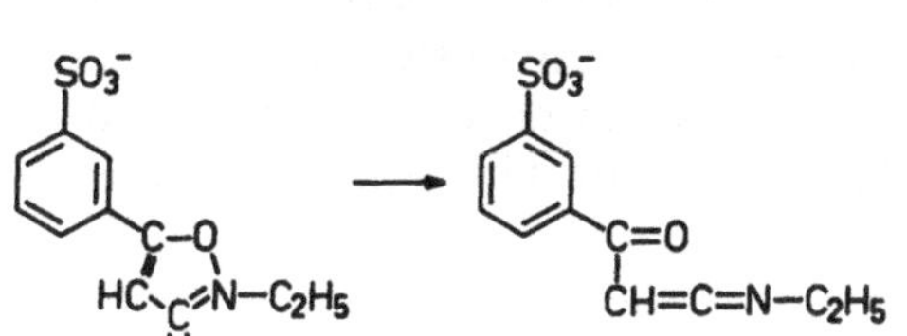

A common shortcoming of the two reagents is the presence of a nucleophile within the molecule of the activated intermediate. Hence, the O → N acyl migration experienced in the DCC method must be expected in the two last mentioned procedures as well. This rearrangement is shown here in connection with diphenylketene imines:

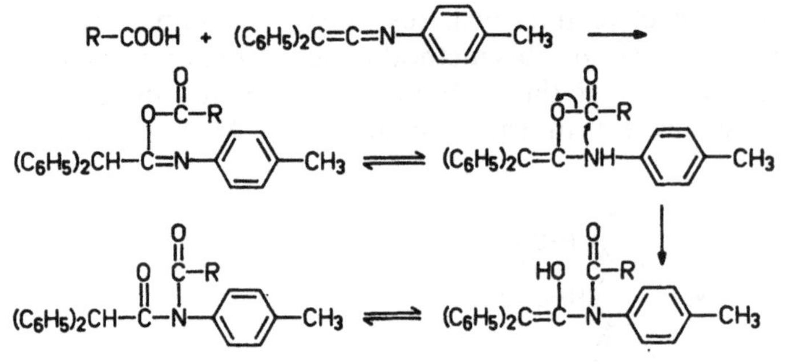

Formation of such poorly reactive by-products was indeed noted in couplings mediated with DCC and with ketenimines. Perhaps a general conclusion can be drawn from these examples, to wit, that the reactive intermediates generated with coupling reagents should not contain nucleophilic centers that can compete with the intended nucleophile, the amino-component. Over and above the production of undesired N-acyl derivatives, the Woodward reagent causes considerable racemization when used in polar solvents such as dimethylformamide [198]. Therefore, the method which was based on the well-known activation of carboxylic acids with isoxazolium salts did not fulfill the expectations attached to it. Subsequent attempts toward the design of better coupling reagents derived from isoxazolium salts are indicated in Table 4.

The reactive intermediates formed in the reaction of coupling reagents with carboxyl-components are not necessarily anhydrides or enol esters, they can be aryl esters or O-acyl-hydroxylamines as well. In connection with the preparation of active esters, we have already mentioned the use of DCC for the esterification of carboxylic acids with phenols. The initial version of this method [180, 181]

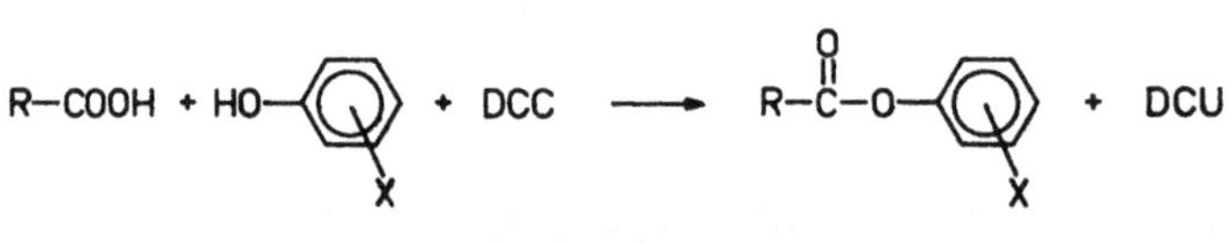

dispensed with the isolation of the active esters and used their solution for acylation. Since the reagent, DCC, is relatively inert toward amines, it is possible to further simplify the process and prepare active esters in the presence of the amino-component. For instance, Kovács and his coworkers proposed [129] complexes of DCC with pentachlorophenol or with pentafluorophenol as coupling reagents:

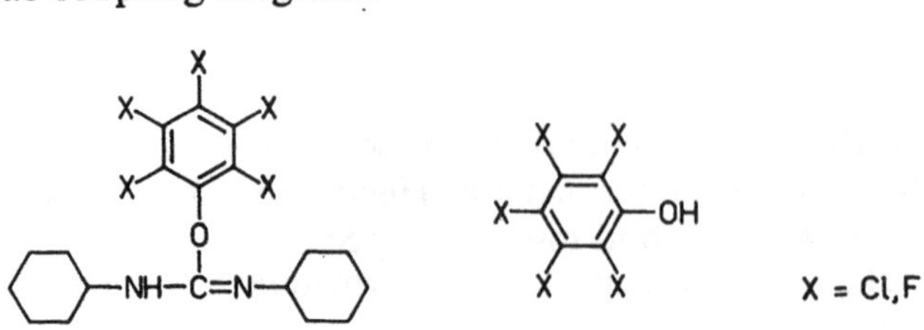

The use of N-hydroxysuccinimide [203, 204] or of 1-hydroxybenzotriazole [169] as additives in couplings with DCC probably belongs to the same category. Active esters are generated *in situ* and are consumed in the acylation of the amino-component already present in the reaction mixture. Of course, acylation of the amine via O-acylisoureas and symmetrical anhydrides will take place concurrently.

The reagents discussed up to this point simplify the execution of the coupling reaction but do not offer entirely novel approaches since the reactive intermediates in the peptide bond forming step are anhydrides or active esters. Yet, the reader should not be left with the impression that nothing new can be invented in the area of coupling reagents. In the remarkably original method of Staab [205], carbonyldiimidazole (CDI) reacts with carboxylic acid to generate acylimidazoles which are powerful acylating agents:

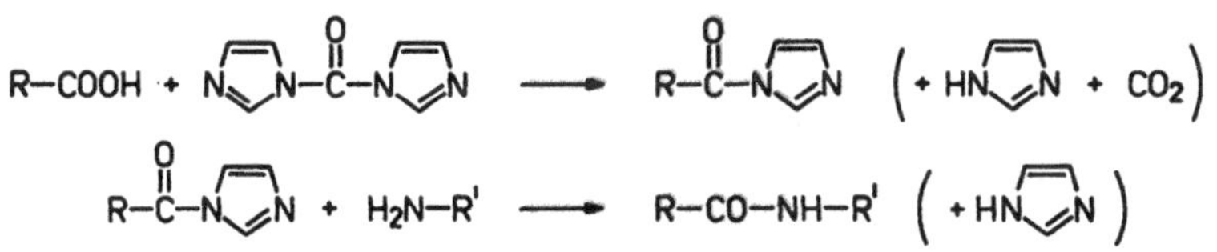

The by-products, imidazole and carbon dioxide, are easily removable from the reaction mixture. The CDI method has been applied [206] in peptide synthesis, including the esterification of protected amino acids with polymeric derivatives of benzyl alcohol [207]. Its popularity, however, could not approach that of DCC, probably because it is fairly expensive to prepare the reagent from phosgene and imidazole and it can be stored only under the rigorous exclusion of moisture.

A whole series of coupling reagents yield acyloxy-phosphonium salts. As an example we show here Bates' reagent [73, 74]:

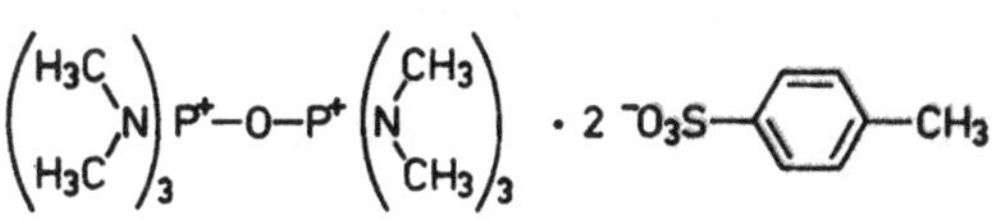

On reaction with carboxylic acids, tri-(dimethylamino)-acylphosphonium salts are produced and these are the acylating intermediates. One drawback

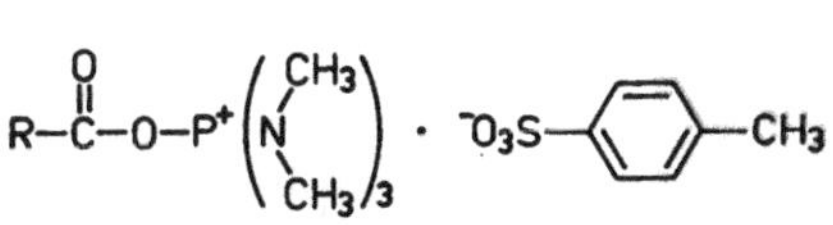

of the method is the need for hexamethylphosphoramide, a carcinogenic material, in the preparation of the reagent. The same material is also released during

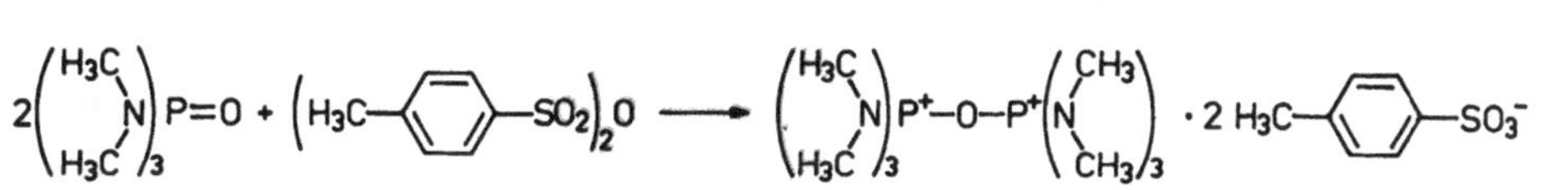

coupling. The significance of the Bates' reagent and of a series of related methods [76–83] based on acyloxyphosphonium ions cannot be stated at this time: it awaits demonstration of applicability in major syntheses.

A particularly independent method for the formation of the peptide bond is the oxidation-reduction process of Mukaiyama and his associates [208]. Of the reagents developed by them, the combination of triphenylphosphine with 2,2′-dipyridyl disulfide might be the best. The phospine is oxidized while the disulfide is reduced in the coupling reaction, an indeed original way of removing water:

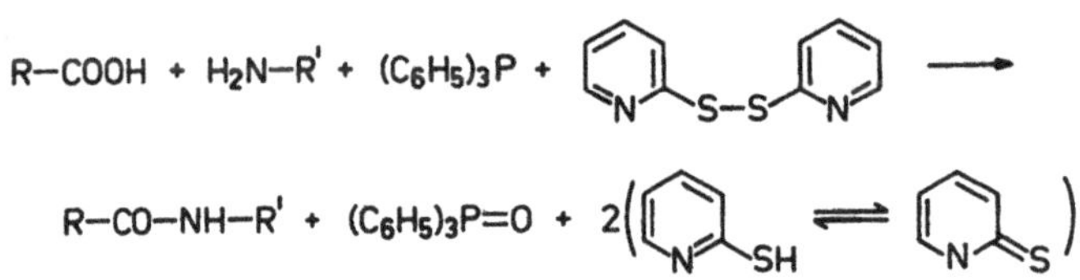

Yet, the complex pathway probably proceeds, also in this case, through acyloxyphosphonium intermediates:

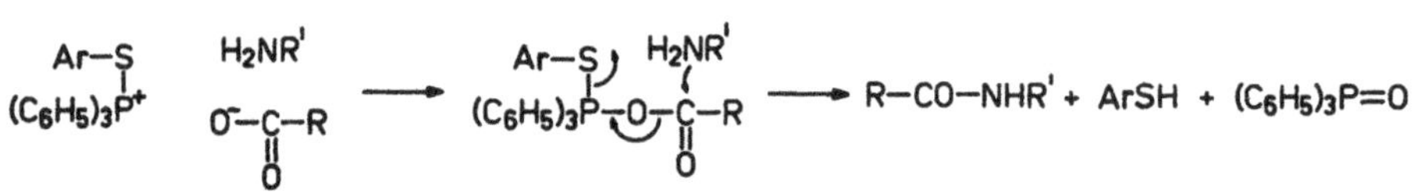

A selection of representative coupling reagents is shown in Table 4. For a review of coupling reagents cf. ref. [209].
In Table 4, several compounds which were mentioned before among mixed anhydride forming reagents are listed a second time. This was done to indicate that the intermediates in acylation with the help of coupling reagents are generally mixed anhydrides, enol esters, O-acyl hydroxylamines or aryl esters. Thus, in most cases the principles involved are not new. In fact, even where N-acyl intermediates form, these are, in turn, converted to mixed anhydrides; e.g.

$$(C_2H_5O)_2PCl + H_2NR \longrightarrow (C_2H_5O)_2P{-}NHR \xrightarrow{R'-COOH}$$

$$R'{-}\overset{\overset{\displaystyle O}{\|}}{C}{-}O{-}P(OC_2H_5)_2 + H_2NR \longrightarrow R'{-}CO{-}NHR + (C_2H_5O)_2P{-}OH$$

It might be also interesting to note that from the array of coupling reagents listed in Table 4, very few were ever used in major syntheses. Among these, dicyclohexylcarbodiimide is still the most popular.

5. Auxiliary Nucleophiles

Some shortcomings of the DCC method can be eliminated by the use of *additives*. Of these, 1-hydroxybenzotriazole [169] is the most commonly

Table 4. Coupling Reagents

Reagent	Reactive Intermediate	Refs.
$C_2H_5O{-}C{\equiv}CH$	$R{-}C(=O){-}O{-}C(=CH_2){-}OC_2H_5$	182,210
$C_2H_5O{-}C(Cl){=}CH_2$	$R{-}C(=O){-}O{-}C(CH_3)(Cl){-}OC_2H_5$	211
$(CH_3)_3C{-}C{\equiv}C{-}N(CH_3)_2$	$R{-}C(=O){-}O{-}C(=CH{-}C(CH_3)_3){-}N(CH_3)_2$	104
$H_3C{-}C(=O){-}C{\equiv}C{-}N(CH_3)_2$	$R{-}C(=O){-}O{-}C(CH_3){=}CH{-}C(=O){-}N(CH_3)_2$	191
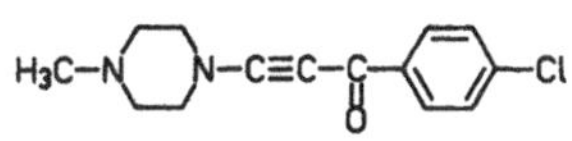	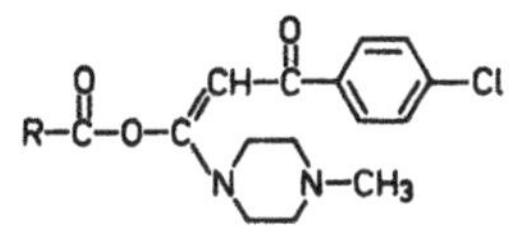	192
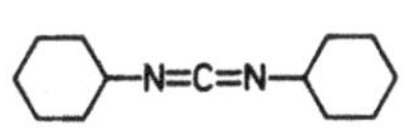	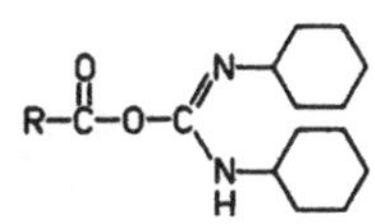	170
$(CH_3)_2CH{-}N{=}C{=}N{-}CH(CH_3)_2$	$R{-}C(=O){-}O{-}C(=N{-}CH(CH_3)_2){-}NH{-}CH(CH_3)_2$	170
$C_2H_5N{=}C{=}N{-}(CH_2)_3{-}N(CH_3)_2$	$R{-}C(=O){-}O{-}C(NHC_2H_5){=}N{-}(CH_2)_3{-}N(CH_3)_2$	107
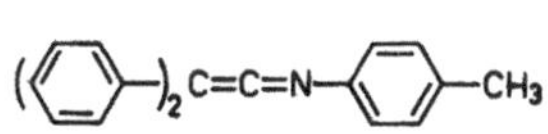	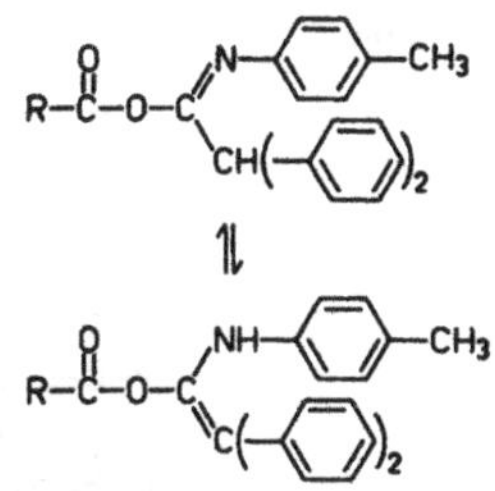	200
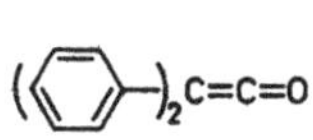	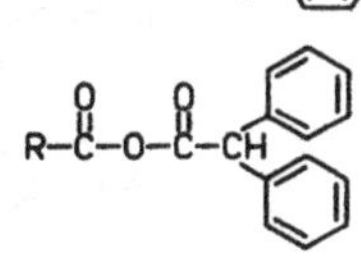	199
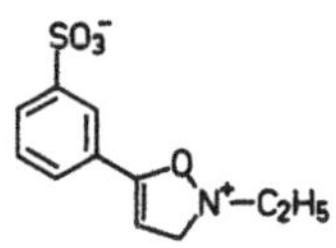	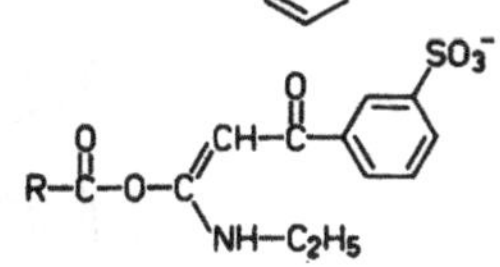	201,202
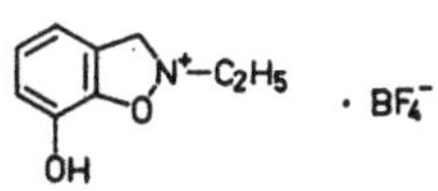	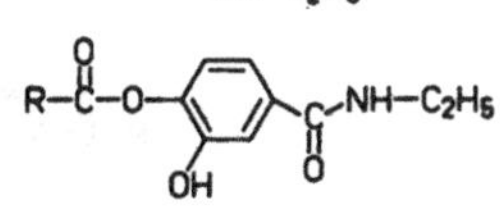	212
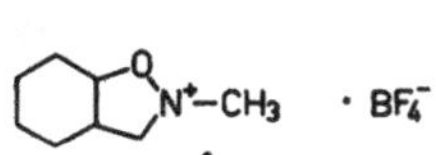	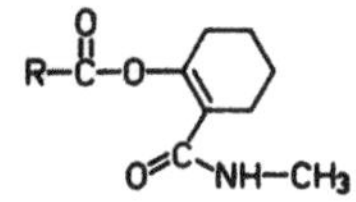	213

Table 4 (continued)

Reagent	Reactive Intermediate	Refs.
$(C_2H_5O)_2P{-}O{-}P(OC_2H_5)_2$	$R{-}C(=O){-}O{-}P(OC_2H_5)_2$	56,58,66
$(C_2H_5O)_2P{-}Cl$	$R{-}C(=O){-}O{-}P(OC_2H_5)_2$	56,58,66
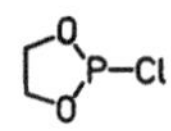	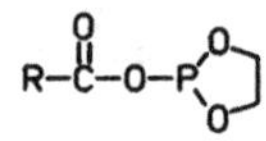	67
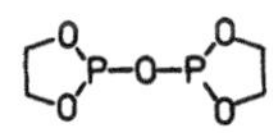	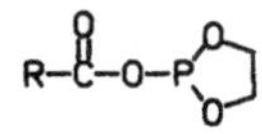	67,68
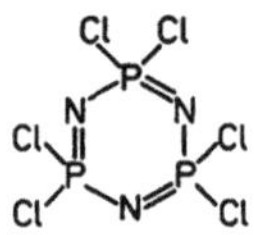	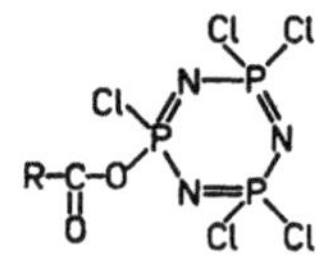	72,188
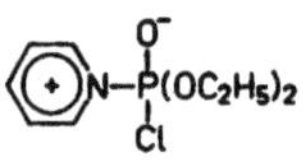	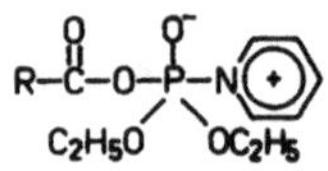	85
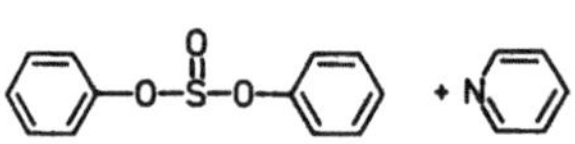	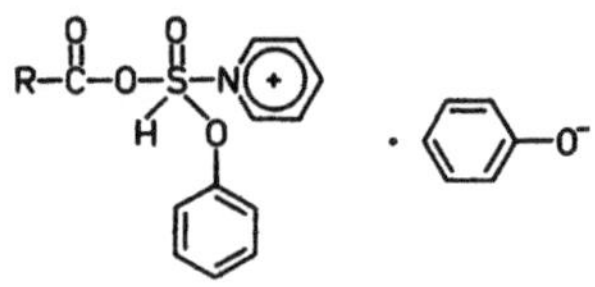	187
$(Me_2N)_3P^+{-}O{-}P^+(NMe_2)_3 \cdot 2\,BF_4^-$	$R{-}C(=O){-}O{-}P(NMe_2)_3 \cdot BF_4^-$	73
$(Me_2N)_3P^+{-}N_3 \cdot PF_6^-$	$R{-}C(=O){-}N_3$	77
$(Me_2N)_3P^+Cl \cdot ClO_4^-$	$R{-}C(=O){-}O{-}P^+(NMe_2)_3 \cdot ClO_4^-$	81
$(Me_2N)_3P + CCl_4$	$R{-}C(=O){-}O{-}P^+(NMe_2)_3\ Cl^-$	76-82
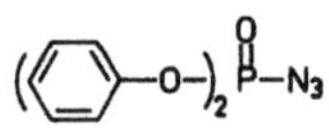	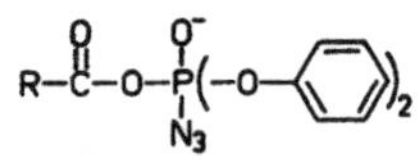	32
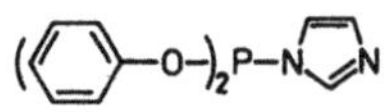	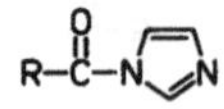	84
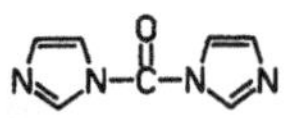	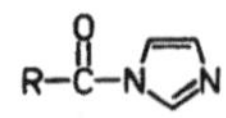	205
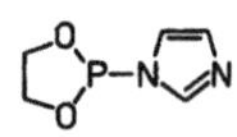	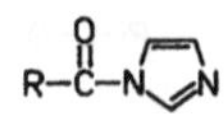	214
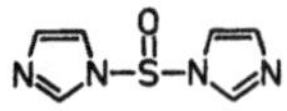	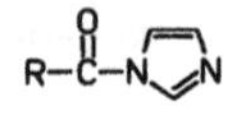	215
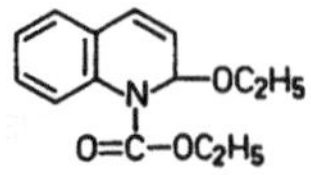	$R{-}C(=O){-}O{-}C(=O){-}OC_2H_5$	195

Table 4 (continued)

Reagent	Reactive Intermediate	Refs.
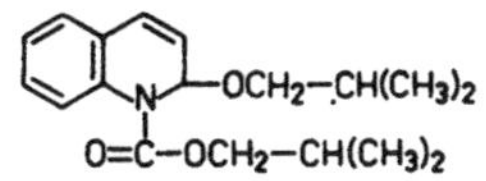	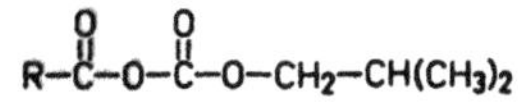	197
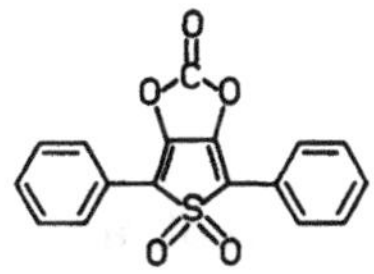	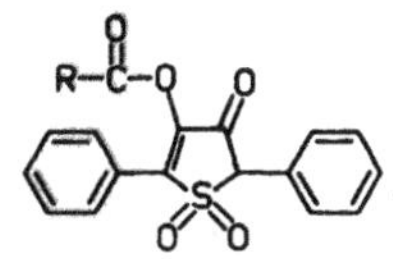	216
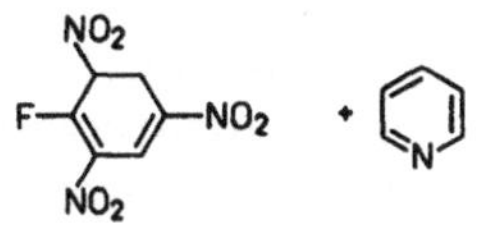	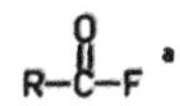 [a]	217
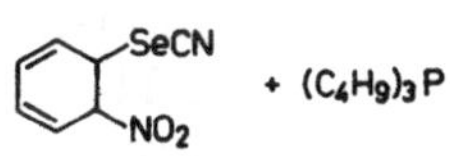	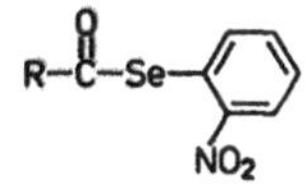	218
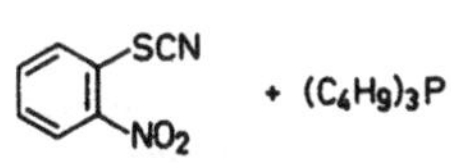	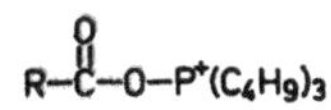	219
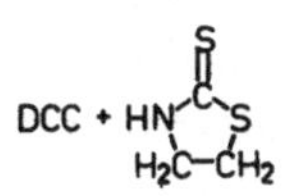	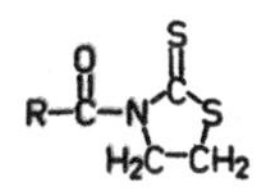	220
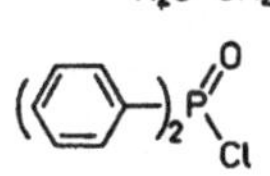	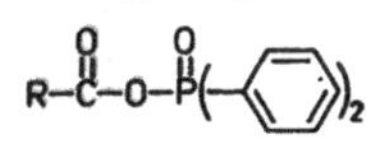	221
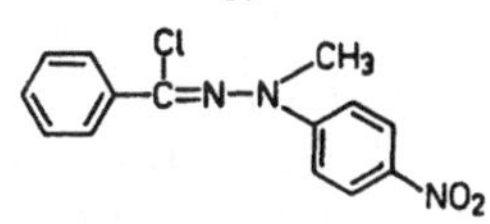	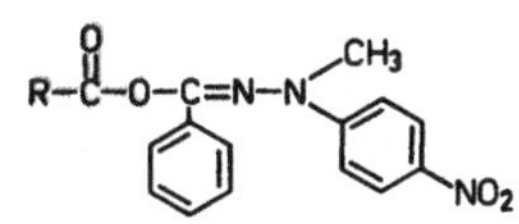	222
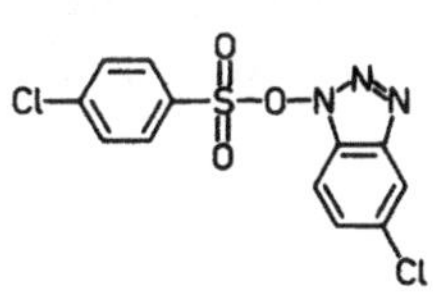	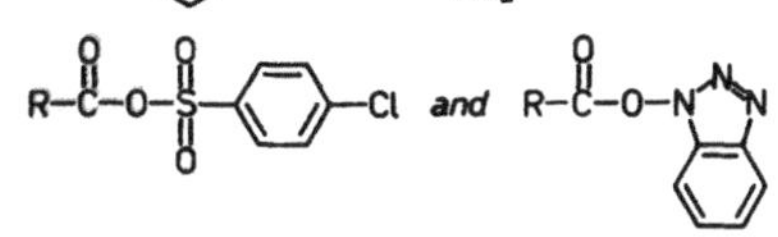	223
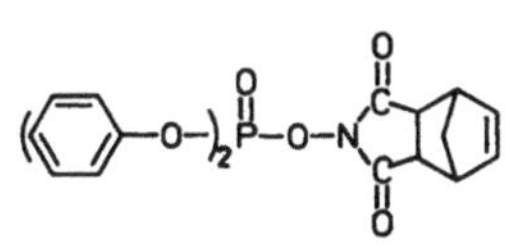	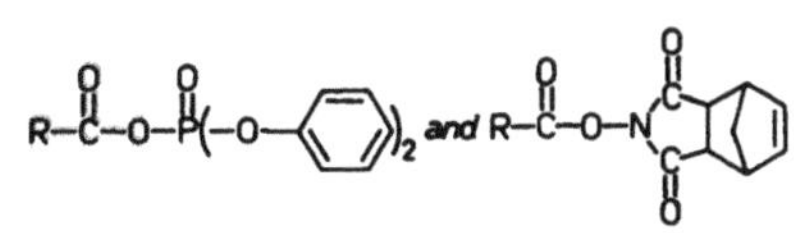	224
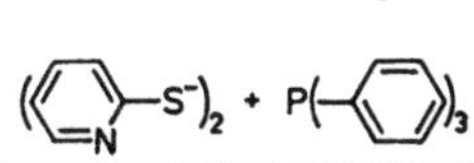	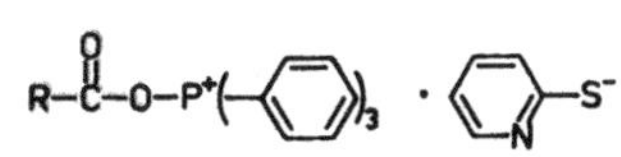	225,226

[a] 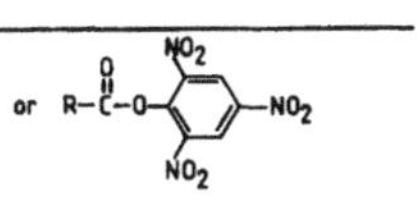

applied, although N-hydroxysuccinimide [203, 204] and 3-hydroxy-3,4-dihydro-1,2,3-benzotriazin-4-one [169] are not less important.

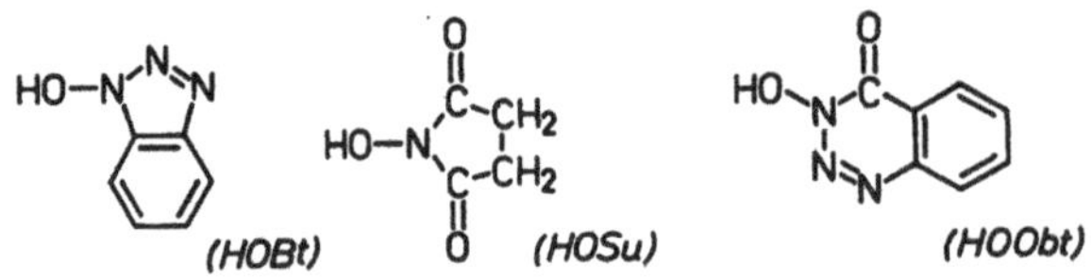

The initial aim in the introduction of these additives, suppression of racemization in the coupling of peptide segments with DCC, will be treated in detail in a later chapter. At this point we wish to stress a more general aspect of the DCC-additive approach. Since the additive, e.g. 1-hydroxybenzotriazole, is usually applied in an amount which is equimolecular with the two components to be coupled, there are two moles of nucleophiles present in the reaction mixture for each mole of carboxyl component or carbodiimide. Therefore, the lifetime of highly reactive intermediates, such as O-acylisoureas, symmetrical anhydrides or azlactones, is considerably reduced. It is particularly noteworthy that the concentration of the additive which acts as a second nucleophile hardly changes during the coupling reaction, because it is continuously regenerated. Hence, there is a significant change in the kinetics of the process. Coupling with DCC involves two or more consecutive reactions, all bimolecular and thus concentration dependent. The second nucleophile, the additive, present in almost constant concentration, will accelerate the entire process, but more importantly it converts the overactivated intermediates of the DCC-reaction to the less reactive esters of 1-hydroxybenzotriazole or N-hydroxysuccinimide. The active ester, formed *in situ*, is less conductive to side reactions such as the rearrangement of the O-acylisourea intermediate to an N-acyl-urea, yet it is sufficiently reactive to ensure satisfactory acylation rates. The ability of additives to provide multiple pathways to the same product is shown in the following scheme:

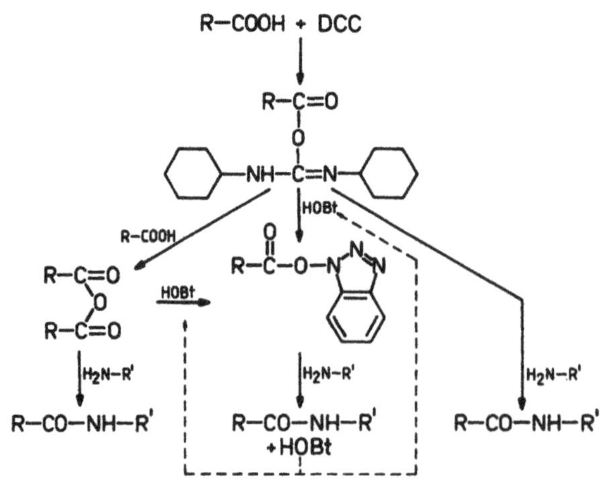

In the introductory part of this chapter, the *principle of excess* was explained and its significance emphasized. While it is possible and even practical to employ the activated derivative of a protected amino acid in excess, the opposite maneuver, the application of the amino-component in excess, is seldom reasonable. The nucleophile is usually a peptide, too valuable to be sacrificed in part, too complex to be recovered from the reaction mixture. The use of additives is a solution for this dilemma. Because of this relationship to the principle of excess, we prefer to designate compounds such as 1-hydroxybenzotriazole or N-hydroxysuccinimide, instead of with the operational term "additives", rather with the more descriptive expression "auxiliary nucleophiles".

The adaptation of auxiliary nucleophiles need not be limited to DCC-mediated coupling reactions. By reducing overactivation, these compounds should improve the performance of other reactive intermediates as well. Also, auxiliary nucleophiles can be generated in the process of activation as shown in the case of the interesting reagent, norborn-5-ene-2,3-dicarboximido diphenyl phosphate [224], which reacts with the carboxyl-component to produce a mixed anhydride with the concomitant liberation of an analog of N-hydroxysuccinimide (or N-hydroxyphthalimide):

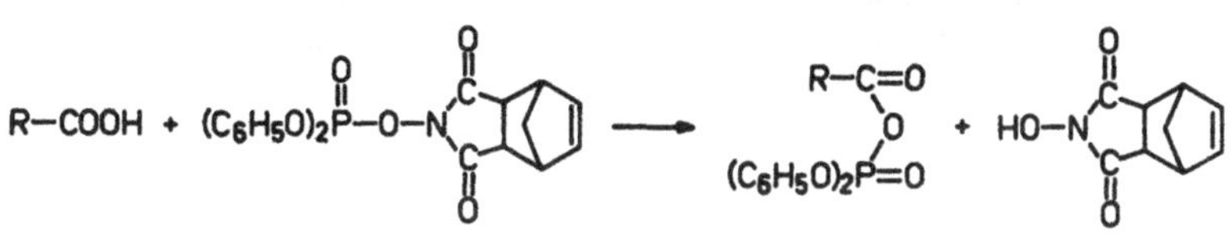

The same principle can be discerned in the application of Itoh's reagent [223] and also in the esterification of protected amino acids with aryl phosphites [126] or aryl sulfites [143].

6. Enzyme-catalyzed Formation of the Peptide Bond

The equilibrium of the reaction

$$R{-}COOH + H_2N{-}R' \rightleftharpoons R{-}CO{-}NH{-}R' + H_2O$$

lies far on the left side; thus the hydrolysis of the peptide bond, but not its synthesis, is a spontaneous process. In the absence of catalysts, the rate of the reaction is extremely low. It is accelerated, however, by astronomical factors, through the intervention of proteolytic enzymes. Such enzymes, as true catalysts, do not affect the equilibrium of the reaction. The latter can be shifted by an increase in the concentration of the reactants or by the removal of a component from the reaction mixture. Hence, the same enzymes which catalyze the hydrolysis of peptide bonds can also be used for the synthesis of peptides. This principle has been applied by Bergmann and Fraenkel-Conrat [227] and also by Bergmann and Fruton [228] and could be used for the resolution of racemic mixtures of amino acids. The proteolytic enzyme papain is added to the mixture of an acylamino acid (DL) and excess aniline in aqueous media. The anilide of the acylamino acid, being insoluble in water, precipitates and this provides the necessary driving force that shifts the equilibrium toward synthesis:

The excess aniline has the same effect. Since the enzyme is specific for L-amino acids, the D-isomer in the starting material remains unchanged.

A more general application of this principle became possible by the recognition of the role of the solvent in Laskowski's laboratory [229]. The addition of organic solvents such as glycerol or acetonitrile, by affecting the dissociation constants of carboxylic acids, also contributes to the right shift of the equilibrium. Through a combination of factors which favor synthesis rather than hydrolysis, practical results can be achieved. The determination of pH regions which are optimal for synthesis but not for hydrolysis further improved the performance of proteolytic enzymes. Thus, papain [230], trypsin [231], and chymotrypsin [232] could be used for the synthesis of selected target peptides. An interesting application of enzyme-catalyzed peptide bond formation is the conversion of pork insulin into human insulin [233–235]. One process [235] is shown in the following scheme:

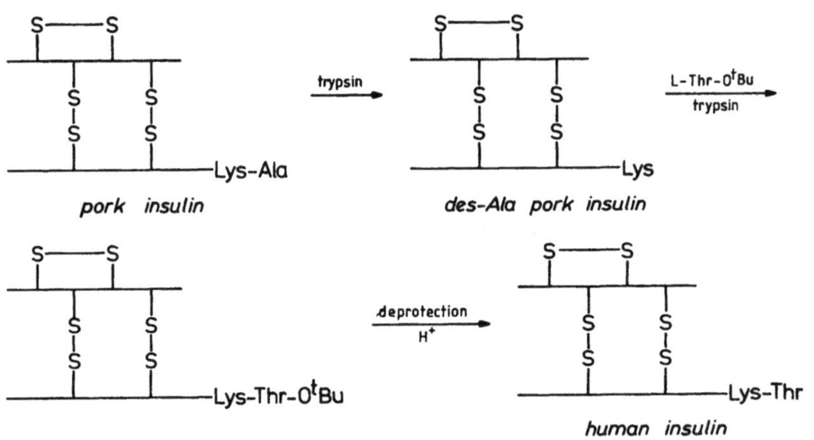

Removal of alanine from the C-terminus of the B-chain by trypsin is fairly selective. Thus, desalanino pork insulin is produced in good yield and can be subjected to the action of trypsin in the presence of a large excess of L-threonine tert. butyl ester. Acidolytic removal of the tert. butyl ester group, e.g. with trifluoroacetic acid, concludes the semi-synthesis of human insulin. The importance of this approach should not be underestimated, but proposals [236, 237] for a systematic synthesis of peptide chains with the help of enzymes have to be regarded as tentative at this time.

7. A Comparison Between Various Methods of Coupling

A common feature of the azide method and of the coupling procedures involving symmetrical or mixed anhydrides is the moderate stability of the reactive intermediates, which can be isolated, but not stored, for longer periods of time. Azides slowly undergo Curtius rearrengement, mixed anhydrides disproportionate to symmetrical anhydrides and even the latter have limited shelf life. In this respect active esters are different. Most of them can be secured in crystalline form and kept intact for years if stored in a refrigerator. The advantage of stability is counterbalanced, of course, by moderate reactivity.

The appeal of coupling reagents is considerable. This is best explained by the simplicity of the execution of coupling reactions, yet the possibility of adding a miraculous reagent to the mixture of the carboxyl- and amino-components to effect coupling also must have some positive psychological influence. Otherwise it is not easy to understand why so many coupling reagents (cf. Table 4) have been proposed in the literature, when most of them merely generate well-known intermediates such as anhydrides or active esters. There is certainly a predilection among practitioners for single-pot procedures for syntheses without the isolation of intermediates. In the synthesis of long chains, where a formidable number of steps may be required, this attitude is understandable and sometimes justified. We would, nevertheless, like to give here some thoughts to the price paid for simplification.

Proceeding from one intermediate to the next without isolation is certainly time saving. One avoids laborious operations and purification steps ranging from simple recrystallization to involved chromatographic procedures or even countercurrent distribution. In the case of intermediates which lack the necessary stability, isolation and attempts at purification are often unrewarding. Acid azides, for instance, are probably more pure if they are used without isolation, since some Curtius rearrangement occurs during handling. In other cases, however, convenience and time saving gained by the omission of isolation of intermediates, might be outweighed by the probability of producing impurities which contaminate the main product of the reaction. Such are, for example, the urea derivatives in azide couplings or the N-acylureas in condensations with DCC. If, in turn, the protected peptide contaminated with such by-products is used in the next step, then the imputities are carried along and will pile up to the point where the final product becomes an intractable mixture or a material from which the target compound must be "fished out" by a series of chromatographic procedures. Such a synthesis cannot serve as proof of structure, since it is not unequivocal. It is also unsuited for the commercial production of peptides needed in medicine, because chromatography on a large scale can be exceedingly complex or prohibitively expensive. Last but not least, if a series of consecutive steps is carried out without isolation of intermediate products, a major deficiency will develop: the loss of analytical information. Both elemental analysis and amino acid analysis are reliable only if they are performed on homogeneous and representative samples. Purification of a small portion of the product for analysis and use of the major portion of the material in the next synthetic step is justified only if the recovery in the purification of the sample is almost quantitative. On the other hand, without analytical control, the investigator loses sight of his products, he works blindfolded.

These considerations motivated the author to advocate [19] the use of active esters as reactive intermediates for the building of peptide chains. In the active ester method, activation and coupling are well separated and the relative stability of protected *and* activated amino acids permits their scrutiny before coupling. Also, the same preparation of the acylating agent can be used in trial runs prior to their application in the final execution of a chain-lengthening step. The relative stability of active esters made it possible that they are available from commercial sources. Hence, the number of steps needed for the synthesis of a complex molecule is not really increased by their preparation. Of course some price has to be paid for these advantages. There are losses in the preparation of active esters and their stability is accompanied by moderate reactivity. The latter shortcoming, however, can be remedied by catalysis of the acylation reaction, e.g. with 1-hydroxybenzotriazole [133].

The arguments supporting the application of active esters are valid mainly for stepwise synthesis, for chain-lengthening through the incorporation of single amino acid residues [19]. In the condensation of peptide segments coupling reagents give the best results. Foremost among them is DCC, best in combination with auxiliary nucleophiles, 1-hydroxybenzotriazole [169], 3-hydroxy-4-keto-3,4-dihydrobenzotriazole [169], or N-hydroxysuccinimide [203, 204]. Good results can be expected in segment condensation also from the use of IIDQ [197]. For the coupling of a peptide with a natural protein or part of it, synthesis catalyzed by proteolytic enzymes holds great promise.

References of Chapter II

1. Rzeszotarska, B., Taschner, E.: Proc. Sixth European Peptide Symposium, Athens, 1963 (Zervas, L. ed.) p. 135, Oxford: Pergamon Press 1966
2. Curtius, Th.: Ber. dtsch. Chem. Ges. *35*, 3226 (1902)
3. Fischer, E.: Ber. dtsch. Chem. Ges. *36*, 2094 (1903)
4. Brenner, M.: in Peptides, Proceedings of the Eighth European Peptide Symposium, Nordwijk, 1966 (Beyerman, H. C., van de Linde, A., Maasen van den Brink, eds.) p. 1, Amsterdam: North Holland Publishing 1967
5. Taschner, E., Rzeszotarska, B., Lubiewska, L.: Chem. Ind. 402 (1967)
6. Heikens, D., Hermans, P. H., van Velden, P. F.: Nature *174*, 1187 (1954)
7. Schwyzer, R., Iselin, B., Feurer, M.: Helv. Chim. Acta *38*, 69 (1955)
8. Taschner, E., Rzeszotarska, B., Kuziel, A.: Acta Chim. Hung. *44*, 67 (1965)
9. Chapman, T. M., Freedman, E. A.: J. Org. Chem. *38*, 3908 (1973)
10. Brenner, M., Zimmermann, J. P., Wehrmüller, Quitt, P., Photaki, I.: Experientia *11*, 397 (1955)
11. Brenner, M., Zimmermann, J. P., Wehrmüller, J., Quitt, P., Hartmann, A., Schneider, W., Beglinger, U.: Helv. Chim. Acta *40*, 1497 (1957)
12. Brenner, M., Hofer, W.: Helv. Chim. Acta *44*, 1794; 1798 (1961)

13. Ugi, I.: Angew. Chem. (Int. Ed. English) *1*, 8 (1962)
14. Ugi, I.: in "The Peptides", Vol. 2, p. 365 (Gross, E., Meienhofer, J., eds.) New York: Academic Press 1980
15. Waki, M., Meienhofer, J.: J. Amer. Chem. Soc. *99*, 6075 (1977)
16. Marquarding, D., Burghard, H., Ugi, I., Urban, R., Klusacek, H.: J. Chem. Res. (M) 0915; (S) p. 82 (1977)
17. Pettit, G. R.: „Synthetic Peptides", Vol. 4, p. 22, Amsterdam: Elsevier 1976
18. Bodanszky, M.: in "Prebiotic and Biochemical Evolution" (Kimball, A. R., Oro, J., eds.) p. 217, Amsterdam: North Holland Publ. 1971
19. Bodanszky, M.: Ann. N.Y. Acad. Sci. *88*, 655 (1960)
20. Merrifield, R. B.: J. Amer. Chem. Soc. *85*, 2149 (1963)
21. Fujii, N., Yajima, H.: J. Chem. Soc. Perkin I, 789; 797; 804; 811; 819; 831 (1981)
22. Sieber, P., Riniker, B., Brugger, M., Kamber, B., Rittel, W.: Helv. Chim. Acta *53*, 2135 (1970)
23. Medzihradszky-Schweiger, H.: Acta Chim. Acad. Sci. Hung. *34*, 213 (1962)
24. Kaufmann, K. D., Raddatz, F., Bauschke, S.: Z. Chem. *15*, 449 (1975)
25. Hofmann, K., Magee, M. Z., Lindenmann, A.: J. Amer. Chem. Soc. *72*, 2814 (1950)
26. Curtius, T., Levy, L.: J. prakt. Chem. *70*, 89 (1904)
27. Cheung, H. T., Blout, E. R.: J. Org. Chem. *30*, 315 (1965)
28. Zahn, H., Schnabel, E.: Liebigs Ann. Chem. *605*, 212 (1957)
29. Guttmann, S., Boissonnas, R. A.: Helv. Chim. Acta *41*, 1852 (1958)
30. Medzihradszky, K., Bruckner, V., Kajtár, M., Löw, M., Bajusz, S., Kisfaludy, L.: Acta Chim. Acad. Sci. Hung. *30*, 105 (1962)
31. Honzl, J., Rudinger, J.: Collect. Czech. Chem. Commun. *26*, 2333 (1961)
32. Shiori, T., Ninomiya, K., Yamada, S.: J. Amer. Chem. Soc. *94*, 6203 (1972)
33. Curtius, T.: Ber. dtsch. Chem. Ges. *23*, 3023 (1890)
34. Inouye, K., Watanabe, K.: J. Chem. Soc. Perkin I, 1911 (1977)
35. Schnabel, E.: Liebigs Ann. Chem. *659*, 168 (1962)
36. Meienhofer, J.: The Peptides, Vol. I, p. 197, New York: Academic Press 1979
37. Klausner, Y., Bodanszky, M.: Synthesis *8*, 549 (1974)
38. Curtius, T.: J. prakt. Chem. *24*, 239 (1881)
39. Chantrenne, H.: Nature *160*, 603 (1947)
40. Chantrenne, H.: Biochim. Biophys. Acta *2*, 286 (1948)
41. Chantrenne, H.: Nature *164*, 576 (1949)
42. Chantrenne, H.: Biochim. Biophys. Acta *4*, 484 (1950)
43. Sheehan, J. C., Frank, V. S.: J. Amer. Chem. Soc. *72*, 1312 (1950)
44. Wieland, T., Kern, W., Sehring, R.: Liebigs Ann. Chem. *569*, 117 (1950); cf. also Wieland, T., Sehring, R.: *ibid. 569*, 122 (1950)
45. Vaughan, J. R., Jr., Osato, R. L.: J. Amer. Chem. Soc. *73*, 5553 (1951)
46. Zaoral, M.: Coll. Czech. Chem. Commun. *27*, 1273 (1962)
47. du Vigneaud, V., Ressler, C., Swan, J. M., Roberts, C. W., Katsoyannis, P. G., Gordon, S.: J. Amer. Chem. Soc. *75*, 4879 (1953)
48. Galpin, I. J., Handa, B. K., Hudson, D., Jackson, A. G., Kenner, G. W., Ohlsen, S. R., Ramage, R., Singh, B., Tyson, R. G.: Peptides 1976 (Loffet, A., ed.) p. 247. Editions de l'Université de Bruxelles, Belgium (1976)
49. Boissonnas, R. A.: Helv. Chem. Acta *34*, 874 (1951)
50. Vaughan, J. R., Jr.: J. Amer. Chem. Soc. *73*, 3547 (1951); Vaughan, J. R., Jr., Osato, R. L.: *ibid. 74*, 676 (1952)
51. Wieland, T., Bernhard, H.: Liebigs Ann. Chem. *572*, 190 (1951)
52. Bodanszky, M., Tolle, J. C.: Int. J. Peptide Protein Research *10*, 380 (1977)

53. Sarges, R., Witkop, B.: J. Amer. Chem. Soc. *87*, 2020 (1965)
54. Tilak, M.: Tetrahedron Lett. 849 (1970)
55. van Zon, A., Beyerman, H. C.: Helv. Chim. Acta *56*, 1729 (1973); *ibid.*, *59*, 1112 (1976)
56. Anderson, G. W., Blodinger, J., Young, R. W., Welcher, A. D.: J. Amer. Chem. Soc. *74*, 5304 (1952)
57. Anderson, G. W., Young, R. W.: J. Amer. Chem. Soc. *74*, 5307 (1952)
58. Anderson, G. W., Blodinger, J., Welcher, A. D.: J. Amer. Chem. Soc. *74*, 5309 (1952)
59. Albertson, N. F.: Org. Reactions, Vol. 12, p. 157. New York: Wiley 1962
60. Tarbell, D. S.: Acct. Chem. Res. *2*, 296 (1969)
61. Meienhofer, J.: The Peptides (Gross, E., Meienhofer, J., eds.) Vol. 1, p. 263. New York: Academic Press 1979
62. E. Wünsch: Synthese von Peptiden. Methoden der Organischen Chemie. Vol. XV2, Houben-Weyl). Stuttgart: Thieme 1974
63. Süs, O.: Liebigs Ann. Chem. *572*, 96 (1951)
64. Young, R. W., Wood, K. H., Joyce, R. J., Anderson, G. W.: J. Amer. Chem. Soc. *78*, 2126 (1956)
65. Goldschmidt, S., Obermeier, F.: Liebigs Ann. Chem. *588*, 24 (1954)
66. Anderson, G. W., Welcher, A. D., Young, R. W.: J. Amer. Chem. Soc. *73*, 501 (1951)
67. Crofts, P. C., Markes, J. H. H., Rydon, H. N.: J. Chem. Soc., 4250 (1958)
68. Crofts, P. C., Markes, J. H. H., Rydon, H. N.: J. Chem. Soc., 3610 (1959)
69. Vaughan, J. R., Jr.: J. Amer. Chem. Soc. *73*, 1389 (1951)
70. Wieland, T., Heinke, B.: Liebigs Ann. Chem. *599*, 70 (1956)
71. Wieland, T., Heinke, B.: Liebigs Ann. Chem. *615*, 184 (1958)
72. Martinez, J., Winternitz, F.: Bull. Soc. Chim. France, 4707 (1972); cf. also Das, K. C., Lin, Y., Weinstein, B.: Experientia *25*, 1238 (1969)
73. Bates, A. J., Galpin, I. J., Hallett, A., Hudson, D., Kenner, G. W., Ramage, R.: Helv. Chim. Acta *58*, 688 (1975)
74. Gawne, G., Kenner, G. W., Sheppard, R. C.: J. Amer. Chem. Soc. *91*, 5669 (1969)
75. Bates, A. J., Kenner, G. W., Ramage, R., Sheppard, R. C.: in "Peptides 1972", p. 124 (Hanson, H., Jakubke, D. eds.). Amsterdam: North Holland Publ. 1973
76. Castro, B., Dormoy, J. R.: Bull. Soc. Chim. France, 3034 (1971)
77. Castro, B., Dormoy, J. R.: Tetrahedron Lett., 3243 (1973)
78. Barstow, L. E., Hruby, V. J.: J. Org. Chem. *36*, 1305 (1971)
79. Wieland, T., Seelinger, A.: Chem. Ber. *104*, 3992 (1971)
80. Yamada, S., Takeuchi, Y.: Tetrahedron Lett., 3595 (1971)
81. Castro, B., Dormoy, J. R.: Tetrahedron Lett., 4747 (1972)
82. Castro, B., Dormoy, J. R.: Bull. Soc. Chim. France, 3359 (1973)
83. Appel, R., Willms, L.: Chem. Ber. *112*, 1057, 1064 (1979)
84. Mitin, Y. V., Glinskaya, O. V.: Tetrahedron Lett., 5267 (1969)
85. Yamazaki, N., Higashi, F.: Tetrahedron Lett., 5047 (1972)
86. Kenner, G. W.: Chem. Ind., 15 (1951)
87. Kenner, G. W., Stedman, R. J.: J. Chem. Soc., 2069 (1952)
88. Clayton, D. W., Farrington, J. A., Kenner, G. W., Turner, J. M.: J. Chem. Soc., 1398 (1957)
89. Theodoropoulos, D., Gazopoulos, J.: J. Org. Chem. *27*, 2091 (1962)
90. Cronyn, M. W., Jiu, J.: J. Amer. Chem. Soc. *74*, 4726 (1952)
91. Sheehan, J. C., Johnson, D. A.: J. Am. Chem. Soc. *74*, 4726 (1952)

92. Leuchs, H.: Ber. dtsch. Chem. Ges. *39*, 857 (1906); Leuchs, H., Geiger, W.: *ibid.* *41*, 1721 (1908)
93. Hunt, M., du Vigneaud, V.: J. Biol. Chem. *124*, 699 (1938)
94. Bailey, J. L.: Nature *164*, 889; J. Chem. Soc., 3461 (1950)
95. Langebeck, W., Kresse, P.: J. prakt. Chem. *2*, 261 (1955)
96. Bartlett, P. D., Jones, R. H.: J. Amer. Chem. Soc. *79*, 2153 (1957); Bartlett, P. D., Dittmer, D. C.: *ibid.* *79*, 2159 (1957)
97. Denkewalter, R. G., Schwam, H., Strachan, R. G., Beesley, T. E., Veber, D. F., Schoenewaldt, E. F., Barkemeyer, H., Paleveda, W. J., Jr., Jacob, T. A., Hirschmann, R.: J. Amer. Chem. Soc. *88*, 3163 (1966)
98. Le Quesne, W. J., Young, G. T.: J. Chem. Soc., 1954, 1959 (1950); 594 (1952)
99. Schröder, E., Klieger, E.: Liebigs Ann. Chem. *673*, 208 (1964)
100. King, F. E., Kidd, D. A. A.: J. Chem. Soc., 3315 (1949); *ibid.* 2976 (1951)
101. Klieger, E., Gibian, H.: Liebigs Ann. Chem. *655*, 195 (1962)
102. Weygand, F., Huber, P., Weiss, K.: Z. Naturf. *22b*, 1084 (1967)
103. Weygand, F., di Bello, C.: Z. Naturf. *24b*, 314 (1969)
104. Buyle, R., Viehe, H. G.: Angew. Chem. *76*, 572 (1964)
105. Wieland, T., Birr, C., Flor, F.: Angew. Chem. *83*, 333 (1971)
106. Khorana, H. G.: Chem. Ind., 1087 (1955)
107. Sheehan, J. C., Hlavka, J. J.: J. Org. Chem. *21*, 439 (1956)
108. Sheehan, J. C., Cruickshank, P. A., Boshart, G. L.: J. Org. Chem. *26*, 2525 (1961)
109. Chen, F. M. F., Kukoda, K., Benoiton, N. L.: Synthesis, 928 (1978)
110. Schüssler, H., Zahn, H.: Chem. Ber. *95*, 1076 (1962)
111. Schwyzer, R., Feurer, M., Iselin, B., Kägi, H.: Helv. Chim. Acta *38*, 80 (1955)
112. Iselin, B., Feurer, M., Schwyzer, R.: Helv. Chim. Acta *38*, 1508 (1955)
113. Gordon, M., Miller, J. G., Day, A. R.: J. Amer. Chem. Soc. *70*, 1946 (1948)
114. Gordon, M., Miller, J. G., Day, A. R.: J. Amer. Chem. Soc. *71*, 1245 (1949)
115. Wieland, T., Schäfer, W., Bokelmann, E.: Liebigs Ann. Chem. *573*, 99 (1951)
116. Lynen, F., Reichert, E.: Angew. Chem. *63*, 47 (1951)
117. Bodanszky, M.: Nature *175*, 685 (1955)
118. Bodanszky, M., Bath, R. J.: Chem. Commun., 1259 (1969)
119. Bodanszky, M., Bath, R. J., Chang, A., Fink, M. L., Funk, K. W., Greenwald, S. M., Klausner, Y. S.: Chemistry and Biology of Peptides (Meienhofer, J., ed.) p. 203. Ann Arbor, Mich.: Ann Arbor Sci. Publ. 1972
120. Bodanszky, M., Funk, K. W., Fink, M. L.: J. Org. Chem. *38*, 3565 (1973)
121. Bodanszky, M., Kondo, M., Yang-Lin, C., Sigler, G. F.: J. Org. Chem. *39*, 444 (1974)
122. Bodanszky, M., Szelke, M., Tömörkény, E., Weisz, E.: Chem. Ind., 1517 (1955); Acta Chim. Acad. Sci. Hung. *11*, 179 (1957)
123. Bodanszky, M.: Acta Chim. Acad. Sci. Hung. *10*, 335 (1957)
124. Bodanszky, M., du Vigneaud, V.: Nature *183*, 1324 (1959)
125. Bodanszky, M., du Vigneaud, V.: J. Amer. Chem. Soc. *81*, 5688 (1959)
126. Farrington, J. A., Kenner, G. W., Turner, J. M.: Chem. Ind., 601 (1955); J. Chem. Soc., 1407 (1957)
127. Kupryszewski, G.: Rocz. Chem. *35*, 595 (1961); Chem. Abstr. *55*, 27121 (1961)
128. Kupryszewski, G., Kaczmarek, M.: Rocz. Chem. *35*, 931 (1961); Chem. Abstr. *56*, 6081 (1962)
129. Kovács, J., Kisfaludy, L., Ceprini, M. Q.: J. Amer. Chem. Soc. *89*, 183 (1967)
130. Kisfaludy, L., Löw, M., Nyéki, O., Szirtes, R., Schön, I.: Liebigs Ann. Chem., 1421 (1973)

131. Kisfaludy, L., Roberts, J. E., Johnson, R. H., Mayers, G. L., Kovács, J.: J. Org. Chem. *35*, 3563 (1970)
132. Pless, J., Boissonnas, R. A.: Helv. Chim. Acta *46*, 1609 (1963)
133. König, W., Geiger, R.: Chem. Ber. *106*, 3626 (1973)
134. Mazur, R. H.: J. Org. Chem. *28*, 2498 (1963)
135. Wieland, T., Vogeler, K.: Liebigs Ann. Chem. *680*, 125 (1964)
136. Wieland, T., Determann, H., Kahle, W.: Angew. Chem. (Int. Ed.) *2*, 154 (1903)
137. Beyerman, H. C., Maassen van den Brink, W.: Proc. Chem. Soc. (London), 266 (1963)
138. Jones, J. H., Young, G. T.: J. Chem. Soc. C, 436 (1968)
139. Trudelle, Y.: Chem. Commun., 639 (1971)
140. Johnson, B. J., Jacobs, P. M.: Chem. Commun., 73 (1968)
141. Johnson, B. J., Jacobs, P. M.: J. Org. Chem. *33*, 4524 (1968)
142. Johnson, B. J., Ruettinger, T. A.: J. Org. Chem. *35*, 255 (1970)
143. Iselin, B., Rittel, W., Sieber, P., Schwyzer, R.: Helv. Chim. Acta *40*, 373 (1957)
144. Wieland, T., Lewalter, J., Birr, C.: Liebigs Ann. Chem. *740*, 31 (1970)
145. Wieland, T., Birr, C., Fleckenstein, P.: Liebigs Ann. Chem. *756*, 14 (1972)
146. Sheehan, J. C., Daves, G. D.: J. Org. Chem. *29*, 2006 (1964)
147. Treiber, R. L.: Experientia *29*, 1335 (1973)
148. Kenner, G. W., Seely, J. H.: J. Amer. Chem. Soc. *94*, 3259 (1972)
149. Martynov, F. V., Samarcev, M. A.: Z. Obsc. Chim. *39*, 940 (1964); C. A. *71*, 405—16 (1969)
150. Bankowski, K., Drabarek, S.: Roczniki Chem. *45*, 1205 (1971)
151. Bankowski, K., Drabarek, S.: Roczniki Chem. *46*, 607 (1972)
152. Stewart, F. H. C.: Austral. J. Chem. *21*, 477; 1639 (1968)
153. Kazmierczak, R., Kupryszewski, G.: Roczniki Chem. *37*, 659 (1963)
154. Jakubke, H. D.: Z. Naturforsch. *20b*, 273 (1965)
155. Taschner, E., Rzeszotarska, B., Lubiewska, L.: Liebigs Ann. Chem. *690*, 177 (1965)
156. Dutta, A. S., Morley, J. S.: J. Chem. Soc. C, 2896 (1971)
157. Mitin, Y. V., Nadezhdina, L. B.: Zh. Org. Khim. *4*, 1181 (1968); Chem. Abstr. *69*, 87227 (1968)
158. Jakubke, H. D.: Z. Chem. *5*, 453 (1965)
159. Lloyd, K., Young, G. T.: Chem. Commun., 1400 (1968; J. Chem. Soc. C, 2890 (1971)
160. Jakubke, H. D.: Z. Chem. *3*, 65 (1963)
161. Jakubke, H. D.: Chem. Ber. *97*, 2816 (1964); Liebigs Ann. Chem. *682*, 244 (1965)
162. Klausner, Y. S., Neiri, T. H., Schneider, E.: in "Peptides. Proc. Fifth Amer. Peptide Symp." (Goodman, M., Meienhofer, J., eds.), p. 536. New York: Wiley 1977
163. Losse, G., Hoffmann, K. H., Hetzer, G.: Liebigs Ann. Chem. *684*, 236 (1965)
164. Weygand, F., Steglich, W.: Angew. Chem. *73*, 757 (1961)
165. Nefkens, G. H. L., Tesser, G. I.: J. Amer. Chem. Soc. *83*, 1263 (1961)
166. Anderson, G. W., Zimmermann, J. E., Callahan, F. M.: J. Amer. Chem. Soc. *85*, 3039 (1963); *ibid.* *86*, 1839 (1964)
167. Bittner, S., Knobler, Y., Frankel, M.: Tetrahedron Lett., 95 (1965)
168. Beaumont, S. M., Hanford, B. O., Jones, J. H., Young, G. T., Chem. Commun., 53 (1965); cf. also Hanford, B. O., Jones, J. H., Young, G. T., Johnson, T. N. F.: J. Chem. Soc., 6814 (1965)
169. König, W., Geiger, R.: Chem. Ber. *103*, 788, 2024 (1970)
170. Sheehan, J. C., Hess, G. P.: J. Amer. Chem. Soc. *77*, 1067 (1955)

171. Savrda, J.: J. Org. Chem. *42*, 3199 (1977)
172. Jeschkeit, H.: Z. Chem. *8*, 20 (1968)
173. Jeschkeit, H.: Z. Chem. *9*, 266 (1969)
174. Lubiewska-Nakonieczna, L., Rzeszotarska, B., Taschner, E.: Justus Liebigs Ann. Chem. *741*, 157 (1970)
175. Govindachari, T. R., Nagarajan, K., Rajappa, S.: Tetrahedron *22*, 3367 (1966)
176. Rajappa, S., Nagarajan, K., Iyer, V. S.: Tetrahedron *22*, 4805 (1967)
177. Paquette, L. A.: J. Amer. Chem. Soc. *87*, 5186 (1965)
178. Sarantakis, D., Sutherland, J. K., Tortorella, C., Tortorella, V.: J. Chem. Soc. C, 72 (1968)
179. Taylor, E. C., Kinzle, F., McKillop, A.: J. Org. Chem. *35*, 1672 (1970)
180. Rothe, M., Kunitz, F. W.: Liebigs Ann. Chem. *609*, 88 (1957)
181. Elliott, D. F., Russell, D. W.: Biochem. J. *66*, 49P (1957)
182. Arens, J. F.: Rec. Trav. Chim. Pay-Bas *74*, 769 (1955)
183. Bodanszky, M., Birkhimer, C. A.: Chem. Ind. (London), 1620 (1962)
184. Iselin, B., Schwyzer, R.: Helv. Chim. Acta *43*, 1760 (1960)
185. Sakakibara, S., Inukai, N.: Bull. Chem. Soc. Jpn. *37*, 1231 (1964); *ibid. 38*, 1979 (1965)
186. Fletcher, G. A., Löw, M., Young, G. T.: J. Chem. Soc. Perkin Trans. I, 1162 (1973)
187. Yamazaki, N., Higashi, F., Niwano, M.: Tetrahedron *30*, 1319 (1974)
188. Martinez, J., Winternitz, F.: Tetrahedron Lett., 2631 (1975)
189. Appel, R., Gläsel, U.: Chem. Ber. *113*, 3511 (1980)
190. Muramatsu, I., Hirabayashi, T., Hagitani, A.: Nippon Kagaku Zasshi *84*, 855 (1963); Chem. Abstr. *60*, 12100c (1964)
191. Gais, H. F.: Angew. Chem. *17*, 597 (1978)
192. Neuenschwander, M., Lienhard, U., Fahrni, H. P., Jurni, B.: Helv. Chim. Acta *61*, 2428 (1978); Neuenschwander, M., Fahrni, H. P., Lienhard, U.: *ibid. 61*, 2437 (1978)
193. Benoiton, N. L., Chen, F. M. F.: Can. J. Chem. *59*, 384 (1981)
194. Bates, H. S., Jones, J. H., Ramage, W. I., Witty, M. J.: in "Peptides 1980" (Brunfeldt, K., ed.), p. 185. Copenhagen: Sriptor 1981
195. Belleau, B., Malek, G.: J. Amer. Chem. Soc. *90*, 1651 (1968)
196. Bodanszky, M., Tolle, J. C., Gardner, J. D., Walker, M. D., Mutt, V.: Int. J. Peptide Protein Res. *16*, 402 (1980)
197. Kiso, Y., Kai, Y., Yajima, H.: Chem. Pharm. Bull. *21*, 2507 (1973)
198. Bodanszky, M., Conklin, L. E.: J. Chem. Soc. Chem. Commun., 773 (1967)
199. Losse, G., Demuth, E.: Chem. Ber. *94*, 1762 (1961)
200. Stevens, C. L., Munk, M. E.: J. Amer. Chem. Soc. *80*, 4065, 4069 (1958)
201. Woodward, R. B., Olofson, R. A.: J. Amer. Chem. Soc. *83*, 1007 (1961)
202. Woodward, R. B., Olofson, R. A., Mayer, H.: J. Amer. Chem. Soc. *83*, 1010 (1961)
203. Weygand, F., Hoffmann, D., Wünsch, E.: Z. Naturforsch. *21b*, 426 (1966)
204. Wünsch, E., Drees, F.: Chem. Ber. *99*, 110 (1966)
205. Staab, H. A.: Liebigs Ann. Chem. *609*, 75 (1957); cf. also Staab, H. A., Wendel, K.: Chem. Ber. *96*, 3374 (1963)
206. Anderson, G. W., Paul, R.: J. Amer. Chem. Soc. *80*, 4423 (1958); *ibid. 82*, 4596 (1960)
207. Bodanszky, M., Sheehan, J. T.: Chem. Ind., 1597 (1966)
208. Mukaiyama, T., Ueki, M., Maruyama, H., Matsueda, R.: J. Amer. Chem. Soc. *90*, 4490 (1968); *ibid. 91*, 1554 (1969)
209. Klausner, Y. S., Bodanszky, M.: Synthesis, 453 (1972)

210. Panneman, H. J., Marx, A. F., Arens, J. F.: Rec. Trav. Chim. Pays-Bas *78*, 487 (1959)
211. Heslinga, L., Arens, J. F.: Rec. Trav. Chim. Pays-Bas *76*, 982 (1957)
212. Kemp, D. S., Chien, S. W.: J. Amer. Chem. Soc. *89*, 2743 (1967)
213. Olofson, R. A., Marino, Y. L.: Tetrahedron *26*, 1779 (1970)
214. Anderson, G. W.: Ann. N.Y. Acad. Sci. *88*, 676 (1960)
215. Staab, H. A., Wendel, K.: Chem. Ber. *93*, 2902 (1960); cf. also Staab, H. A.: Angew. Chem. *74*, 407 (1962)
216. Hollitzer, O., Seewald, A., Steglich, W.: Angew. Chem. *15*, 444 (1976)
217. Kinoshita, H., Inomata, K., Miyano, O., Kotake, H.: Bull. Chem. Soc. Jpn. *52*, 2619 (1979)
218. Grieco, P. A., Yokoyama, Y., Williams, E.: J. Org. Chem. *43*, 1283 (1978)
219. Grieco, P. A., Clark, D. S., Withers, G. P.: J. Org. Chem. *44*, 2945 (1979)
220. Yajima, H., Akaji, K., Hirota, Y., Fujii, N.: Chem. Pharm. Bull. *28*, 3140 (1980)
221. Jackson, A. G., Kenner, G. W., Moore, G. A., Ramage, R., Thorpe, W. D.: Tetrahedron Lett., 3627 (1976)
222. Hegarty, A. F., McCarthy, D. G.: J. Amer. Chem. Soc. *102*, 4537 (1980)
223. Itoh, M., Nojima, H., Notani, J., Hagiwara, D., Takai, K.: Tetrahedron Lett., 3089 (1974)
224. Kiso, Y., Miyazaki, T., Satomi, M., Hiraiwa, H., Akita, T.: J. Chem. Soc. Chem. Commun., 1029 (1980)
225. Mukaiyama, T., Matsueda, R., Suzuki, M.: Tetrahedron Lett., 1901 (1970)
226. Mukaiyama, T., Matsueda, R., Maruyama, H.: Bull. Chem. Soc. Jpn. *43*, 1271 (1970)
227. Bergmann, M., Fraenkel-Conrat, H.: J. Biol. Chem. *119*, 707 (1937)
228. Bergmann, M., Fruton, J. S.: J. Biol. Chem. *124*, 321 (1938)
229. Homandberg, G. A., Mattis, J. A., Laskowski, M., Jr.: Biochemistry *17*, 5220 (1978)
230. Isowa, Y., Ohmori, M., Ichikawa, T., Kurita, H., Sato, M., Mori, K.: Bull. Chem. Soc. Jpn. *50*, 2762 (1977)
231. Oka, T., Morihara, K.: J. Biochem. *84*, 1277 (1978)
232. Morihara, K., Oka, T.: Biochem. J. *163*, 531 (1977)
233. Bodanszky, M., Fried, J.: U.S. Pat. 3,276,961 (1966)
234. Inouye, K., Watanabe, K., Morihara, K., Tochino, Y., Kanaka, T., Emura, J., Sakakibara, S.: J. Amer. Chem. Soc. *101*, 751 (1979)
235. Morihara, K., Oka, T., Tsuzuki, H.: Nature *280*, 412 (1979)
236. Breddam, K., Widmer, F., Johansen, J. T.: Carlsberg Res. Commun. *45*, 237 (1980)
237. Takai, H., Sakato, K., Nakamizo, N., Isowa, Y.: in "Peptide Chemistry 1980" (Okawa, K., ed.), p. 213. Protein Research Foundation, Osaka (1981)

III. Reversible Blocking of Amino and Carboxyl Groups

A. General Aspects

1. The Need for Protecting Groups

In the preparation of even the smallest peptide, it becomes obvious that certain functional groups must be blocked. If amino acid $H_2N{-}CHR{-}COOH$ (A) has to be coupled with amino acid $H_2N{-}CHR'{-}COOH$ (B) to produce the dipeptide $H_2N{-}CHR{-}CO{-}NH{-}CHR'{-}COOH$ (AB), then, in order to acylate the amino group of the amino-component (B), we have to activate the carboxyl group of the carboxyl-component (A). The activated carboxyl-component, $H_2N{-}CHR{-}CO{-}X$, can acylate, however, not only the amino-component (B), but also some still unreacted molecules of A, to yield, instead of the desired compound AB, rather a derivative of AA:

$$H_2N{-}CHR{-}CO{-}X + H_2N{-}CHR{-}CO{-}X \longrightarrow H_2N{-}CHR{-}CO{-}NH{-}CHR{-}CO{-}X$$

In addition to such a (still reactive) derivative of AA, formation of the cyclic dipeptide ⌞AA⌟ and peptides with sequences AAB, AAAB, etc. can also be expected. Thus, for an unequivocal course of the coupling reaction, it is necessary that only a single nucleophile, the amino group of B, should be available for acylation. This, in turn, requires the masking of the amino group of the carboxyl-component. Similarly, the amino groups in the side chains of lysine residues, if present in either component, must be blocked, otherwise a branching of the chain will occur:

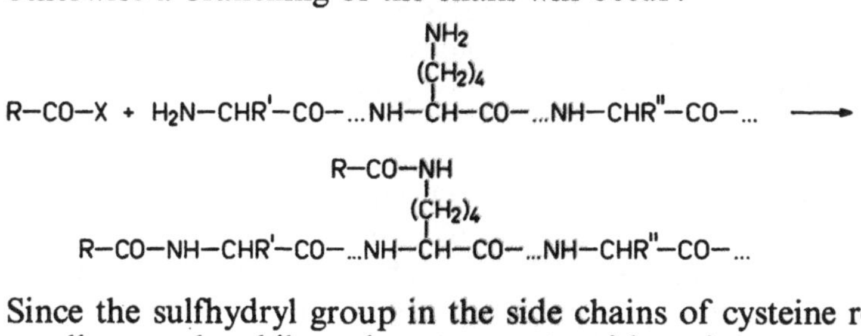

Since the sulfhydryl group in the side chains of cysteine residues is also an excellent nucleophile and can compete with amino groups for the acylating agent, its protection is mandatory. Masking of the SH-group also prevents its oxidation to the disulfide, a reaction which occurs even on exposure to air. Carboxyl groups are often left unprotected, although this can be the source of serious complications. The latter will be discussed in a separate chapter dedicated to side reactions.

2. Minimal Versus Global Protection

Protection of the functional groups which occur in the side chains of amino acids is, at least to some extent, optional. In fact, there are two almost diagonally opposed opinions in this matter. According to one view, the unequivocal execution of a synthesis requires *global protection*, because some participation of the various functional groups in coupling reactions or in the steps needed for the introduction or removal of protecting groups cannot be completely excluded. The alternative possibility, however, synthesis with *minimal protection*, also has numerous advocates, who can point to the better solubility of only partially protected intermediates in solvents used in peptide synthesis and to the problems created by the frequently experienced insolubility of fully protected peptides in the same solvents. It is often practical to incorporate serine, threonine and tyrosine residues without protecting their side chain hydroxyls, since the latter will suffer acylation only in the presence of bases or when overactivated acylating agents (cf. previous chapter) are used in the coupling reaction. The thioether in methionine residues, the imidazole in histidine and the indole in tryptophan are sufficiently inert to permit their presence in unmasked form. The guanidino group in arginine will be acylated only under extreme conditions [1] which generally do not occur in peptide synthesis. On the other hand, removal of protecting groups by acidolysis is accompanied by the formation of alkylating agents and the thioether in methionine, the indole of tryptophan and the phenol in tyrosine side chains are all quite sensitive to alkylation. Also, even if the unprotected guanidinine in arginine remains protonated under most conditions and does not act as a nucleophile, its basic character should not be forgotten: one has to pay attention to the counterion attached to it. An unsubstituted imidazole in the side chain of histidine can catalyze O-acylation of hydroxy-amino acid residues [2]. Thus, there are several good reasons for caution in the application of the minimal protection principle. Synthesis with minimal protection requires mild acylating agents and considerable care in the selection of the methods of protection and deprotection. It is understandable, therefore, that a compromise between the two principles is practiced in many syntheses. For instance, in addition to the mandatory blocking of amino and sulfhydryl groups, some investigators protect the hydroxyl group of serine and tyrosine but leave the secondary alcoholic hydroxyl in threonine side chains unprotected, since it is hindered and not readily acylated. Many practitioners dispense with the masking of methionine, tryptophan or arginine side chains and the imidazole in histidine is left without blocking in numerous syntheses. In fact, only a few peptide chemists insist on global protection to the extent of protecting the carboxamide groups in asparagine and glutamine. On the other hand, selective activation of the α-carboxyl groups of aspartic acid and glutamic acid renders some kind of blocking of the side chain carboxyls unavoidable.

The problems of protection are limited by the fortunate circumstance that of the twenty amino acid constituents of proteins seven have no functional groups in their side chains. Yet, the decision between minimal and global protection and the selection of the most appropriate blocking group remain important steps in the planning of the synthesis of a peptide. Because of the individuality of the target compounds, it is impossible to suggest general guidelines which could help in such considerations. The final choice must depend primarily upon the sequence of the peptide to be synthesized, but also upon its length, the amounts needed and the criteria of purity.

3. Easily Removable Protecting Groups and Methods Used for Their Removal

The most obvious masking of the amino function by acetylation or benzoylation cannot be applied in peptide synthesis because the acid or base catalyzed hydrolysis needed for the removal of an N-acetyl or N-benzoyl group would affect the newly formed peptide bonds as well:

$$CH_3{-}CO{-}NH{-}CHR{-}CO{-}NH{-}CHR'{-}CO{-}NH{-}CHR''{-}CO\ldots \xrightarrow[(H^+ \text{ or } OH^-)]{HOH}$$
$$CH_3COOH + H_2N{-}CHR{-}COOH + H_2N{-}CHR'{-}COOH + H_2N{-}CHR''{-}COOH + \ldots$$

Protection of carboxyl groups is less demanding in this respect. Esters, the derivatives of carboxylic acids which first come to mind, can be hydrolyzed, e.g. by alkali, under conditions which are mild enough to leave the peptide bonds intact:

$$\ldots{-}NH{-}CHR{-}CO{-}NH{-}CHR'{-}CO{-}NH{-}CHR''{-}CO{-}OC_2H_5 \xrightarrow[(HO^-)]{HOH}$$
$$\ldots{-}NH{-}CHR{-}CO{-}NH{-}CHR'{-}CO{-}NH{-}CHR''{-}COOH + C_2H_5OH$$

It was logical, therefore, to transform [3] the free amino group of an amino acid to a substituted urethane, which, in turn, is hydrolyzed to yield an unstable carbamoic acid. Decarboxylation of the latter regenerates the amine:

$$C_2H_5O{-}CO{-}NH{-}CHR{-}CO{-}\ldots \xrightarrow{HOH} C_2H_5OH + HO{-}CO{-}NH{-}CHR{-}CO{-}\ldots$$
$$\downarrow$$
$$CO_2 + H_2N{-}CHR{-}CO{-}\ldots$$

Hence, masking of amino groups in the form of urethanes seemed to be a promising approach.

Unfortunately, the first attempts in this direction failed to provide the expected results. The carbonyl group in urethanes is flanked on both sides by atoms with unshared pairs of electrons and this probably explains the resistance of the ester bond in these compounds to saponification by alkali. This bond still can be cleaved, but elimination of the alcohol occurs as a result of intramolecular nucleophilic attack by the (negatively charged) nitrogen atom of the second amino acid in the sequence:

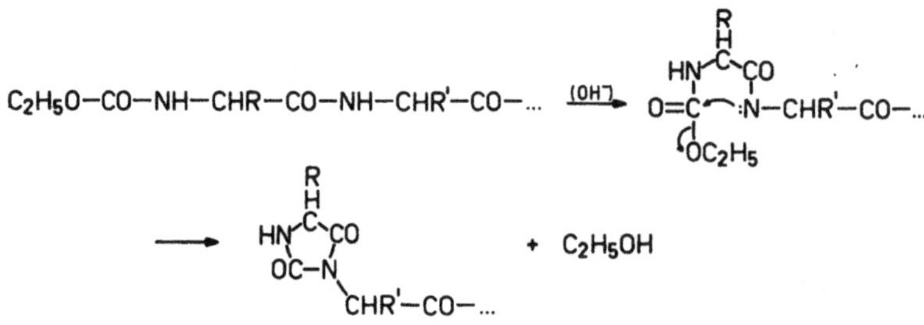

The hydantoin thus formed opens up under the influence of excess alkali but with the production of a urea derivative rather than the desired amine:

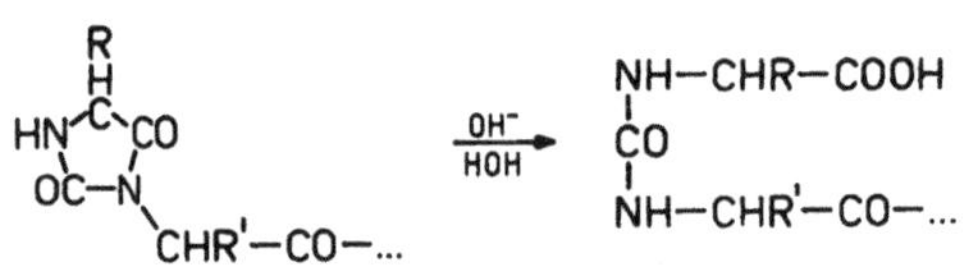

The ring-closure-elimination reaction [4] occurs particularly readily [4] if glycine is the second amino acid in the sequence [5], as was the case in the experiments of Emil Fischer [3]. The failure of this attempt considerably delayed the development of a truly applicable amino protecting group. The long awaited breakthrough occurred in 1932 when Bergmann and Zervas proposed [6] the now classical and yet in many respects still unsurpassed benzyloxycarbonyl group for the protection of the amino function.

a. Reduction and Oxidation

The benzyloxycarbonyl group is also of the urethane type but, instead of methyl or ethyl esters, it was based on *benzyl esters* of carbamoic acids. This allows the cleavage of the ester bond by catalytic hydrogenation [7], an elegantly simple procedure which can be executed without special equipment and in quantitative yield. Most importantly, removal of the benzyloxycarbonyl protection by catalytic reduction does not affect the peptide bonds.

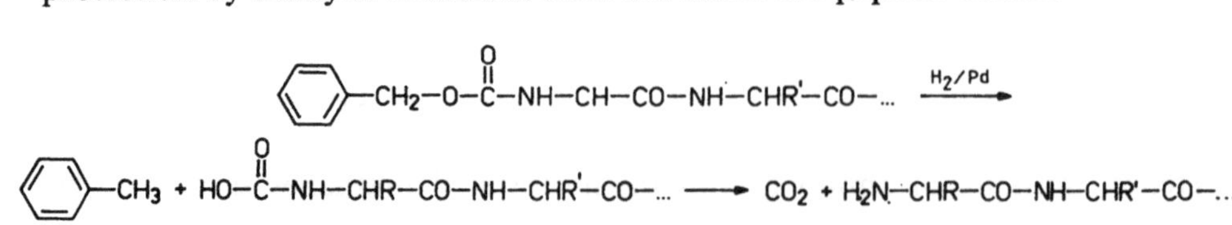

There are also some additional benefits in blocking the amino function with the benzyloxycarbonyl (formerly "carbobenzyloxy" or "carbobenzoxy") group, designated in contemporary literature with the letter "Z". The reagent most commonly used for the introduction of the Z group, benzyl chlorocarbonate[4], is easily prepared from inexpensive starting materials, benzyl alcohol and phosgene,

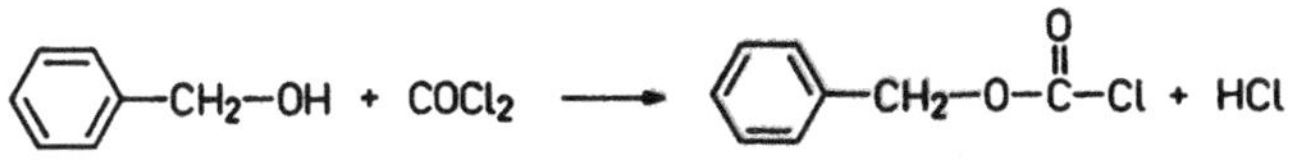

and the by-products of the deprotection reaction, toluene and carbon dioxide, are readily removed from the mixture. A further significant advantage of the Z group lies in the fact that over and above the protection it provides against undesired acylation, it also prevents — with few exceptions — the racemization of the amino acid to which it is attached. This ability of the Z group to counter losses in chiral purity during activation and coupling is shared also by other urethane-type blocking groups and will be discussed in more detail in the section dealing with racemization.

Deblocking by *reduction* is not limited to the Z group. The p-toluenesulfonyl (or "tosyl" or Tos) group, applied quite early in peptide synthesis, was cleaved by the action of hydrogen iodide-phosphonium iodide [8]. The discovery of reduction with sodium in liquid ammonia by du Vigneaud [9] greatly improved [10] the usefulness of the tosyl group in masking amino functions. The Na/NH_3 approach could be extended to the reductive removal of benzyl groups, including cleavage of the benzyloxycarbonyl group [11]. Alternative means for deblocking via reduction are catalytic hydrogenation in liquid ammonia [12] and transfer hydrogenation with donors such as cyclohexene [13], cyclohexadiene [14], hydrazine [15], or formic acid [16]. Thus, reduction remains an attractive method for the removal of amino-protecting groups and can be recognized as a stimulating idea in the many reductive pathways advanced for the deblocking of various side chain functions, e.g. for the removal of the nitro-group from nitroarginine residues. For the latter purpose, reduction with zinc in acetic acid [17], with stannous chloride [18], titanium trichloride [19], as well as electrolytic reduction [20], were recommended.

Interestingly, *oxidation* as a method for the removal of protecting groups was proposed only exceptionally, e.g. for the unmasking of amines blocked with the butylthiocarbonyl group [21]. Oxidative removal of sulfhydryl protecting groups, such as the S-trityl [22] or the S-acetamidomethyl [23] group, entails also the conversion of the regenerated SH groups to disulfides.

b. Acidolysis — Carbocation Formation

A new era was initiated by Ben Ishai and Berger who found [24] that the benzyloxycarbonyl group, in addition to reduction, can also be cleaved by strong acids in anhydrous media. Hydrobromc acid in glacial acetic acid

[4] In order to emphasize the application of carbonic acid chemistry, we prefer the expression "benzyl chlorocarbonate" instead of the formally correct "benzyl chloroformate" or the somewhat archaic "carbobenzoxy chloride".

seemed to be the most suitable for this purpose. It must be emphasized that this method relies on *acidolysis* and not on hydrolysis. Breaking of the ester bond in urethanes requires the presence of a group that can give rise to stable carbocations:

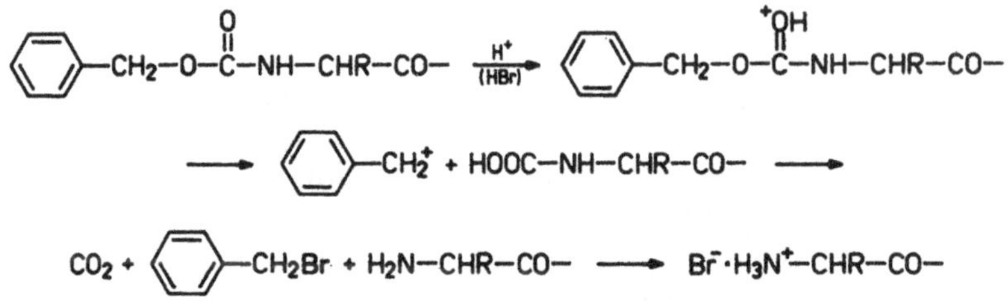

Delocalization of the positive charge produced by protonation of the carbonyl oxygen will be productive, in terms of heterolytic fission of the ester bond, only if this charge can be accommodated by the carbon atom attached to the ester oxygen. The moderate stability of the benzyl cation allows the process to go to completion, but strong acids are required for practical rates in the deblocking step. While in the absence of water, no hydrolysis of peptide bonds has to be feared, strong acids can produce N → O shifts in serine-containing peptides, and they also might damage the sensitive indole nucleus of tryptophan side chains. It is obvious that deprotection with less strong acids is highly desirable. A logical solution to this problem can be found in the application of urethanes in which the benzyl group is replaced by groups that can generate more stable carbocations. Substitution of the aromatic ring of the benzyl group with electron-releasing substituents, such as the methoxy group, or the complete replacement of the benzyl group by the tert. butyl group are equally effective. Both the p-methoxybenzyloxycarbonyl group [25] and the tert. butyloxycarbonyl (t.Boc or Boc) group [26–28] are cleaved with moderately

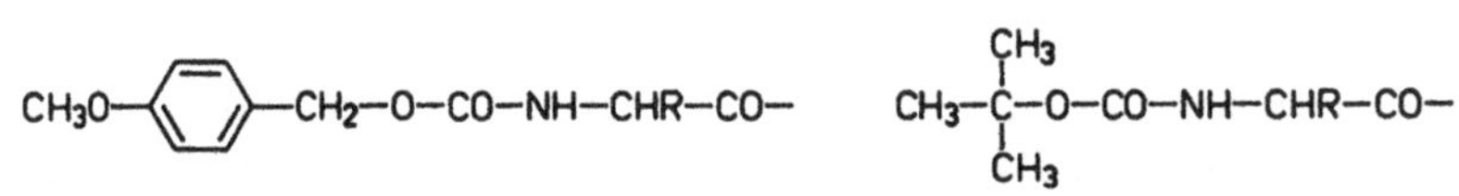

strong acids. They are sensitive to dilute solutions of hydrochloric acid in organic solvents. Moreover, they can be removed with trifluoroacetic acid, a reagent that can also play the role of *solvent*, an important consideration in the synthesis of longer peptide chains. From the two alternatives, the Boc group was more widely accepted by the practitioners and became also the starting point of further developments. A systematic and extensive search for protective groups with enhanced acid sensitivity culminated in the discovery of the biphenylisopropyloxycarbonyl (Bpoc) group [29].

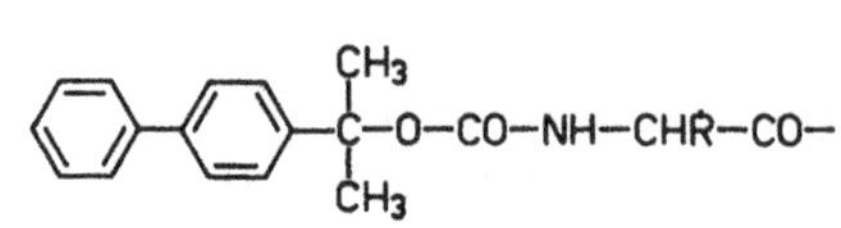

A general shortcoming of most acid sensitive protecting groups is the possible alkylation of amino acid side chains. The alkyl cations generated in the process of deprotection can cause alkylation by electrophilic aromatic substitution of the phenol in tyrosine or the indole in tryptophan. Alternatively, the products formed in the deprotection step, e.g. benzyl bromide or tert-butyl trifluoroacetate, act as alkylating agents. These can also convert the thioether in methionine side chains to ternary sulfonium salts. The frequently applied remedy, addition of cation-scavengers, is sometimes helpful but can also backfire: the most popular additive, anisole, might itself serve as a source of methyl groups and cause alkylation during deprotection with strong acids, such as methanesulfonic acid [30]. The danger of alkylation could be circumvented by the use of acid labile blocking groups which do not produce cations during deprotection. For instance, diphenylphosphinamides [31] are cleaved because of their inherent sensitivity to acids:

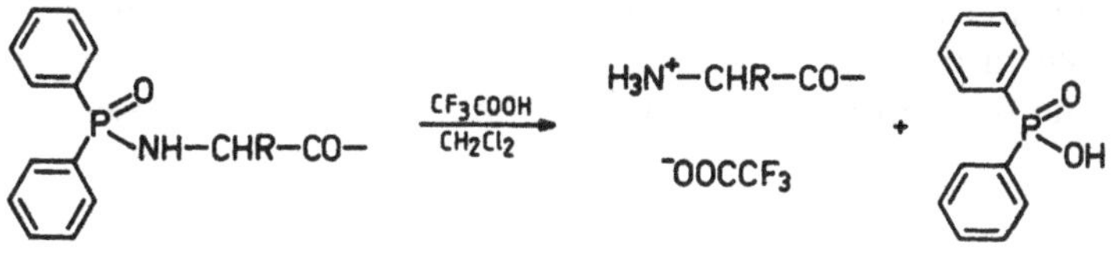

c. Proton Abstraction (Carbanion Formation)

Alkylation of readily substituted amino acid side chains and possible damage to the peptide by strong acids create problems which are serious enough to warrant the exploration of deblocking by pathways other than acidolysis. A logical counterpart for masking groups based on carbocations is the generation of *carbanions* in the deblocking process. For instance, the protons on the β-carbon atom of ethyl esters can be rendered acidic by an electron-withdrawing substituent on the same carbon atom. Sulfones are quite effective in this respect:

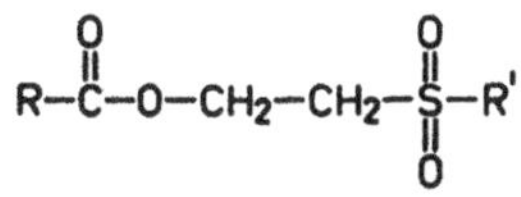

Under the influence of bases, proton abstraction occurs and the resulting carbanion is stabilized by the elimination of a vinylsulfone and regeneration of the carboxyl function:

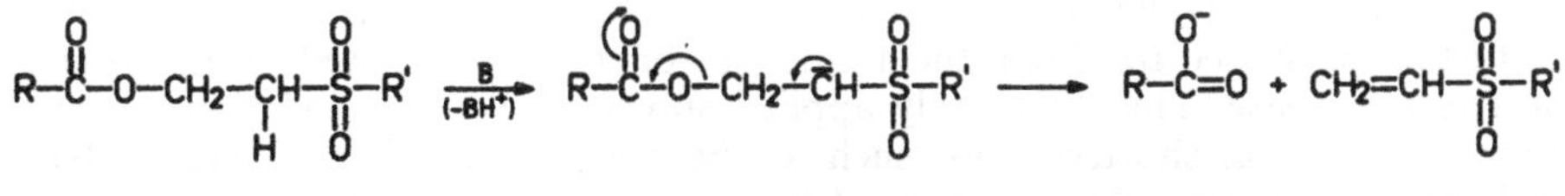

The development of protecting groups removable by β-elimination started with carboxyl-protection by ethyl esters having electron withdrawing substituents at their β-carbon [32, 33]. The subsequent adaptation of this principle for amino-protection by the use of urethanes with similarly designed ester groups culminated in the introduction of the 9-fluorenylmethyloxycarbonyl (Fmoc) group [34, 35] in which the β-carbon atom of the ester is part of the fluorenyl system:

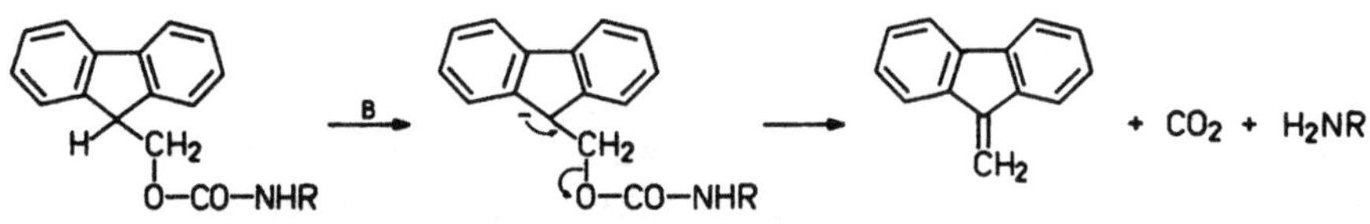

For the abstraction of the acidic proton on carbon 9 of fluorene, weak bases, such as piperidine, are sufficient. The ensuing elimination yields dibenzofulvene and a carbamoic acid which in turn loses carbon dioxide and regenerates the free amine. Because of the absence of acids, this amino group is indeed "free", unlike in deprotection by acidolysis where it is recovered in protonated form. The free amine need not be "liberated" for the subsequent acylation.[5]

d. Nucleophilic Displacement

A further alternative to acidolysis is the nucleophilic displacement of masking groups. As an example, we mention here the *o*-nitrophenylsulfenyl (Nps) group, which can be cleaved not only with acids, as originally proposed [36, 37], but also by nucleophiles [38–41], particularly by thiols [42, 43]:

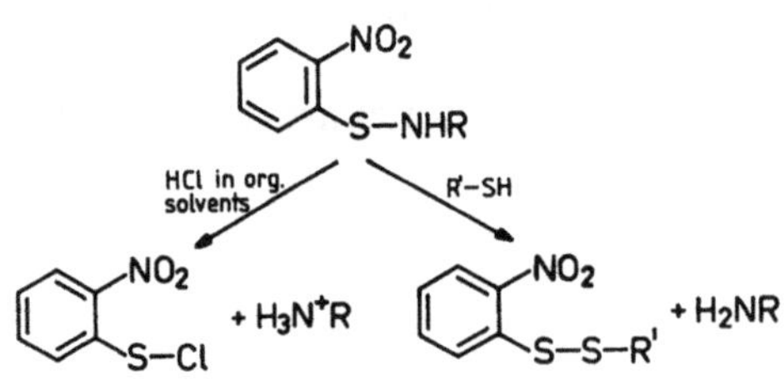

[5] A not unimportant aspect of deprotection is the difference between regeneration of the amino group in protonated form or as the free amine. In the latter case, its entire amount is available for acylation, while the protonated form either must be further processed to obtain the amine free or it has to be used in the presence of a tertiary base. This commonly applied substitute for the acylation of a free amine has several shortcomings, such as the racemizing effect of tertiary bases and the more general disadvantage of having, at any given time, only a fraction of the amino-component available for acylation according to the equilibrium:

$$R{-}N^{+}H_3 + NR'_3 \rightleftharpoons R{-}NH_2 + HN^{+}R'_3$$

A remarkably selective way of deprotection is possible with the 2-trimethylsilylethyloxycarbonyl group [44] which is cleaved by fluoride anions:

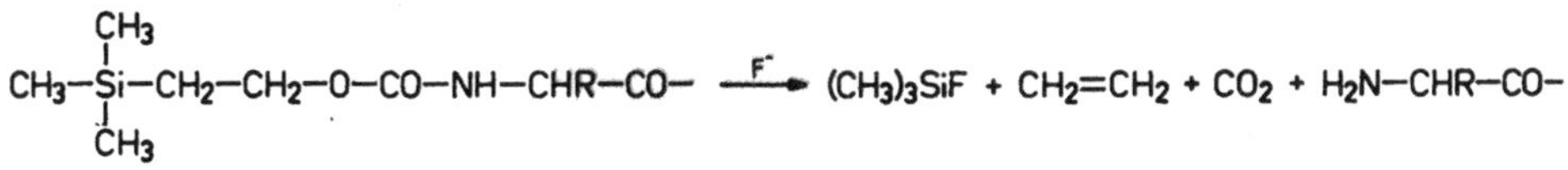

e. Photolysis

In the practical synthesis of peptides, so far only reduction, oxidation, acidolysis, proton abstraction and nucleophilic displacement were commonly applied for the removal of protecting groups, but sometimes other quite promising approaches were also explored. Of these, photolysis is particularly attractive since it can be performed without the use of reagents which subsequently have to be disposed. Yet, photolytic reactions are not always quantitative. For instance, the photolytic cleavage of sulfonamides [45] liberates the amines in low yield. Other photolytic reactions require irradiation at a wavelength which coincides with the absorption bands of tryptophan or tyrosine. Still, a few proposals are promising enough to expect their application in the future. Such are some groups advanced for the protection of the amino function, e.g. the 3,5-dimethoxybenzyloxycarbonyl [46, 47] and the 2-nitro-4,5-dimethoxybenzyloxycarbonyl [48] group,

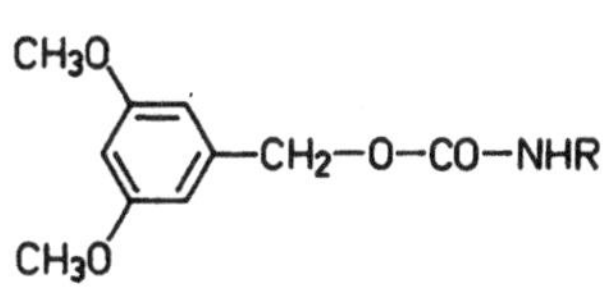

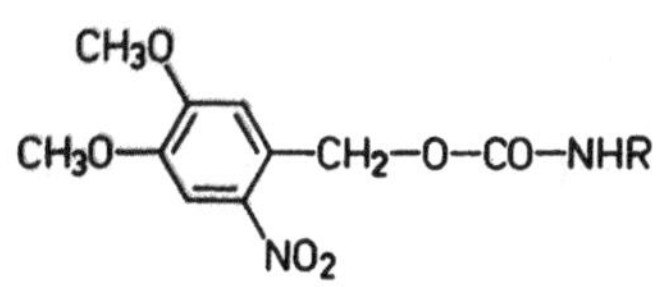

or for the blocking of carboxyls 2,2′-dinitrodiphenylmethyl esters [48], α-methylphenacyl esters [49]

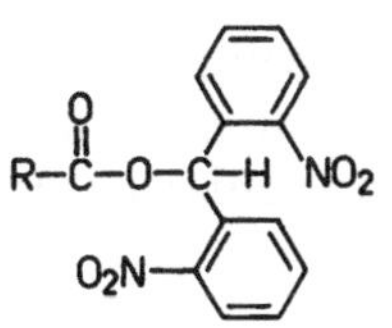

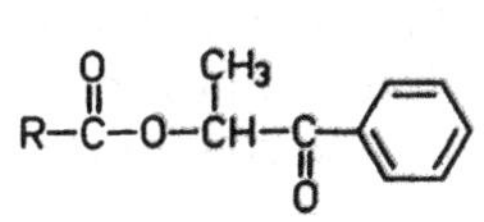

and the esters of dimethoxybenzoin [50]:

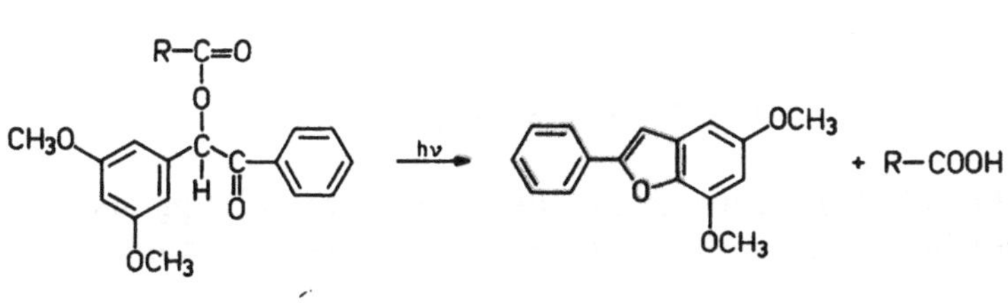

Finally, even substituted benzyl esters which serve as points of attachment between peptides and insoluble polymeric supports can be cleaved by irradiation at 3500 Å if the benzyl ester type resin carries an o-nitro substituent [51]:

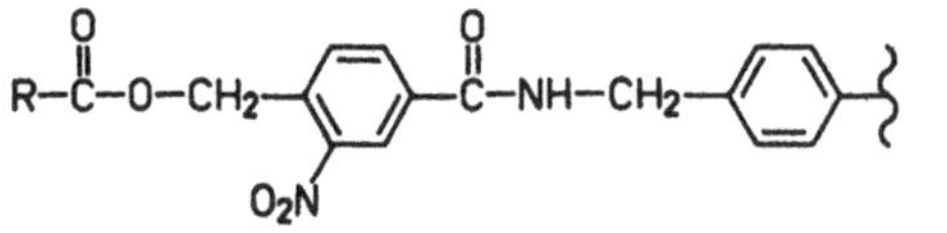

f. Enzyme Catalyzed Hydrolysis

The remarkable selectivity of enzyme-catalyzed reactions suggests that they could be applied for the removal of protecting groups. For instance, an acyl-lysine deacylase [52] will effect the cleavage of acyl groups from the ε-amino group in the side chain of lysine moieties, but leaves the α-amides in the peptide backbone intact. Thus, it seems to be feasible to use the small and easily introduced acetyl group for the protection of lysine side chains:

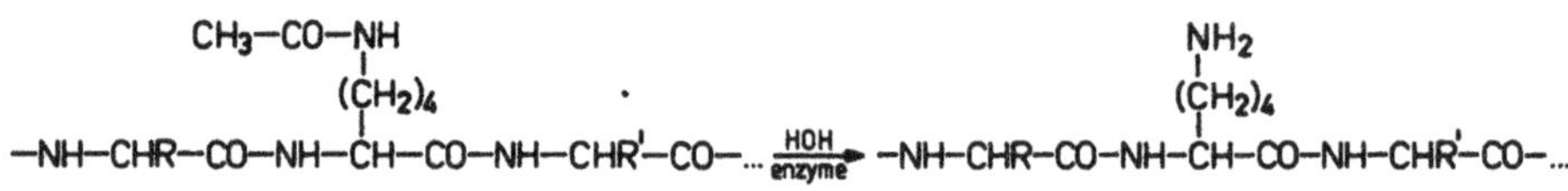

Similarly, carboxyls protected in the form of alkyl esters can be hydrolyzed, without the risks that accompany saponification with alkali, simply by exposure to the action of proteolytic enzymes [53]. In a selected p_H range, carboxypeptidase Y will catalyze ester hydrolysis without affecting peptide bonds [54].

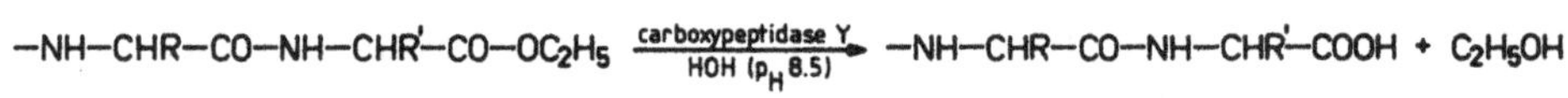

The essentially intramolecular catalysis characteristic for enzymes is simulated in the sophisticated cleavage of benzyl esters (used as links between peptides and insoluble polymers) through the intervention of dimethylaminoethanol [55] which after transesterification plays the role of the intramolecular base:

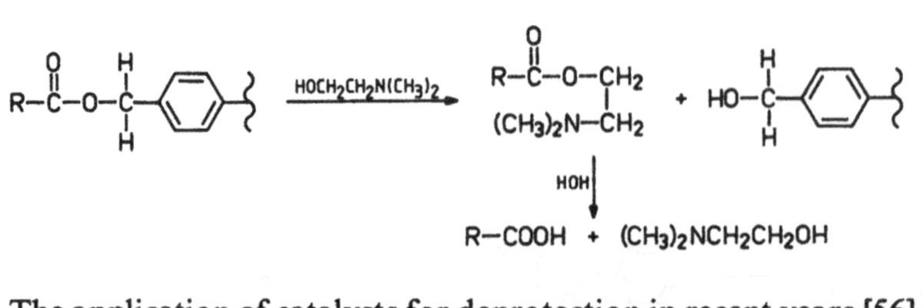

The application of catalysts for deprotection in recent years [56] suggests their more extensive use in future syntheses. In any case, this list, enumerating the principles which are operative at this time, is obviously incomplete[6] and will also increase by the addition of new methods of peptide chemists challenged by the difficulties of their endeavors. As an example of stimulating ideas, we mention here the use of *precursors* of amino acids, which are transformed at a preselected stage of the synthesis to the amino acid residues. Such precursors are α-azido-acids [58] which yield α-amino acids and 2-amino-6-nitro-caproic acid [59] which forms a lysine residue after incorporation and reduction.

B. Protection of the Carboxyl Group

Although carboxylates are excellent nucleophiles which are certainly able to react, in competition with amino groups, with acylating agents, their protection is not absolutely necessary. The products of such reactions are readily hydrolyzed, if water is present, and the carboxyl group is regenerated:

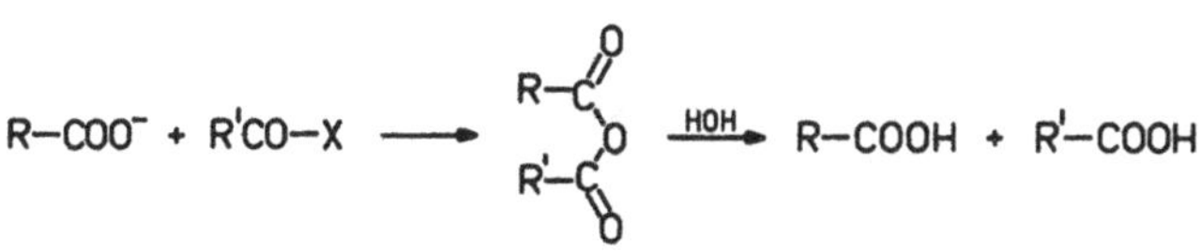

Nevertheless, there are several good reasons for the masking of carboxyl groups. Solubility is one of these reasons. For instance, acylation of the sodium salt of an amino acid requires aqueous media and, in order to accomodate the acylating agent, one has to resort to mixtures of organic solvents with water. This leads, of course, to a competition between aminolysis and hydrolysis. Furthermore, in the above sketched process, the mixed anhydride intermediate can cause [60] ring closure in aspartyl residues with unprotected carboxyls in their side chains. Even the transient formation of a reactive intermediate might lead [61] to a loss of chiral purity in C-terminal residues if these are applied without carboxyl protection. Thus, carboxyl groups are masked in most syntheses. Semipermanent protecting groups are preferred: these can be kept intact through the chainbuilding process and are removed at its completion or prior to the coupling of segments in the final deprotection.

It is rather fortunate that quite a few biologically active peptides have a carboxamide group rather than a free carboxyl at their C-termini. This might be a device of Nature to protect the molecules of hormones such as oxytocin, vasopressin, gastrin, cholecystokinin, secretin, etc. from degradation by carboxypeptidases. In the synthesis of these peptides, the problem

[6] Because of their inherent limitations, we did not include methods which are based on the specific cleavage at certain selected amino acid residues, e.g. the use of arginine residues as "protecting groups" to be hydrolyzed later by trypsin or the incorporation of methionines (cf. for instance ref. [54] which are subsequently cleaved at their carboxyl side with cyanogen bromide [57].

of finding a suitable protecting group for the α-carboxyl of the C-terminal residue is absent, but the blocking of the side chain carboxyls of aspartyl and glutamyl residues remains to be considered.

A review on recent development in methods for the esterification and protection of the carboxyl group was written by Haslam [62] and a more peptide oriented article by Roeske [63].

1. Benzyl Esters and Substituted Benzyl Esters

The already discussed possibilities inherent in benzyl esters, their cleavage both by reduction and by acidolysis, rendered them very popular tools for the masking of the carboxyl function:

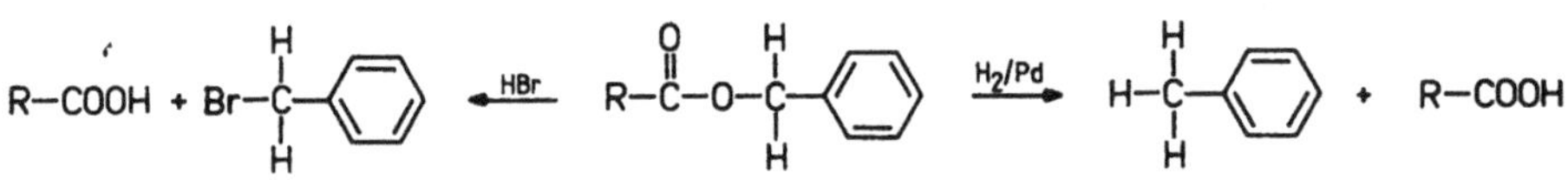

The semipermanent blocking group character of benzyl esters is based on the rapid removal of the benzyloxycarbonyl group by HBr in acetic acid and the relative stability of benzyl esters toward this reagent. This useful selectivity can be further enhanced by destabilizing the intermediate benzyl cation with the help of electron withdrawing substituents. The nitro group is quite efficient in this respect and *p*-nitrobenzyl esters [64, 65] turned out to be very practical for the blocking of carboxyl functions, since they remain intact during the acidolytic removal of benzyloxycarbonyl groups, yet are readily cleaved by reduction when they are no more needed. The similar affect of halogen substituents [66, 67] or of the cyano group [68] attached to the aromatic nucleus of benzyl esters has not been utilized so far in major syntheses.

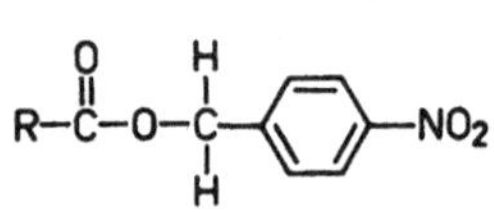

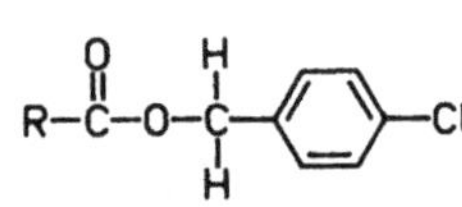

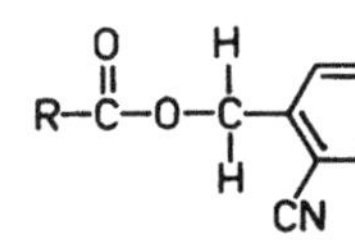

An increase in sensitivity toward acids can readily be achieved by electron releasing substituents, such as the methoxy group [25], or the combined effect of three or five methyl groups [69, 70]:

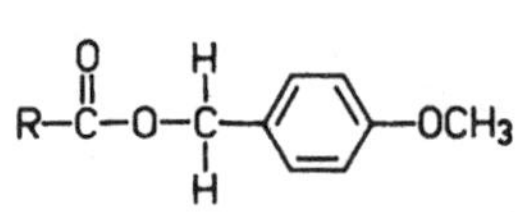

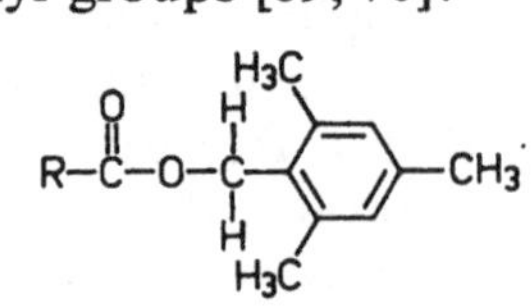

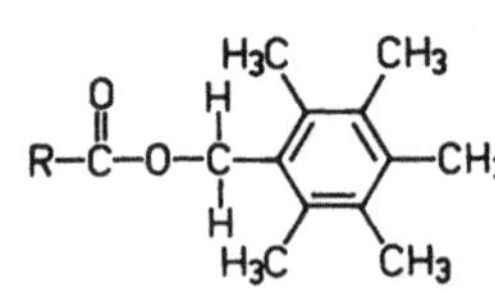

Esters of 3,4-methylene-dihydroxybenzyl alcohol [70] are similarly acid labile and the logical extension of this thought, acid sensitive protection in the form of 3,4- and 2,4-dimethoxybenzyl esters, has also been investigated [71].

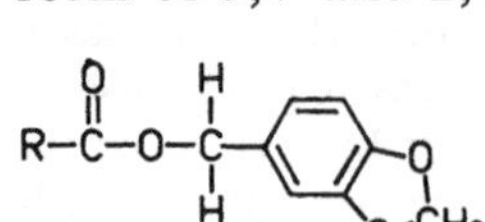

An interesting modification of the benzyl ester group is the introduction of chromophores which absorb in the visible region of the spectrum and thus facilitate the isolation or purification of protected intermediates, because extraction, chromatography, etc. can be followed with the naked eye. One such group, esters of 4-dimethylamino-4′-hydroxymethylazobenzene [72],

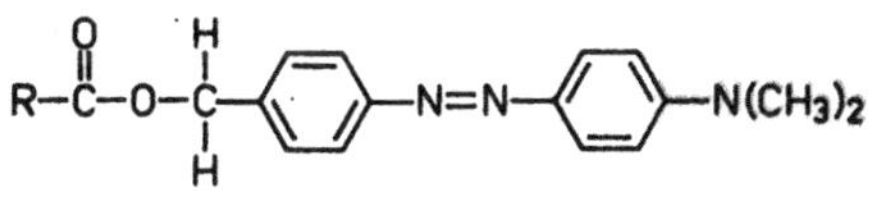

has also a basic substituent which makes it possible to attach the intermediates to cation exchangers and thereby helps in the removal of starting materials and by-products.

By analogy, two more protecting groups can be mentioned here: the esters of 9-hydroxymethylanthracene [73] and acyl derivatives of 4-hydroxymethylpyridine [74]. The latter are very resistant to acids because

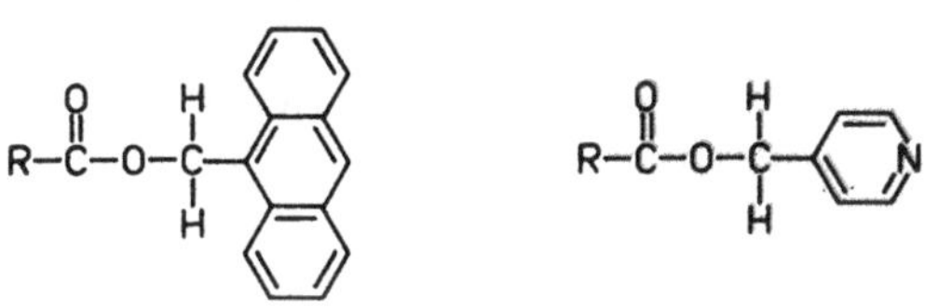

protonation of the pyridine nitrogen prevents the formation of a second cationic center and the benzylic carbon cannot be the charged atom of a carbo-cation. Yet, just because of this resistance to acidolysis and the basic character of the protecting group, picolyl esters form a "handle" on the peptide intermediates by which they can be selectively [75] adsorbed to ion-exchange resins. The handle method is probably one of the most important approaches to a facilitation of peptide synthesis.

In connection with benzyl esters, one should not forget that the polymeric support invented by Merrifield [76] for solid phase peptide synthesis is a derivative of benzyl chloride; the attachment between the C-terminal residue and the insoluble support is a form of benzyl esters. Separation of the already assembled peptide from the resin is carried out mostly by acidolysis as in the case of other benzyl esters.

In a sense, the esters of diphenylmethanol or benzhydrol [77, 78] also can be looked upon as substituted benzyl esters and are, indeed, removable by reduction and also by saponification

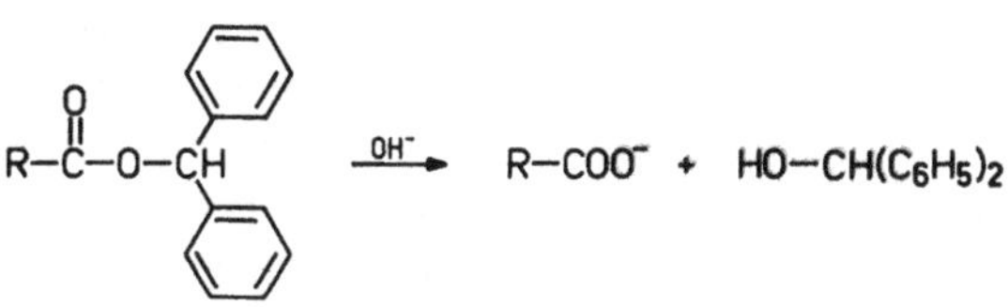

with alkali. They are more sensitive to acids than unsubstituted benzyl esters and are cleaved by trifluoroacetic acid in the cold.

The *preparation* of benzyl esters and substituted benzyl esters usually presents no major difficulties. The acid catalyzed esterification of carboxylic acids with benzyl alcohol is suitable also for the conversion of free amino acids to their benzyl esters. The initially used hydrochloric acid [79] can be advantageously replaced by benzenesulfonic [80] or *p*-toluenesulfonic acid [81] and the ester equilibrium shifted by the removal of water through azeotropic distillation with the solvent, benzene or toluene:

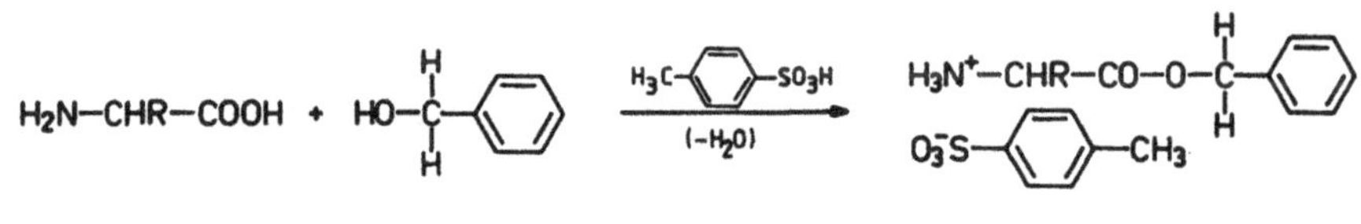

This simple esterification method can be extended for protected amino acids, for instance for N-benzyloxycarbonyl amino acids [24], but less firmly protected derivatives or sensitive peptides are seldom exposed to elevated temperatures used in the process; milder procedures are needed for their esterification. Of these, transesterification with benzyl borate [82] and esterification with the help of dineopentyl acetal [83] could be mentioned.

Preparation of a substituted benzyl ester is usually the first step in solid phase peptide synthesis because this anchors the C-terminal residue to the insoluble polymeric support. In most cases a reaction between a salt of the protected amino acid and the chloromethyl group on an aromatic nucleus of the polymer provides the ester bond sought for the purpose of attachment [76]. An improved incorporation was attained by the use of cesium salts [84] instead of the initially applied triethylammonium salts:

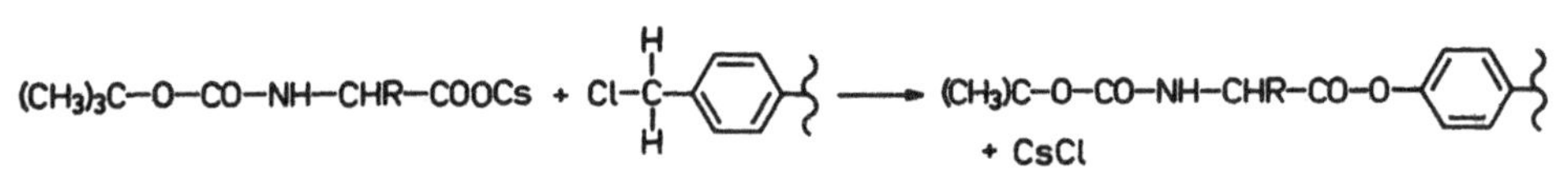

An alternative solution is the imidazole-catalyzed transesterification of active esters of protected amino acids with benzyl alcohol or substituted benzyl alcohols [85], an approach that can be adapted [86] also for the anchoring of the first amino acid to hydroxymethyl type polymers.

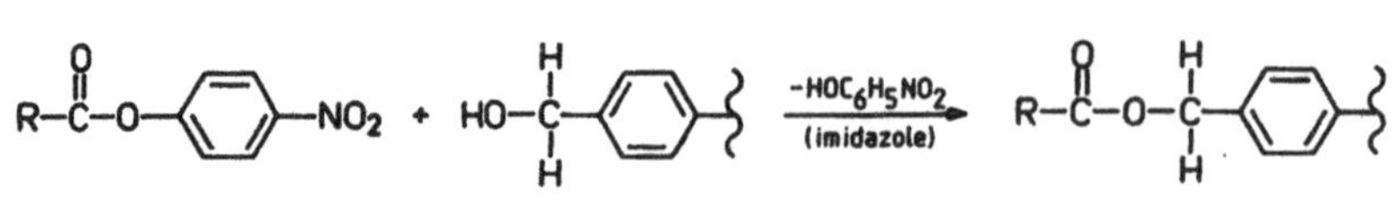

Active esters are by no means the only reactive derivatives of carboxylic acids that can be applied for the preparation of benzyl esters or substituted benzyl esters. Carboxyl-activation with carbodiimides, carbonyldiimidazole (cf. e.g. ref. [86]) or in the form of mixed anhydrides is equally possible. In fact, all that could be said in the section on activation and coupling remains valid for the cases in which alcohols rather than amines are the nucleophiles to be acylated.

In the preparation of benzhydryl esters of free amino acids, the use of a reactive derivative of the alcohol component gives good results [77, 78]. The reaction between a salt of the amino acid and diphenyldiazomethane in dimethylformamide

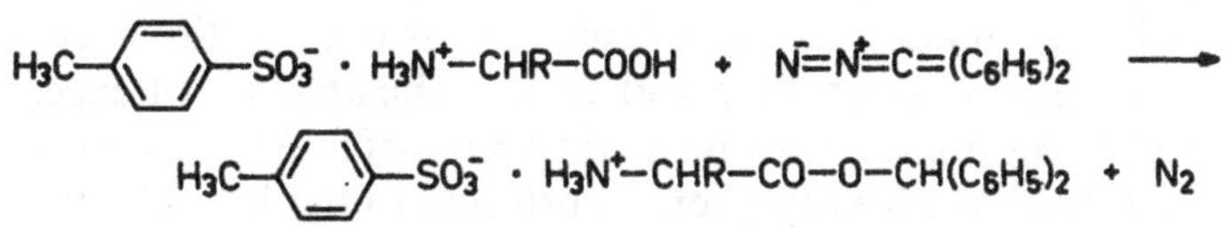

provides the desired amino acid benzhydryl esters in high yield and it is also possible to generate the reagent, diphenyldiazomethane, from benzophenone hydrazone [87] in the presence of the amino acid salt. With protected amino acids or peptides, more conventional methods of esterification, e.g. with the help of diphenylchloromethane [88] or diphenylmethanol [89], are also feasible.

Deprotection of carboxyl groups, blocked in the form of their benzyl esters, by reduction (H_2—Pd, Na—NH_3) was discussed earlier in this chapter as was the possible saponification with alkali. Hydrolysis [55] via transesterification with dimethylaminoethanol and intramolecular catalysis proposed for solid phase synthesis should have broader application, but the most general approach for the removal of benzyl groups is acidolysis. Modifications of the aromatic ring were introduced mainly to render the blocking group more sensitive or less sensitive to acids. The relationship between this acid sensitivity and the stability of the intermediate carbocations has a so been pointed out. We have to mention here, however, a more recently proposed [90, 91] alternative, the cleavage of benzyl esters through transesterification with iodotrimethylsilane. The trimethylsilyl esters thus formed are readily hydrolyzed by water. The conditions described for this ester-cleavage are not mild enough for the general praxis of peptide synthesis, but the new approach might be amenable to further improvements. The rapid reaction of 9-anthranylmethyl esters [73] with sodium methylmercaptide leading to the formation of the sodium salt of the acid and 9-anthranylmethyl methyl thioether [92] is similarly stimulating and should initiate broader exploration.

2. Methyl Esters and Substituted Methyl Esters

It is fairly simple to prepare methyl esters of amino acids. In addition to the classical method of esterification, introduction of HCl gas into a suspension of the amino acid in methanol [93, 94], a whole series of variations of this approach appeared in the literature, e.g. the addition of the amino acid to a cold mixture of thionyl chloride and methanol [95], or to acetyl chloride-methanol [96], the treatment of amino acids with acetone dimethyl ketal and aqueous hydrochloric acid [97] and transesterification of methyl acetate with sulfuryl chloride as catalyst [98], etc. Methylation of protected amino acids and peptides with diazomethane is also used occasionally. Methyl esters are excellent blocking groups if they need not be removed after chain building because they are subsequently converted to amides by ammonolysis or treated with hydrazine to form hydrazides for coupling by the azide method. If, however, methyl esters were used simply as semipermanent blocking of a carboxyl function, their removal is fraught with difficulties. The simplest and most practiced method for cleaving methyl esters is saponification with aqueous alkali, mostly in the presence of organic solvents such as methanol, acetone or dioxane. Although usually considerable care is used in the execution of the reaction, even at low temperature and even in the absence of excess alkali, side reactions might accompany saponification. One of these, hydantoin formation at the N-terminus, has already been mentioned in the introduction of this chapter. Peptides which contain serine residues suffer some decomposition [99], perhaps due to the elimination of water or formaldehyde while C-terminal S-benzylcysteine can lose chiral purity to an unacceptable extent [100]. With other amino acids in the same position, some racemization was noted in one laboratory [101] but not in another [102]. Probably the solvents used or subtle differences in the conditions applied account for such discrepancies. These difficulties notwithstanding, methyl esters were used for carboxyl protection in numerous instances and were removed mostly by saponification. Yet, alternative methods of deblocking are obviously needed. Unfortunately, transesterification with dimethylaminoethanol followed by hydrolysis, a method which serves so well in the case of benzyl esters, can be used for methyl esters only with thallium salts as catalysts [55]. An approach involving the use of iodotrimethylsilane [90, 91] or the selective cleavage of methyl esters with the help of lithium *n*-propyl mercaptide in hexamethylphosphoramide [103] have not found their way into peptide synthesis so far. Thus, modified methyl esters might be the answer to the questions raised in this paragraph.

In a formal sense, benzyl esters, benzhydryl esters, and esters of hydroxymethylpyridine or 9-hydroxymethylanthracene are modified methyl esters. Because of their specific properties, such as removability by reduction at a benzylic carbon atom, we discussed them under the heading of benzyl esters. At this point we try to emphasize a different kind of modification and call the attention to phenacyl (Pa) esters [104]:

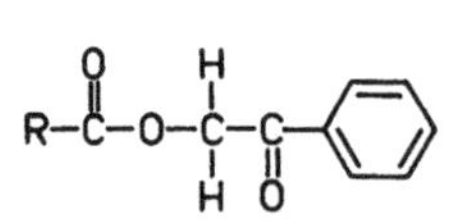

Preparation of phenacyl esters through the reaction of salts of carboxylic acids with ω-bromoacetophenone (a lachrymator!) proceeds smoothly and the masking group is stable toward acids, including liquid HF. It can be removed by nucleophiles, particularly with sodium thiophenoxide [104] which cleaves the alkyl-oxygen bond in phenacyl esters and also in phthalimidomethyl esters [105], but leaves benzyl esters,

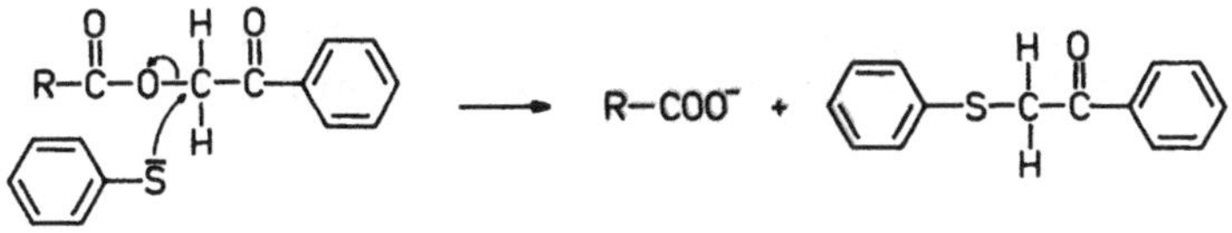

e.g., unaffected. Reduction with zinc in acetic acid also removes the masking group, but catalytic hydrogenation does not give satisfactory results because reduction of the keto group accompanies the process of hydrogenolysis [106]. If one looks upon phenacyl esters as substituted methyl esters, it becomes obvious that substitution of a hydrogen atom of the methyl group by a benzoyl group must increase the reactivity of the esters toward nucleophiles. Therefore, it is not surprising that diketopiperazine formation occurs as a side reaction in dipeptide phenacyl esters [107]

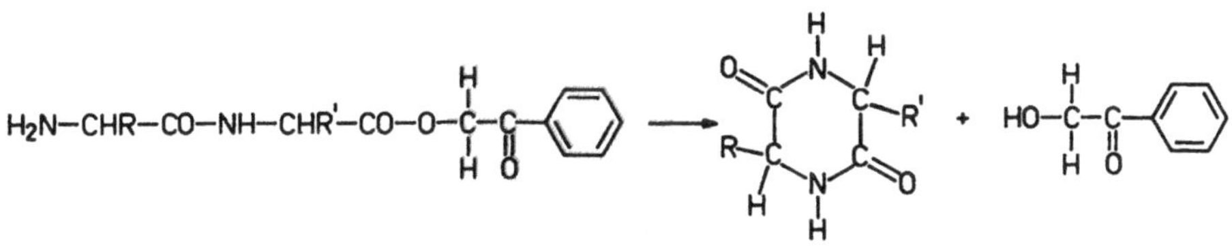

or that a similar intramolecular attack converts aspartyl residues blocked at their β-carboxyl by the phenacyl group into undesired succinimide derivatives [108].

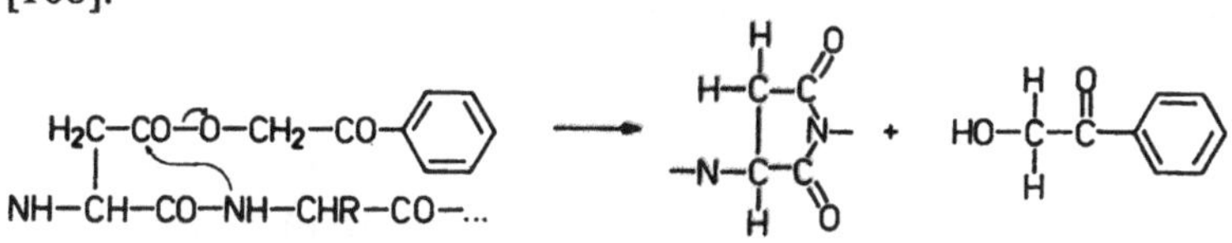

The concern felt about the activated methyl ester character of phenacyl esters is even more justified if the aromatic ring carries an electron withdrawing substituent in ortho or para position. Therefore, the readiness to form crystalline derivatives, shown by 4-bromophenacyl esters [109, 110], is probably not sufficient reason for their

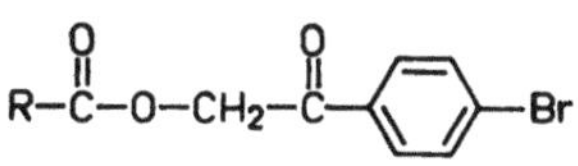

general use in peptide synthesis. This example demonstrates that the borderline between carboxyl protecting groups and carboxyl activating groups is not easily drawn and substituted methyl esters must be carefully scrutinized in this respect. The pronounced electronic effects seen in phthalimidomethyl esters [105] might be harmless on account of the bulkiness of this protecting group, but less hindered esters

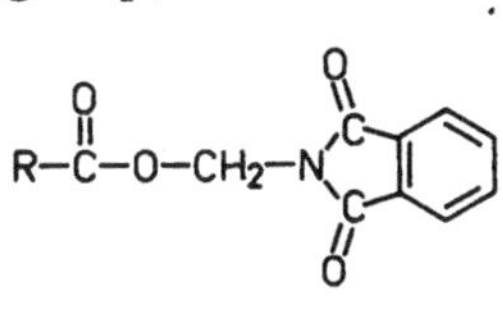

such as 2,2,2-trichloroethyl esters [111, 112] of protected amino acids, which can be regarded as methyl esters substituted with a trichloromethyl group, are

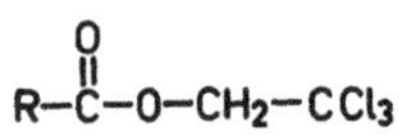

certainly not sufficiently inert toward nucleophiles[7] to provide unequivocal protection for the carboxyl group. In fact, the related 2,2,2-trifluoroethyl esters [113] and hexafluoro-2-propyl esters [114] are acylating agents recommended [114] for the synthesis of polypeptides.

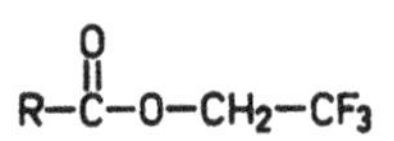

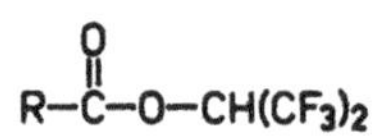

3. Ethyl Esters and Substituted Ethyl Esters

Amino acid ethyl esters are readily available through direct esterification, that is by the introduction of HCl into an alcoholic suspension of the amino acids [93] or by alkylation, e.g. with ethyl *p*-toluenesulfonate [115]. Their usefulness in peptide synthesis is still limited because of the difficulties encountered when carboxyl groups protected in the form of ethyl esters are to be unmasked. Saponification with alkali has been applied and often with satisfactory results, yet, the side reactions mentioned in connection with methyl esters are due to occur in the case of ethyl esters as well. The fact that ethyl esters are less sensitive to nucleophilic attacks than methyl esters renders them less vulnerable, but also slows down the process of hydrolysis or the conversion of the esters to amides or hydrazides. Interestingly, not only the bond between carbonyl carbon and oxygen is more resistant, but certain methods used for the fission of the alkyl-oxygen bond work better with methyl than with ethyl esters. For instance, splitting of methyl esters with lithium iodide in pyridine [116] proceeds several times faster than the similar fission of ethyl esters:

[7] For the removal of the masking provided by trichloroethyl esters, not a displacement by nucleophiles is used, but an elimination reaction [112] induced by zinc (in acetic acid):

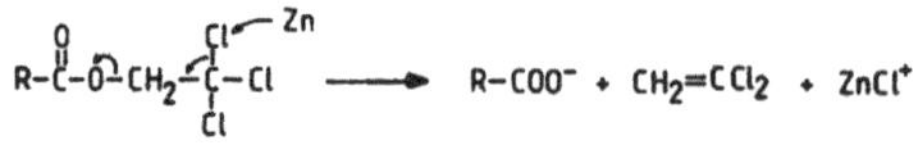

Enzyme-catalyzed hydrolysis remains an attractive possibility and the recommended enzymes, thermolysine [117] and carboxypeptidase Y [54], could turn out to be important tools in the hands of peptide chemists.

The situation is quite different with ethyl esters substituted on their β-carbon atom. Appropriate substituents can mobilize the hydrogen atom(s) of the β-carbon and facilitate proton abstraction by bases. The resulting carbanion is then stabilized by elimination:

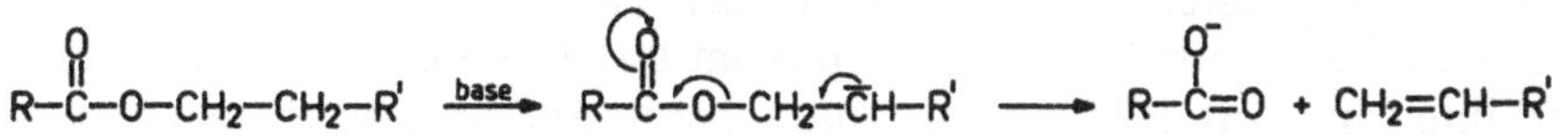

Of the various electron withdrawing R′ groups, dimethylsulfonium salts were first proposed [32, 33] and their alkali lability stimulated further developments, e.g. the introduction of sulfones [118] obtained from the intermediate methylthioethyl esters used in the synthesis of the sulfonium salts:

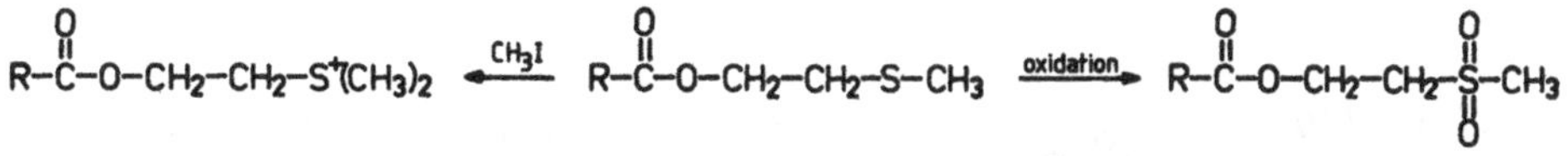

Subsequently, even more potent electron withdrawing substituents were recommended, such as the *p*-toluenesulfonyl group [119] and the *p*-nitrophenylsulfonyl groups [120]:

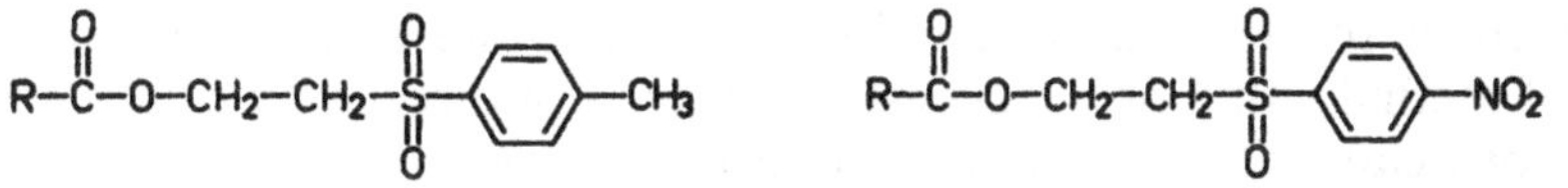

Beta elimination could also be applied in combination with a chromophore [121] that renders the peptide intermediates visible to the naked eye.

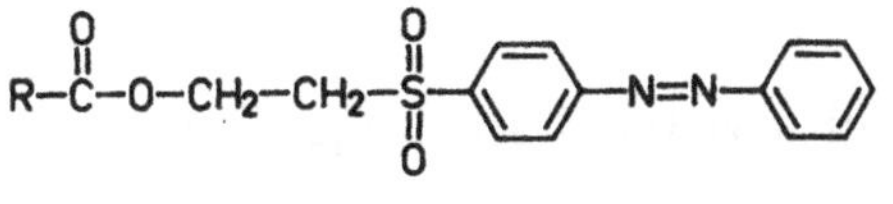

A most interesting contribution is the use of β-trimethylsilylethyl esters for carboxyl protection [122]. The new blocking group is selectively removable with fluoride ions, preferably with tetraalkylammonium fluorides.

The usefulness of the trimethylsily ethyl group has already been demonstrated in practical syntheses [123].

4. Tert.Butyl Esters and Related Compounds

In the introductory section of this chapter we have already sketched the logical pathway in the development of acid labile protecting groups. The role of the stability of the intermediate carbocations also has been pointed out. Since the moderately stable benzyl cation forms only in strong acids such as hydrobromic acid (in acetic acid) or liquid HF, the need for carboxyl protecting groups which are cleaved selectively by less strong acids led to the introduction of tert. butyl esters in peptide synthesis. The stability of the tert. butyl cation permits the removal of protection from tert. butyl esters of carboxylic acids by dilute solutions of HCl in organic solvents and also by trifluoroacetic acid. The same electron distribution renders these esters resistant to base catalyzed hydrolysis and more generally against the attack of nucleophiles. Yet, tert. butyl esters do not resist the similar attack of *intramolecular* nucleophiles. Thus, carboxylic acids are liberated from their tert. butyl esters even during brief exposure to aqueous NaOH in methanol, if the carboxyl belongs to an asparaginyl residue [124]:

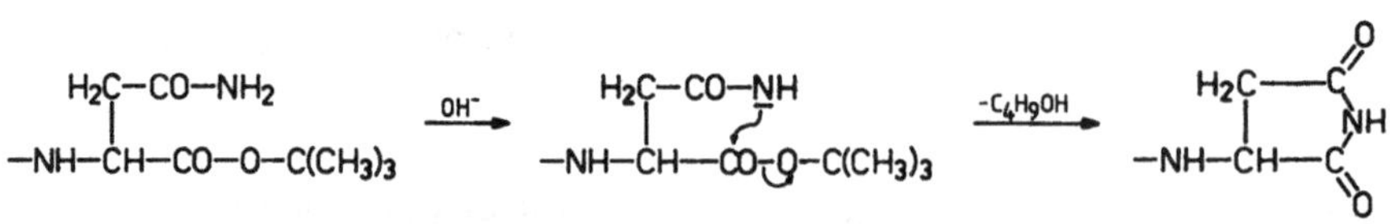

In spite of such incomplete selectivity, protection of carboxyl groups in the form of their tert. butyl esters remains an important method in peptide synthesis.

Introduction of the tert. butyl blocking group is more demanding than preparation of methyl, ethyl or benzyl esters. Instead of direct acid catalyzed esterification of acids with tert. butanol, one has to resort to the addition of the carboxyl group to isobutene [125].

$$R-COOH + CH_2{=}\underset{|}{\overset{CH_3}{C}}-CH_3 \xrightarrow[(H_2SO_4)]{H^+} R-CO-O-C(CH_3)$$

Transesterification of carboxylic acids with tert. butyl acetate [126] was also

$$R-COOH + CH_3-CO-O-C(CH_3)_3 \xrightarrow{HClO_4} R-CO-O-C(CH_3)_3 + CH_3COOH$$

recommended.

While tert. butyl groups are removable with moderately strong acids, they are sufficiently resistant to weak acids to allow the safe handling of intermediates protected by them, e.g. the washing of organic solutions of the peptides with dilute aqueous acids or conversion of hydrazides to azides with nitrites in the presence of a moderate excess of hydrochloric acid. This relative stability is missing in compounds blocked by the triphenylmethyl (trityl) group [88]. Trityl esters are cleaved even by acetic acid at room temperature and within a short time [127]. The even more labile trimethylsilyl esters [128]

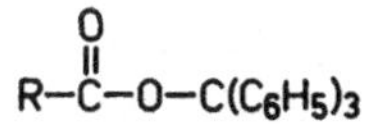

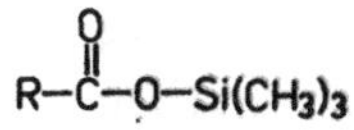

are hydrolyzed by water and can be applied merely for the transient protection of carboxyl groups [129]. A more practical enhancement of the acid lability of tert. butyl esters was sought by Blotny and Taschner [130] who, after a detailed comparison of various substituted methyl esters, found that the replacement of one methyl group of tert. butyl esters by a phenyl group yields a masking group about as sensitive

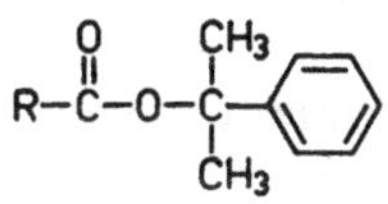

to acids as the benzhydryl esters mentioned earlier in this section.

5. Aryl Esters

The marked reactivity of phenyl esters in ammonolysis [131] suggests that they are not best suited for the protection of the carboxyl function. Nevertheless, the studies of Kenner [132] demonstrated that it is possible to use unsubstituted phenyl esters for the semipermanent masking of carboxyl groups and also that they are smoothly removed, under mildly alkaline conditions, if the hydrolysis is catalyzed by the peroxide anion. This method was shown to

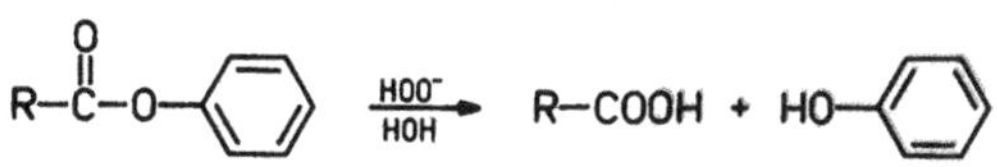

be useful even in major syntheses [133, 134]. A certain reservation still must be felt about the protection of carboxyls in the form of their phenyl esters if the activated derivatives of the carboxyl component are not sufficiently potent to successfully compete with phenyl esters which have weak, but not negligible, active ester character. Also, the results with highly activated carboxyl groups still could be less than perfect and the potential acylation by phenyl esters could cause serious complications if the reaction of the latter is catalyzed, e.g. by one mole of acetic acid [135]. Similar concerns must accompany the proposed use of 4-chlorophenylazophenyl esters [136] or *m*-nitrophenyl esters [137]. The latter can be equally well regarded as

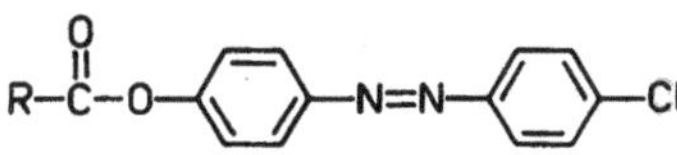

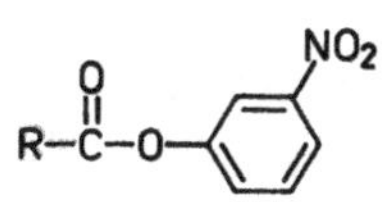

active esters [138] of protected amino acids or peptides. On the other hand, even *p*-nitrophenyl esters, acylating agents in peptide synthesis, could occasionally be used for carboxyl protection [139, 140]. The so-called "backing off" approach, however, is limited to syntheses in which highly reactive intermediates are used for acylation and even then the outcome of the procedure

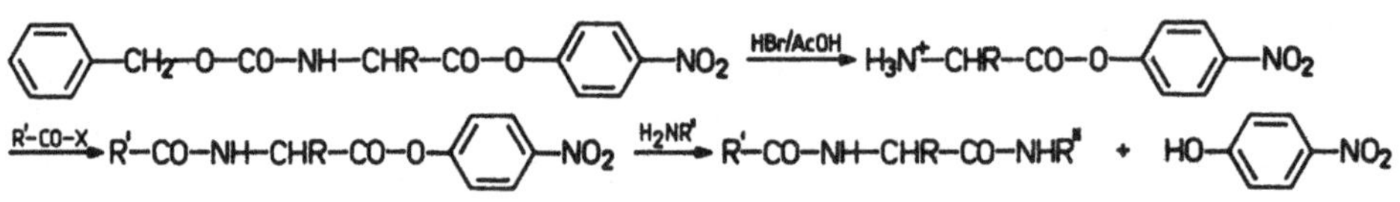

is ambiguous [141] since it must greatly depend on the method of activation and the nature of the amino acids involved.

Aryl esters with electron releasing substituents in the aromatic nucleus could better approach the ideal carboxyl protecting group since their reactivity toward nucleophiles might be sufficiently reduced. Therefore, more unequivocal masking is possible with 2-benzyloxyphenyl esters [142] or with 4-methylthiophenyl esters [143, 144]

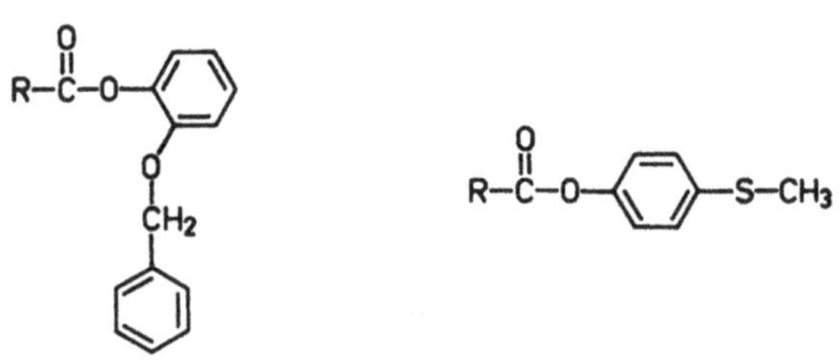

than with the above mentioned reactive aryl esters. As discussed in the chapter on activation, these esters can be converted, one by reduction and the other by oxidation, into reactive acylating agents.

6. Hydrazides

Amino acid and peptide hydrazides, in addition to being intermediates in activation via carboxylic acid azides, can also serve for the semipermanent blocking of the carboxyl function [145]. The protection, however, provided by simple hydrazides is limited to syntheses in which only moderately active acylating agents are used and in not large excess. With more potent derivatives of carboxyl components, one must expect substitution at the second hydrogen as well:

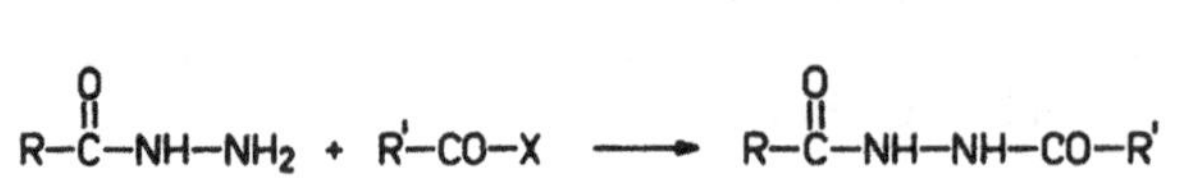

In this respect N'-substituted monoacyl hydrazines, such as the enzymatically prepared amino acid or peptide phenylhydrazides [146—150], are more auspicious because of the further reduced nucleophilicity of the second nitrogen atom. In N′-isopropylhydrazides [151], the bulkiness of the substituent hinders unwanted acylation.

The hydrazides mentioned so far can be oxidized to acyldiimides, which, in turn, are reactive enough to be aminolyzed by nucleophiles or hydrolyzed by water [152–154]:

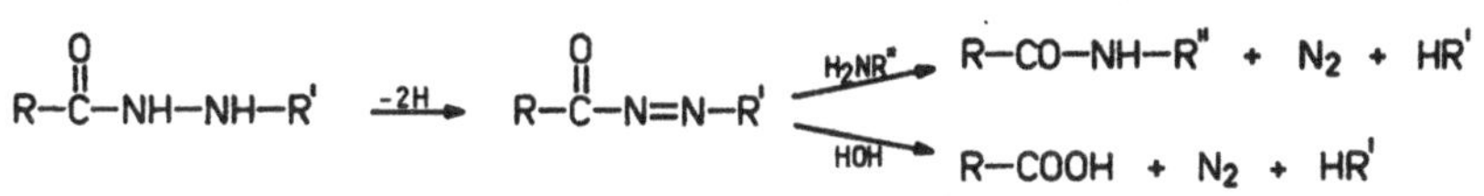

Both nitrogen atoms are substituted in N,N′-diphenylhydrazides [155] and N,N′-diisopropylhydrazides [156]. These too can be oxidized to carboxylic acids, but their usefulness in peptide synthesis has not yet been established.

An entire series of substituted hydrazides was designed for the protection of the carboxyl group of the C-terminal residue in peptides which serve later as segments in the preparation of a still larger chain. Selective removal of the N′-substituent unblocks the hydrazide and opens the way for the preparation of an azide. The first such combination [157] consists of protected amino acid N′-benzyloxycarbonyl hydrazides

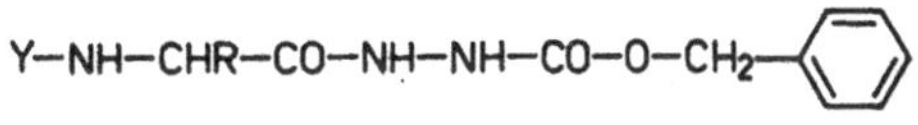

in which Y is so chosen that its removal does not affect the benzyloxycarbonyl group on N′. The partially deprotected amino acid is then acylated with the penultimate residue in the sequence and the chain is lengthened to the desired size. At this point the benzyloxycarbonyl group is removed, e.g. by catalytic hydrogenation, and the peptide hydrazide treated with nitrous acid or an alkyl nitrite to form the azide. This interesting and often practical concept led to new combinations of protecting groups, e.g. to the masking of hydrazides with the tert.butyloxycarbonyl [158], trifluoroacetyl [159], trityl [160], picolyloxycarbonyl [161], formyl [162], and trichloroethyloxycarbonyl [163] groups. Obviously, almost all the blocking groups discussed in the next section, dealing with amine protection, can be applied for the masking of the second nitrogen in the hydrazides of amino acids or peptides. Blocking the carboxyl function in the form of the hydrazide has already been discussed in the preceding chapter, in connection with the azide method.

C. Protection of the Amino Group

1. Alkyl and Alkylidene Protecting Groups

Monoalkylation of primary amines does not seriously interfere with their acylation and only bulky substituents can provide sufficient protection. (Of course a single alkyl group can block the secondary amine proline.) Furthermore, simple alkyl groups require drastic conditions for their removal from nitrogen atoms, conditions which are not compatible with the sensitivity of peptides to high temperature or to strong acids. These limitations, however, do not extend to the N-benzyl group, which can be cleaved by acidolysis or more readily by catalytic hydrogenation or reduction with sodium in liquid ammonia. In spite of their removability, N-benzyl and N-dibenzyl amino acids were only occasionally used [164–167].

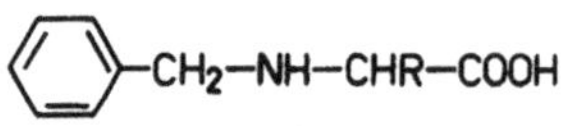

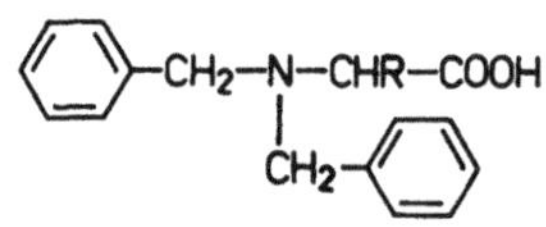

Monobenzyl amino acids were used for the preparation of N-methyl amino acids [168] but seldom in peptide synthesis because the benzyl group does not remove the basic character of the amino group and can cause steric hindrance in coupling. The steric factor is even more noticeable in the acid labile methoxy derivatives of the diphenylmethyl group [169]

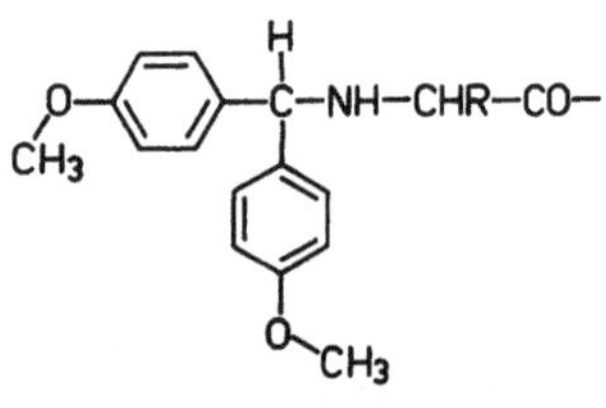

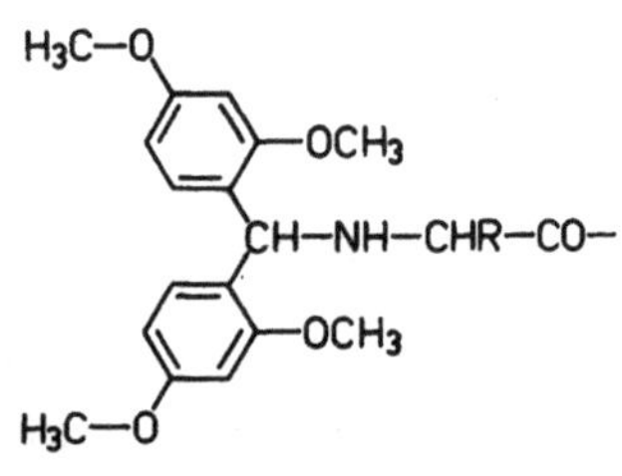

and is quite pronounced in the triphenylmethyl (trityl) amino acids [170–177]

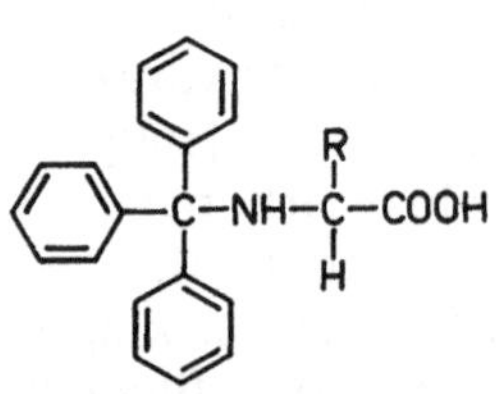

which are readily cleaved by weak acids such as 80% acetic acid and also, albeit slowly, by catalytic hydrogenation. Protection of the amino group by tritylation is a tempting idea and experiments in this direction have been reported for more than half a century [170]. Yet, the difficulties experienced in the preparation of trityl amino acids required continued attention [178] and the poor coupling rates caused by the bulk of the triphenylmethyl group had to be overcome by carefully chosen reaction conditions or by catalysis. It is understandable, therefore, that the trityl group has not been generally used for the protection of the amino function. A notable application should be mentioned here, a synthesis of insulin [179] in which the N-trityl groups were selectively removed by HCl in 90% trifluoroethanol in the presence of biphenylisopropyloxycarbonyl (Bpoc) and tert.butyloxycarbonyl (Boc) groups which could be kept intact under these conditions. It would not be surprising if the masking of amino groups by tritylation would gain more significance in the future.

Several attempts were made to take advantage of the readiness of amino acids to form well defined Schiff bases with benzaldehyde [180–182], 4-methoxysalicylaldehyde, 5-chlorosalicylaldehyde, or 2-hydroxynaphthaldehyde [183, 184]. These aralkylidene derivatives are potentially useful in synthesis because

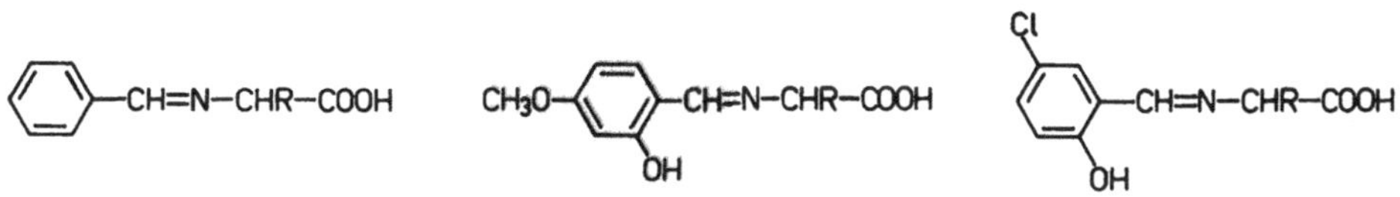

Schiff bases are hydrolyzed under mildly acidic conditions. As shown in the example of the aldimine obtained from 2-hydroxy-1-naphthaldehyde, intramolecular hydrogen bonds stabilize the Schiff base and make it somewhat more

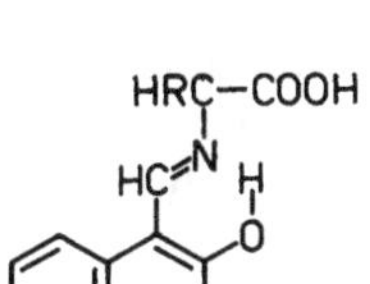

resistant to acids and hence less delicate and more tractable during the operations of peptide synthesis. Yet, even with such improvements, blocking of the amino function by condensation with aldehydes did not gain general acceptance and the same can be said about the use of a ketone, 2-hydroxy-5-chlorobenzophenone [185, 186], proposed

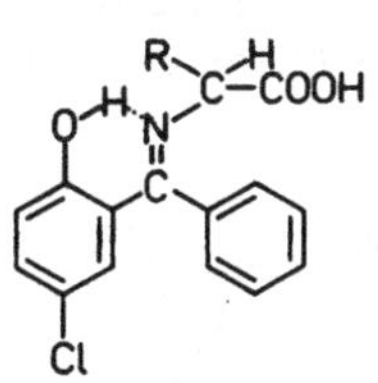

for the same purpose.

Application of the analytical reagent 1,1-dimethylcyclohexane-3,5-dione (dimedone) for the protection of the amino function [187, 188] is interesting, but has led, so far, to no further development, probably because the reagents proposed for deprotection, bromine-water or nitrous acid, can harm the side chain of certain amino acids, particularly those of tyrosine, tryptophan and methionine. The dimedone-protecteced amino acids

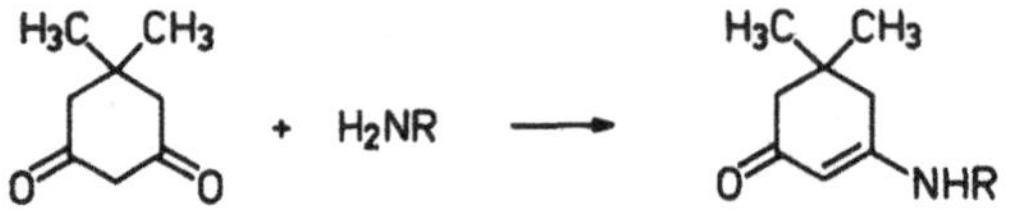

are enamines rather than Schiff bases, yet the geometry of the molecule does not provide additional stabilization similar to the one found in the enamines formed between amino acids and 1,3-dicarbonyl compounds [189, 190] such as pentane-2,4-dione (acetylacetone), benzoylacetone or esters of acetoacetic acid. The enamines obtained in the reaction of these diketones with amino acids are cleaved by weak acids, e.g. aqueous acetic acid, but intramolecular hydrogen bonds lend

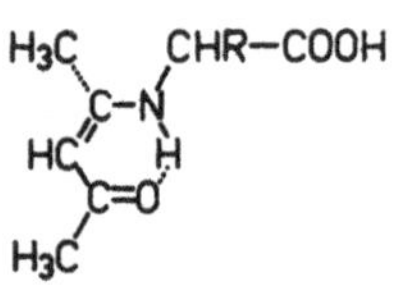

them sufficient stability to be useful in practical operations like the preparation of new penicillins through the acylation of 6-amino-penicillanic acid with enamine-protected amino acids [191] and also in the execution of peptide synthesis [192].

While dicarbonyl compounds can produce a two-point attachment with amino acids, simple monoketones, such as acetone, are able to engage two amino groups in a peptide [193, 194]. Dipeptides protected in the form of such imidazolidinones

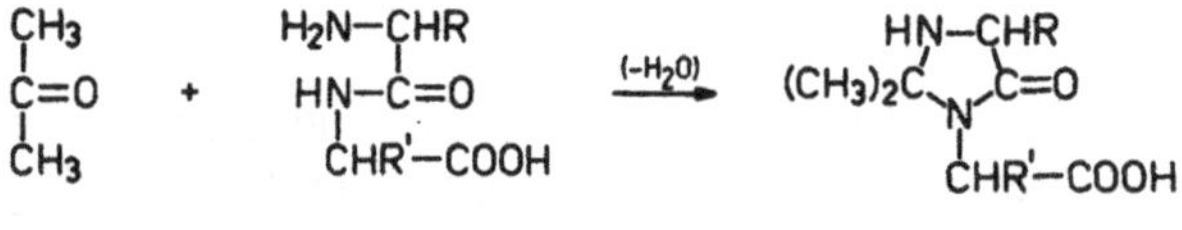

can be used [195] for the construction of peptide chains. Deblocking, removal of the isopropylidene group by hydrolysis, requires only weak acids, but under fairly drastic conditions.

The hydroxyketone benzoin can be converted to an enol carbonate that, in turn, reacts with amino acids to form oxazolinones. The latter can be coupled to other amino acids or peptides [196].

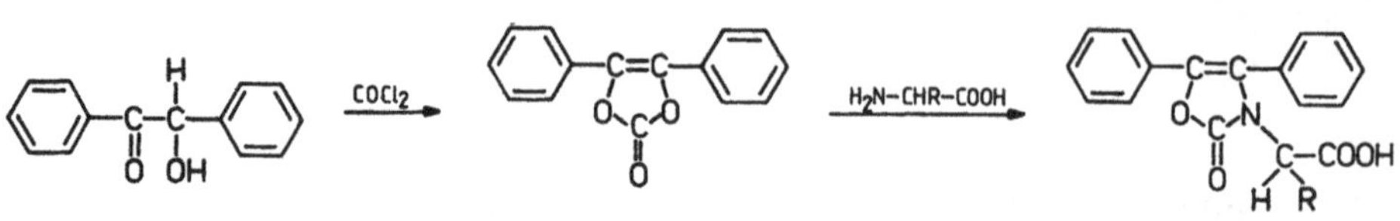

Removal of the fluorescent masking group requires reduction with sodium in liquid ammonia or catalytic hydrogenation in alcohol or dimethyl formamide.

2. Protection by Acylation

The smallest carboxylic acid, formic acid, offers an obvious solution for the protection of the amino function. Introduction of the *formyl group* presents no serious difficulty: formylamino acids were first obtained simply by heating amino acids with formic acid [197]. Later, several reagents were found which can formylate under milder conditions and without the loss of chiral purity. Thus, formic acid-acetic acid mixed anhydride [198, 199], alkyl- [200] or *p*-nitrophenyl formate [201] are equally suitable for this purpose, as is activation of formic acid with the help of coupling reagents, e.g. dicyclohexylcarbodiimide [202, 203].

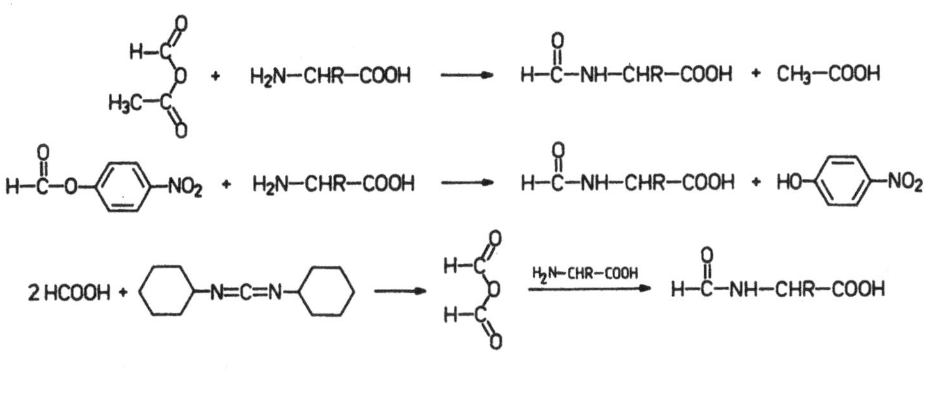

One can choose between several procedures for the removal of the formyl group. The initially used acid catalyzed hydrolysis [197] and alcoholysis [204] were replaced by oxidation with hydrogen peroxide [206] or by the application of nucleophilic reagents such as hydrazine acetate [206], aniline [207] or substituted anilines [207]. Catalytic hydrogenation was also suggested [208]. Yet, in spite of these many choices both for the introduction and for the removal of the formyl group, it was used only occasionally (e.g., ref. [209]) in the synthesis of peptides. For reasons not fully understood, formyl amino acids and formyl peptides tend to yield products in less than satisfactory yield and purity. A partial explanation for this may lie in the lack of protection by the formyl group against racemization.

The protection provided by the formyl group is imperfect only with respect to the conservation of chiral purity. Masking of the amino function by acylation with *carbonic acid*, however, provides less than sufficient blocking. In N-carboxyamino acid anhydrides, a carbonic acid residue serves both for protection and for activation and carbamoic acids are formed on reaction with nucleophiles:

Yet, the spontaneous decarboxylation of the protected intermediates is too rapid to prevent the incorporation of a second or even third amino acid residue through the reaction of the newly formed nucleophile with still unreacted N-carboxyanhydride [210–212]:

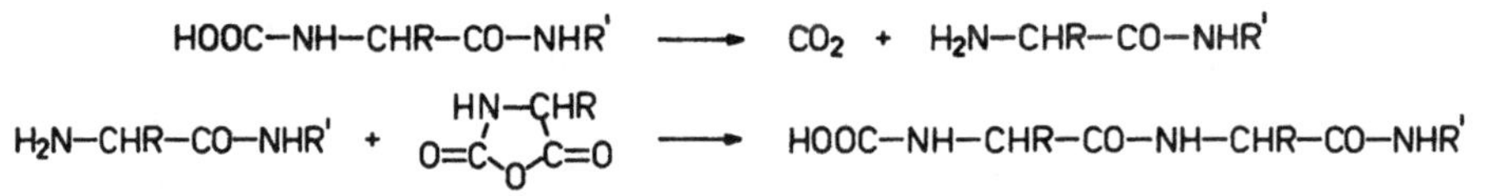

Under carefully selected and maintained basic conditions [213], premature decarboxylation can be prevented and double incorporation avoided. Nevertheless, the area for the general application of N-carboxyanhydrides remains the preparation of polyamino acids. The ambiguity inherent in N-carboxyanhydrides is absent from half esters of carbonic acid, which, in the form of urethanes, provide unequivocal protection for the amino group. Because of the numerous

$$R{-}O{-}\overset{O}{\overset{\|}{C}}{-}X \; + \; H_2NR' \longrightarrow R{-}O{-}\overset{O}{\overset{\|}{C}}{-}NHR' \quad (+ \; HX)$$

variations of the R group in such urethanes, a separate section will be dedicated to them.

Acetylation is not useful as a general method of amine-protection because the acetyl group does not prevent the racemization of activated amino acids, but it can be considered for the blocking of the side chain amino group of lysine residues from which the acetyl group can be selectively cleaved with acyl-lysine deacylase [52]. More options are available for the removal of substituted acetyl groups. Even substitution by a single chlorine atom opens up several possibilities. Thus, *chloroacetyl* peptides can be deblocked through the action of *o*-phenylenediamine [214] in water and also with

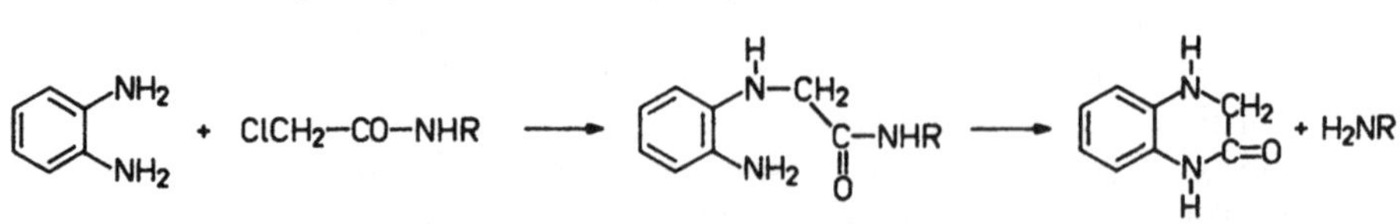

thiourea [215, 216] or substituted thioureas [217] such as 1-piperidinethiocarboxamide

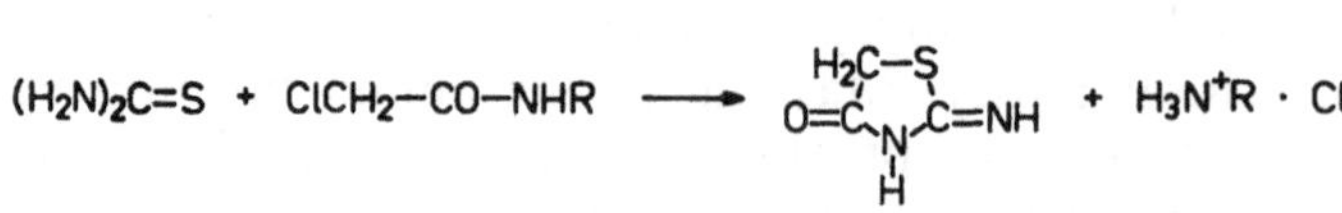

or with the aid of 2-aminothiophenol [218]. Removal of the chloroacetyl group with these reagents occurs, however, at elevated temperature, at the boiling point of ethanol. The related process [219], the reaction of the chloroacetyl group with pyridine-2-thione or 3-nitro-pyridine-2-thione, takes place in aqueous solutions of sodium bicarbonate and requires only cold trifluoroacetic acid for cyclization.

The principle of amine-protection by acyl groups which are removed by cyclization can also be discerned in the proposed application of homologs [220] and analogs [221] of the *o*-nitrophenoxyacetyl group

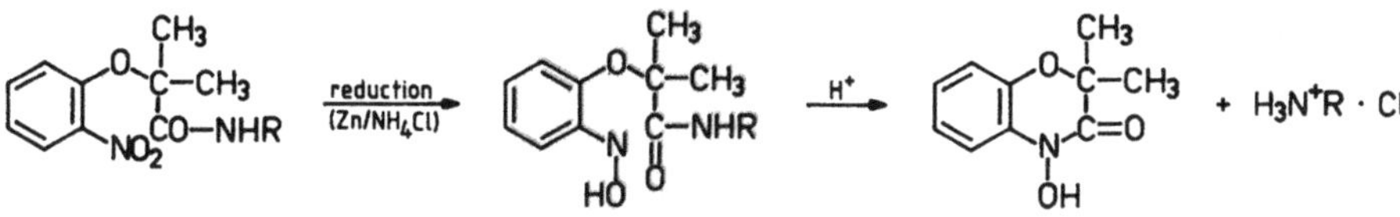

and also in the use of 4-hydroxy-4-methyl-pentanoic acid [222]. The latter can be introduced via the lactone while under acidic conditions (e.g., in aqueous trifluoroacetic acid) the protecting group is removed with the concomitant regeneration of the lactone:

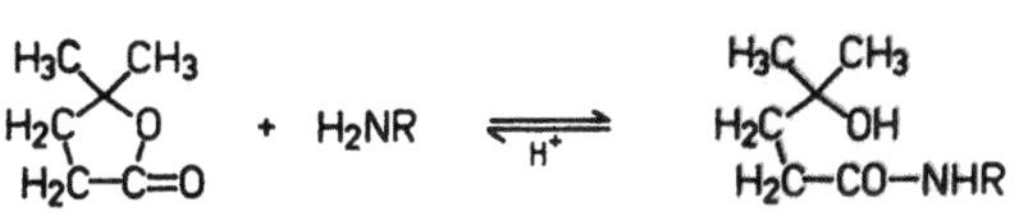

Similarly, cyclic reagents are used in the introduction of the *aceto-acetyl group* [223] and cyclization is the driving force in its removal by hydrazine, phenylhydrazine or hydroxylamine [224, 225]:

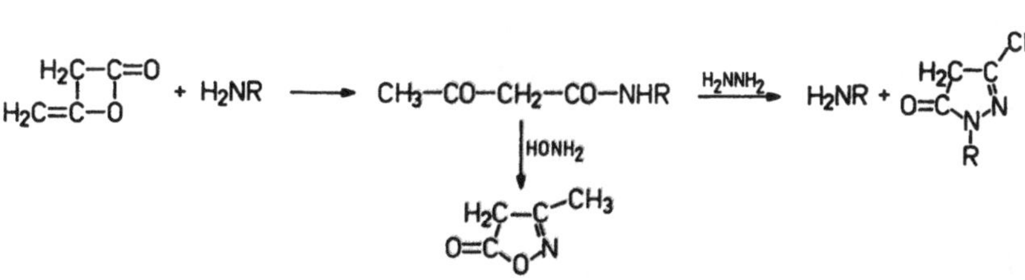

A definite practical significance has to be assigned to the *trifluoroacetyl* group [226]. It is readily introduced by acylation with trifluoroacetic anhydride in trifluoroacetic acid [227] or through reactive esters [228, 229] such as phenyl trifluoroacetate [203, 231]

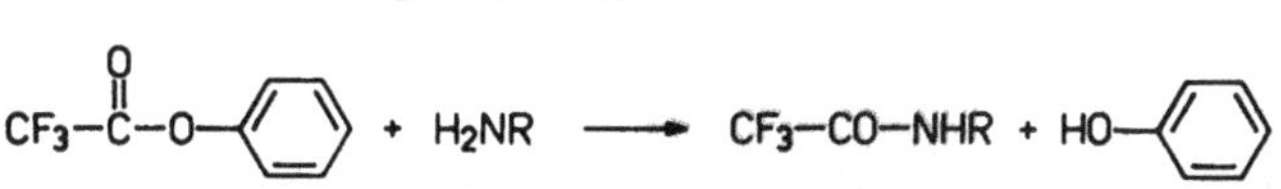

and also with trichlorotrifluoroacetone [232]. Removal of the trifluoroacetyl protection is equally simple because the highly reactive group is sensitive to alkali and to organic bases such as aqueous piperidine [233]. Reduction with sodium borohydride has also been recommended [234] for unmasking trifluoroacetylated amines. Racemization, however, is not reduced but rather enhanced in amino acids carrying a trifluoroacetyl group on their α-amino group. Therefore, this method of protection is usually limited to the blocking of the side chain function in lysine residues.

From the attempts to protect the amino function by simple acylation, the interesting concept of using cysteic acid as blocking agent [235] must be mentioned. The highly ionized sulfo group lends higher solubility, a much desired feature, to peptide intermediates. Removal of the blocking group by Edman degradation requires, however, that side chain amino groups remain masked, otherwise certain residues, e.g. lysine, are converted to phenylthiocarbamoyl derivatives.

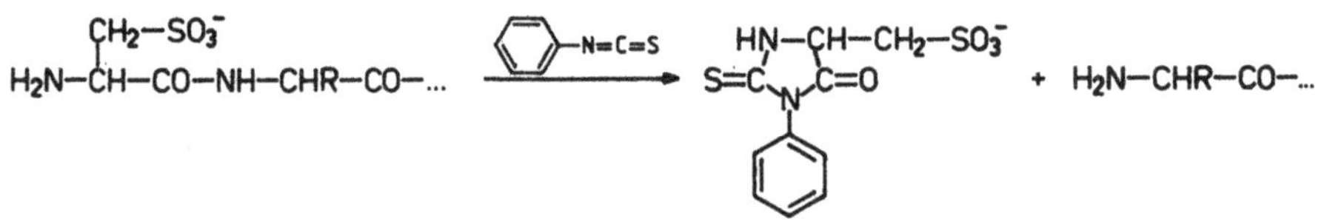

The possibility of selective removal led to proposals in which N-acyl-arginine [236] or acylmethionine [237] residues play the role of blocking groups. The former can be split off by trypsin catalyzed hydrolysis, the latter by reaction with cyanogen bromide [238]. The obvious limitations, however, caused by the presence of additional basic residues or methionine moieties respectively cannot be overlooked.

Diacylamides ensure a complete blocking of primary amines and, therefore, the *phthalyl group* (or *phthaloyl group*) looked promising from the start [239, 240]. Introduction of the protecting group was greatly improved when, instead of acylation with molten phthalic anhydride [241, 242], the reagent of Nefkens [243], N-ethoxycarbonylphthalimide, was applied [244] for phthalylation.

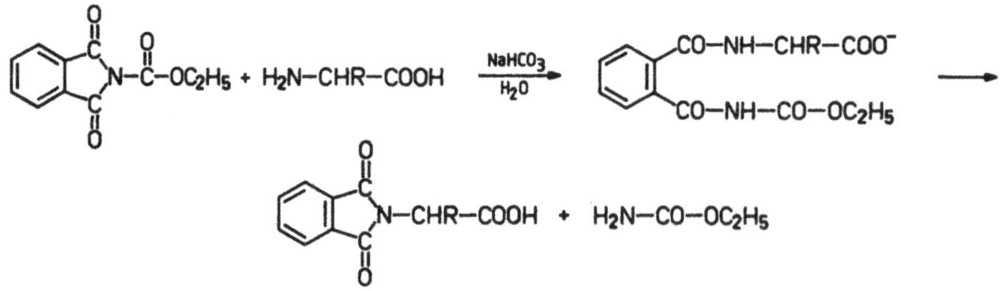

The reaction is carried out in aqueous sodium bicarbonate solution, at room temperature, and does not affect the chiral purity of the amino acids which are blocked in this way.

Removal of the phthalyl group is also simple: hydrazine attacks the protecting group in phthalyl peptides to form the cyclic, rather insoluble hydrazide of phthalic acid. In the presence of excess hydrazine, the more soluble hydrazinium salt of the byproduct is generated [245]

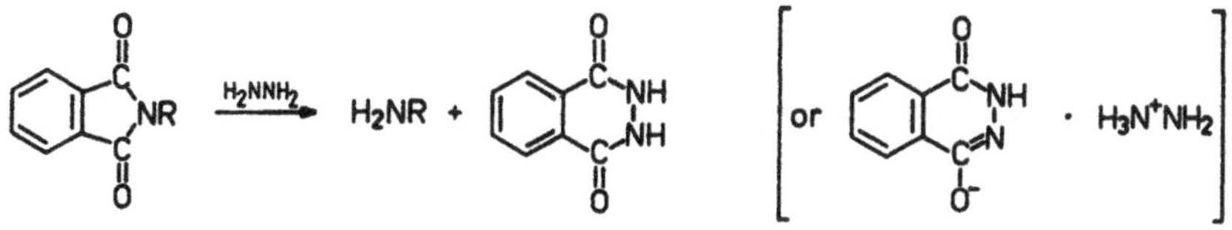

Improvements in the deblocking procedure, e.g. the use of phenylhydrazine [246] instead of hydrazine, were also recommended. Since both the introduction and the removal of the blocking can be carried out under favorable conditions and the phthalyl group is quite resistant to acids, the lack of continued popularity of the method must be sought in the imperfect stability of the phthalyl group toward nucleophilic attack, particularly against alkaline conditions. The phthalimide ring is readily opened up by bases; even aqueous sodium bicarbonate attacks it slowly.

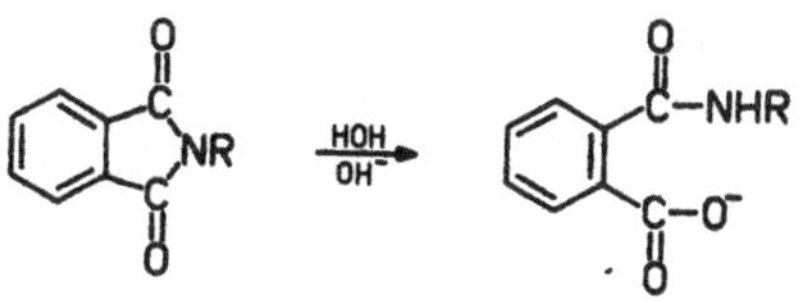

The analogous diacylamides, maleoyl derivatives [247] of amino acids or peptides have not gained practical significance in synthesis thus far. A relatively recent proposal of Barany and Merrifield, however, the use of the dithiasuccinyl group [248], could become important in the preparation of complex molecules because the new amine protecting (Dts) group

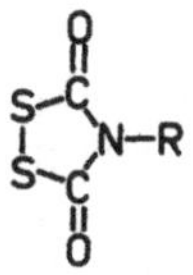

is resistant to acids and also to photolysis, while it is readily cleaved by thiols, e.g. by mercaptoethanol in the presence of tertiary bases. At this time the preparation of the protected amino acids could cause some difficulties, since it proceeds through several steps:

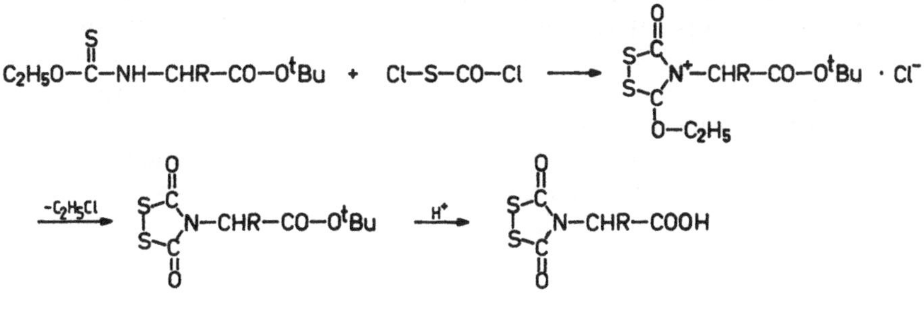

Also, in the thiolytic removal [249] of the Dts group, carbonyl sulfide is produced, a compound with rather unattractive properties:

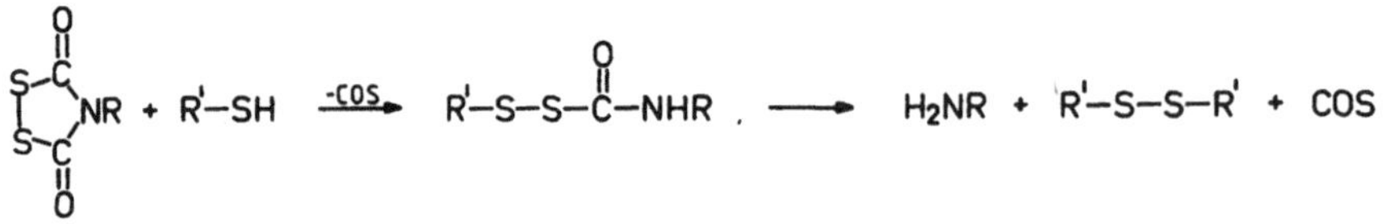

3. Protection of the Amino Group in the Form of Urethanes

a. Urethane Type Protecting Groups

The principle of using carboxyl-protecting groups for the blocking of amino groups through the conversion of the latter into urethanes was the most successful innovation in peptide synthesis. The significance of the benzyloxycarbonyl group (formerly carbobenzoxy or Cbz, now abbreviated as Z group) transcends its continued use over 50 years [6] and must be recognized in the development leading to a vast array of urethane-type protecting groups. Removal of the protection provided by these groups starts with the cleavage of the urethane ester bond

$$R{-}O{-}\overset{\displaystyle O}{\overset{\|}{C}}{-}NH{-}CHR{-}CO{-}\ldots \longrightarrow HOOC{-}NH{-}CHR{-}CO{-}\ldots$$

and is concluded with the loss of carbon dioxide from the resulting carbamoic acid. This spontaneous decarboxylation regenerates the unblocked amine.

$$HOOC{-}NH{-}CHR{-}CO{-}\ldots \longrightarrow CO_2 + H_2N{-}CHR{-}CO{-}\ldots$$

Many different solutions were found for the implementation of the first step. Esters were designed which render the urethane more resistant toward strong acids and others which provide the urethane with higher sensitivity to weak acids. Esters were proposed which are cleaved by elimination rather than by acidolysis or hydrogenolysis, etc. This extensive development and the confinements of a small volume force us to condense the urethane-type protecting groups in a few tables. Even in this way we do not aim at a complete listing of all urethanes proposed for amine-protection in peptide synthesis. The tables demonstrate many suggested variations, most of which were never tried in demanding preparative work. Also, just because of their large number, no direct comparisons between the results obtainable with individual methods of protection are available. This lack of information leads to a situation in which most laboratories stick to a few well established groups and ignore the vast array of alternative possibilities. The only way out of this dilemma seems to be a demonstration of the advantages of a new protecting group by those who recommend its use and the demonstration must extend to the synthesis of complex molecules.

A thorough and systematic treatment of urethanes in peptide synthesis can be found in a review article by Geiger and König [250] and an even more comprehensive account was written by Wünsch [251]. Here we show, in Table 5, those protecting groups which were developed from the benzyloxycarbonyl group [6]. Because of the presence of a *benzylic* carbon atom, these groups are removable both by hydrogenolysis and by acidolysis.

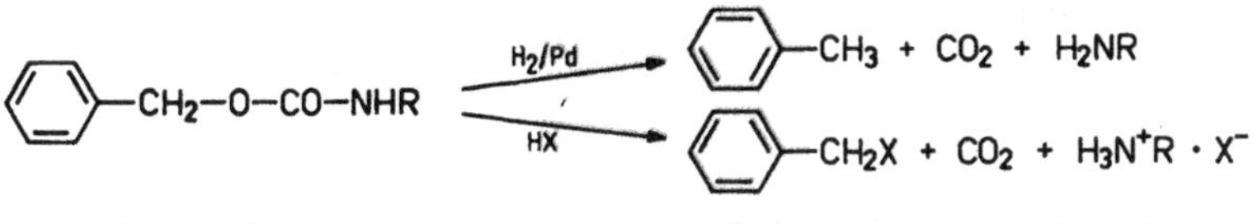

In Table 6, blocking groups derived from the tert. butyloxycarbonyl group [26–28] demonstrate the almost unlimited possibilities to form protecting groups removable on account of the generation of a stable carbo-cation.

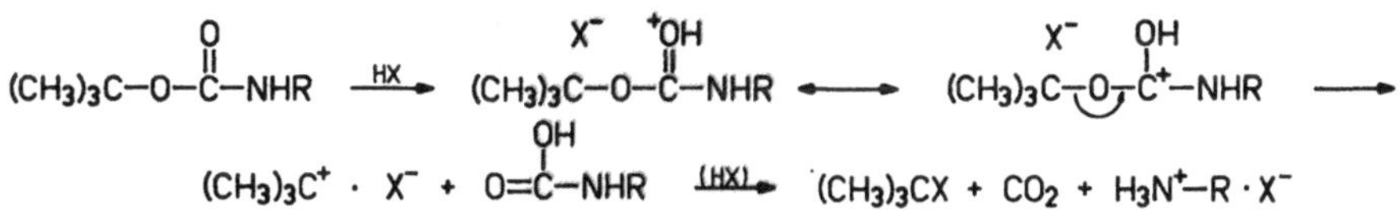

In Table 7, urethane-type blocking groups which are removable by base-induced β-elimination are listed:

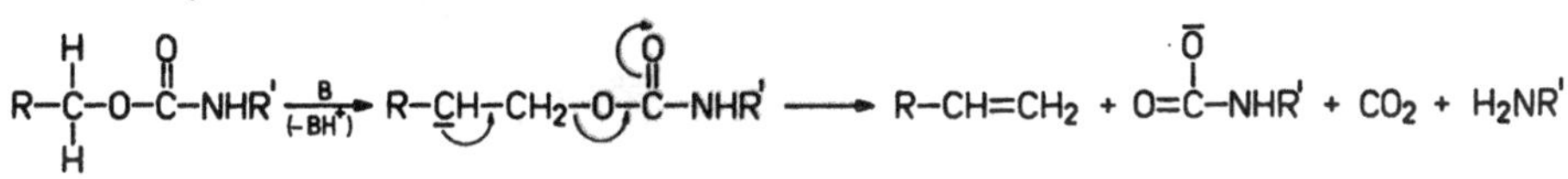

Table 5. "Benzyl"-components in Urethane-type Protecting Groups

„Benzyl" Component	Refs.	„Benzyl" Component	Refs.	„Benzyl" Component	Refs.
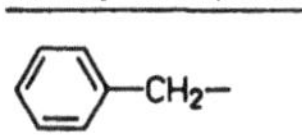	6	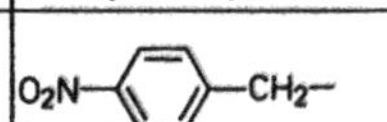	259,260	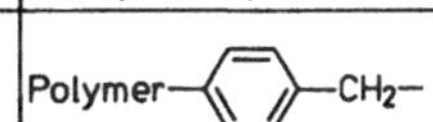	263
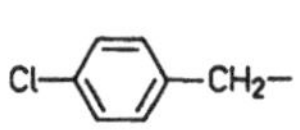	252,253	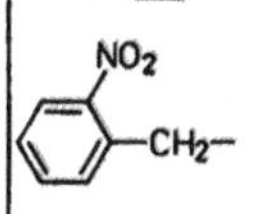	48,261	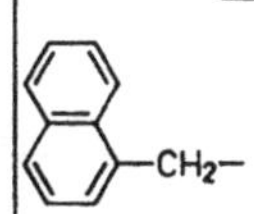	257
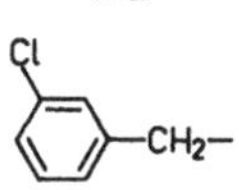	254	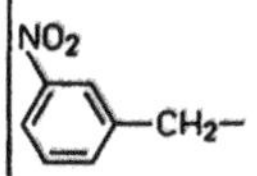	255	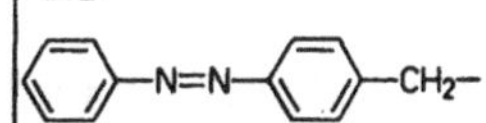	158,264
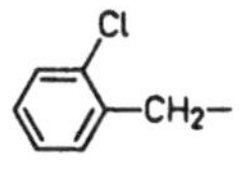	255,256	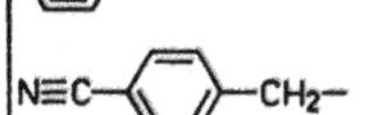	255,262	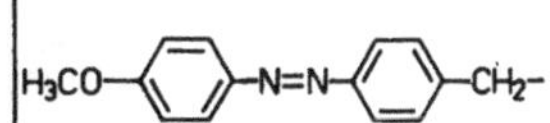	264
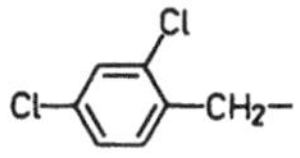	256	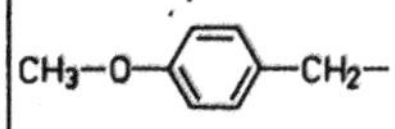	25,27	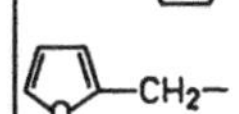	265,266
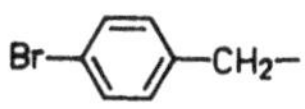	257	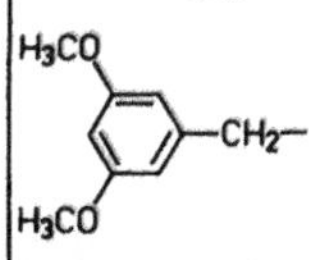	46,47	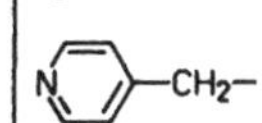	267
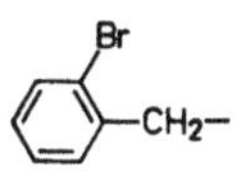	258	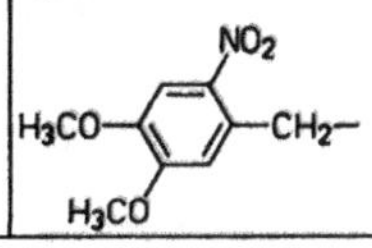	48,261	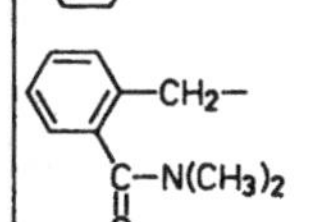	268

Table 6. Carbocations Formed in Deprotection by Acidolysis

Cations	Refs.	Cations	Refs.	Cations	Refs.
a Secondary carbocations		b Tertiary carbocations			
$(CH_3)_2CH-\overset{+}{C}H-CH(CH_3)_2$	27,269	$CH_3-C^+(CH_3)_2$	26,27	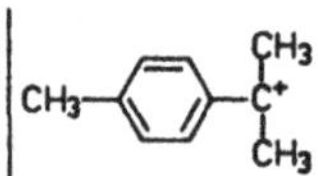	29
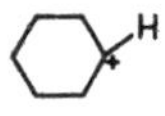	27	$CH_3CH_2-C^+(CH_3)_2$	273,274	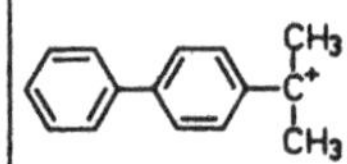	29
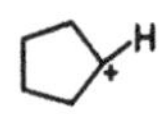	27	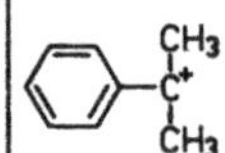	29	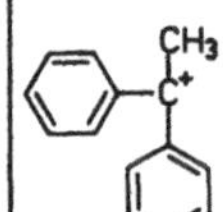	29
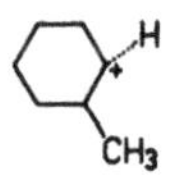	254	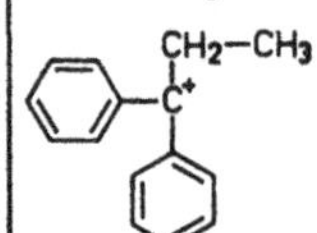	29	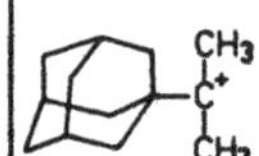	277
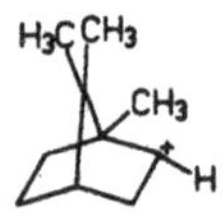	270,271	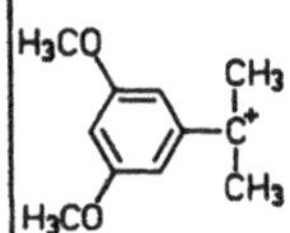	275	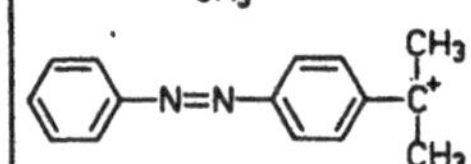	278
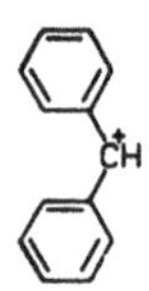	89,272	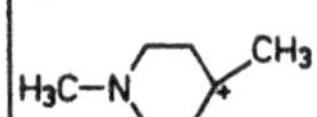	268,276	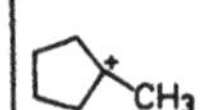	279
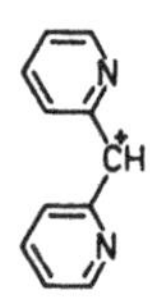	268	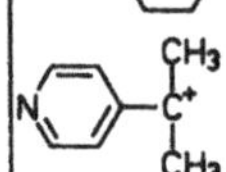	268,276	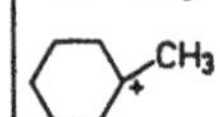	279
		$(CH_3)_2N-CO-CH_2-CH_2-C^+(CH_3)_2$	268,276	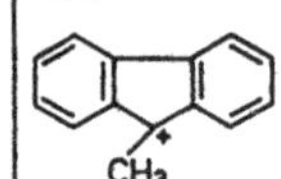	280
		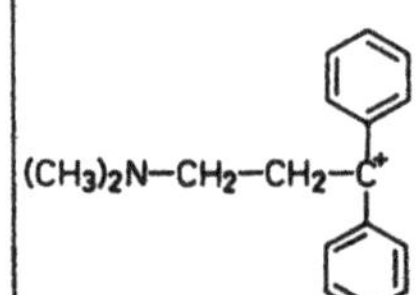	268,276		281

Table 7. Urethanes Cleaved by β-elimination

Alcohol component	Refs.	Alcohol component	Refs.	Alcohol component	Refs.
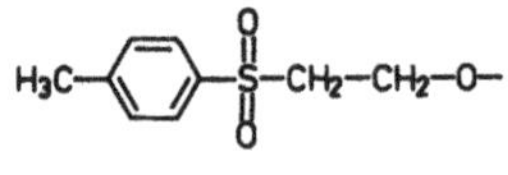	282	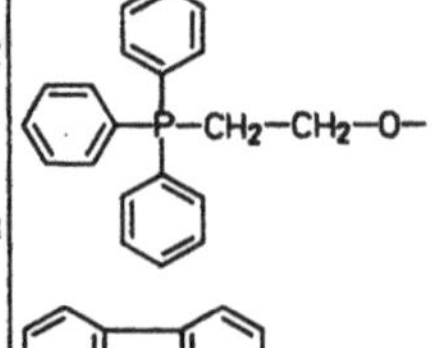	284	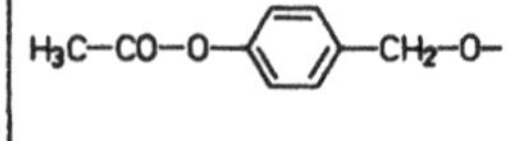	286
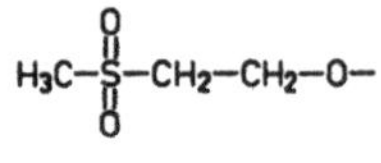	283			Cl_3C-CH_2-O-	287
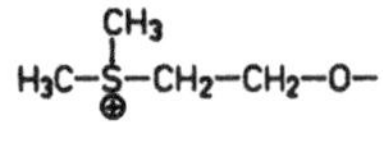	284	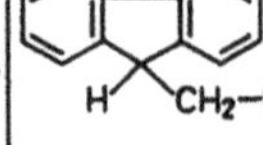	34,35	$N{\equiv}C-CH_2-C(CH_3)_2-O-$	288
		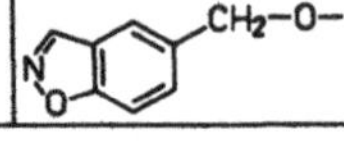	285		

An attempt to rigidly classify the urethane-type protecting groups which appeared in the literature cannot be entirely successful. There are several among the suggested methods of protection which do not fit in the categories of Tables 5, 6 or 7. Because of novel or unconventional concepts in some of these methods, they should be discussed here separately. For instance, the allyloxycarbonyl group [289], removable by acidolysis and also

$$CH_2{=}CH{-}CH_2{-}O{-}CO{-}NHR$$

by catalytic hydrogenation (albeit incompletely, because of competing saturation of the double bond), or the homologous vinyloxycarbonyl group [290] which is cleaved by addition of HCl followed by a reaction with ethanol at somewhat elevated temperature:

$$CH_2{=}CH{-}O{-}CO{-}NHR \xrightarrow[\text{(dioxane)}]{HCl} CH_3{-}CHCl{-}O{-}CO{-}NHR \xrightarrow{C_2H_5OH}$$

$$CH_3{-}CH(OC_2H_5){-}O{-}CO{-}NHR \xrightarrow{C_2H_5OH} CH_3{-}CH(OC_2H_5)_2 + CO_2 + Cl^-N^+RH_3$$

A certain revival of the allyl group idea can be recognized in the 2-ethynyl-2-propyloxycarbonyl group [291, 292] cleaved by catalytic hydrogenation[8],

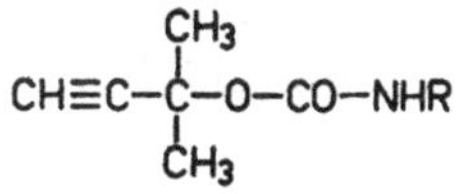

although acid hydrolysis also can split the masking group with the formation of a cyclic carbonate [293]:

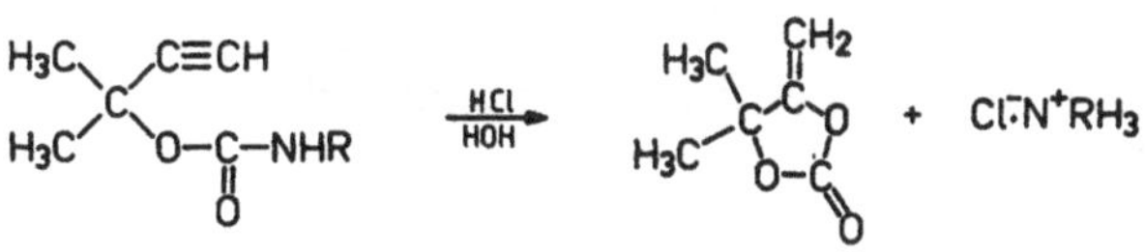

Urethanes constructed from cyclopropylcarbinol and 1-cylopropylethanol [279]

are acid labile, but offer no particular advantage over other acid sensitive protecting groups. On the other hand, masking groups that contain a hydrophobic moiety can lend favorable solubility properties to the protected intermediates. In this respect the very acid sensitive 4-decyloxy-benzyloxy-

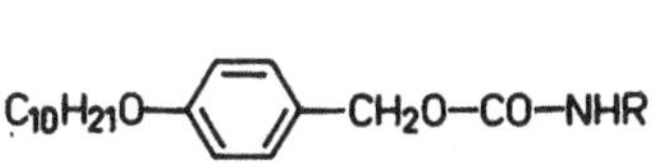

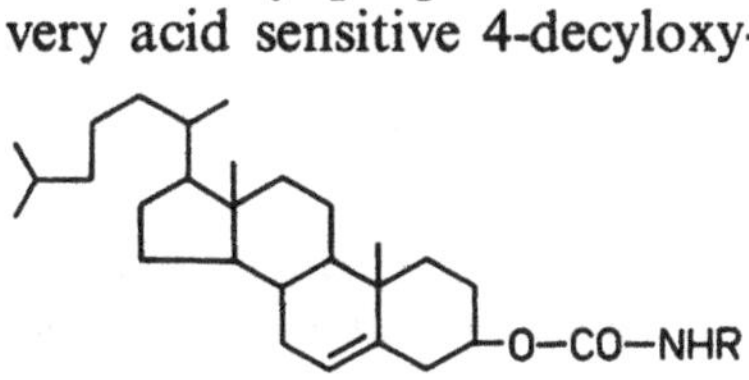

[8] This group can be removed by hydrogenation (with a Lindlar catalyst) even from methionine containing peptides.

carbonyl group [294] is quite promising. The further enhanced lipid character of peptides protected by the cholesteryloxycarbonyl group [295] is somewhat counterbalanced by the need of relatively strong acids for its removal. A wide range of solubility properties can be selected for peptides protected by the 4-dihydroxyboronato-benzyloxycarbonyl group [296] since the dihydroxyboron function allows reversible reactions with a variety of

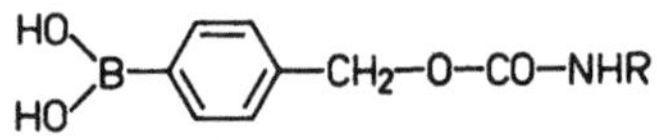

vicinal diols. Favorable modifications in solubility should be provided also by groups in which dialkylhydroxylamines play the role of the alcohol component of the urethane. Both 1-hydroxypiperidine [297] and N,N-dimethylhydroxylamine [298] can serve for this purpose.

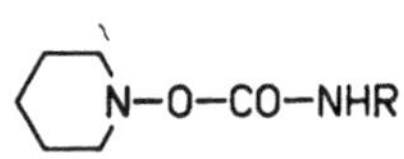

$(CH_3)_2N{-}O{-}CO{-}NHR$

The tempting replacement of alcohols by thiols is less practical. The phenylthiocarbamoyl group [299] retains some of the reactivity of thiophenyl esters in spite of deactivation by the urethane grouping [300]. The acid stable alkylthiocarbonyl peptides [21] require oxidation

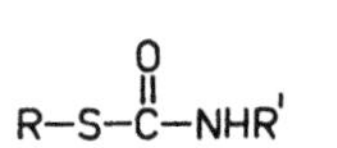

$(R = CH_3{-}, C_2H_5{-}, C_4H_9{-}, C_6H_5{-}CH_2{-})$

with ozone or peracids for deprotection. The more attractive reductive cleavage becomes possible in aryl or alkyl dithiocarbamoyl derivatives or carbamoyl disulfides [301, 302].

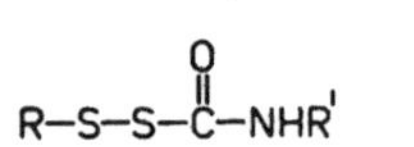

$(R = CH_3{-}, C_2H_5{-}, (CH_3)_2CH{-}, (CH_3)_3C{-}, C_6H_5{-})$

b. Introduction of Urethane-type Protecting Groups

The preparation of benzyloxycarbonylamino acids [6] presents no serious problems. Benzyl chlorocarbonate, the reagent generally used for this purpose, is readily obtained through the reaction of phosgene with benzyl alcohol, with [303] or without [304] a diluent. This half ester, half

chloride of carbonic acid disproportionates, however, on heating or storage, to form symmetrical derivatives of the acid, phosgene and dibenzyl carbonate:

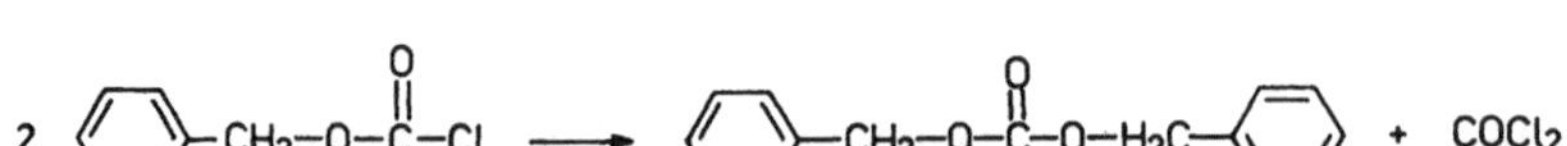

Therefore, it is advisable to store the reagent in the cold and distil it, if necessary, in high vacuum at moderate temperature, certainly not exceeding 60 °C. Also, preparations stored for prolonged periods and commercially obtained samples should be freed from phosgene by passing a stream of nitrogen through them (hood!). The remaining dibenzyl carbonate is relatively harmless in the ensuing Schotten-Baumann acylation of an amino acid since, if it reacts at all, it will yield the same Z-amino acid with the elimination of benzyl alcohol.

In spite of the simplicity of the procedure and the low cost of the chlorocarbonate, attempts were made to find reagents which are more stable and can produce Z-amino acids under more carefully controlled conditions. An early solution for this problem is the adaptation of the active aryl ester procedure [138] for the preparation of Z-amino acids with aryl benzyl carbonates as acylating reagents [305]:

The principle of using mixed carbonates for the introduction of urethane-type amine-protecting groups into amino acids found many followers and, in addition to various substituted aryl carbonates, reagents with 1-hydroxysuccinimide and 1-hydroxypiperidine as activating "alcohol"-components were also recommended for the same purpose. In the absence of direct comparisons between the efficacies of diverse mixed carbonates, it is difficult to choose between them. This might be one of the reasons for the undiminished popularity of benzyl chlorocarbonate.

In the preparation of p-methoxybenzyloxycarbonyl amino acids, the freshly prepared acid chloride can be used for acylation of the amino group [306–308], but it is probably more practical to secure a storable intermediate, such as the carbazate [307, 308] which is converted to the azide just prior to acylation:

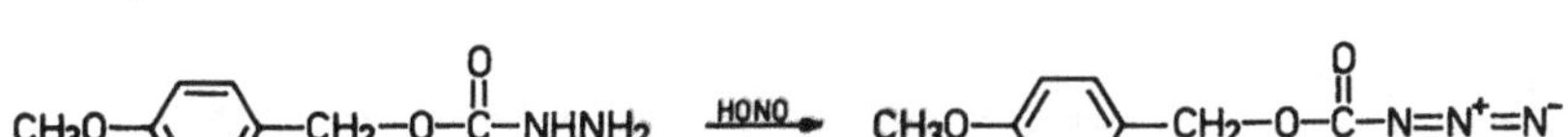

For the same reason, the tert. butyloxycarbonyl and the tert. amyloxycarbonyl groups were introduced [309, 310], initially with the aid of the half ester-half azides (tert. butyl azidoformate) or as the mixed carbonates [28].

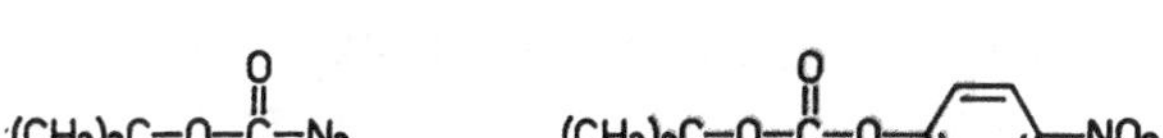

Acylation with alkyl carbonic acid *p*-nitrophenyl esters could be extended to the incorporation of the *p*-methoxybenzyloxycarbonyl group as well [307]. These reactions can be carried out in aqueous-organic media and high yields were achieved in acylation with azides at a constant p_H [311], but Boc-amino acids could be secured also in non-aqueous media, e.g. in dimethylformamide in the presence of tertiary amines [312, 313]. The azide procedure was applied in the case of more complex protecting groups as well, e.g. for the incorporation of the benzhydryloxycarbonyl group [89]. In connection with highly acid-sensitive protecting groups, the active ester method is usually preferred. Thus, the biphenylisopropyloxycarbonyl group is attached to amino acids via the carbonic acid phenyl ester [315] or substituted phenyl esters [316]. Yet, while the chlorides of *p*-methoxybenzyl carbonic acid and of tert.butyl carbonic acid are unstable, the corresponding fluorides could be secured and used for acylation [317].

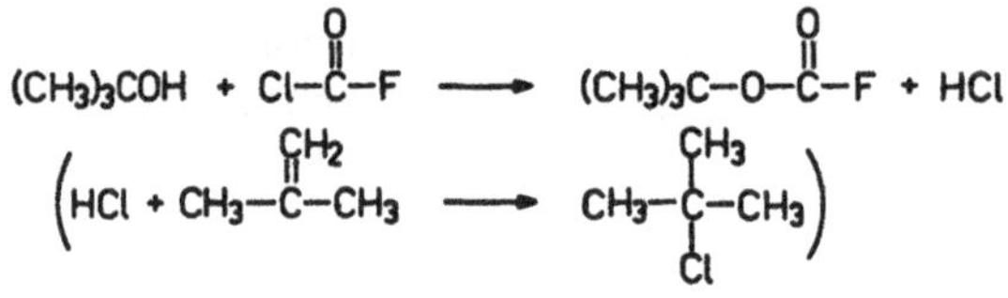

In an analogous manner, fluoroformates could be obtained from anisalcohol, 2-furfurylcarbinol and 3,4,5-trimethoxybenzylalcohol [317] and later also from 1-adamantanol [318].

In addition to activation in the form of acid halogenides, azides or active esters, alkyl carbonic acids can be converted to reactive intermediates such as mixed anhydrides [319] or symmetrical anhydrides (pyrocarbonates) [320, 321]:

An interesting alternative for the incorporation of urethane-type masking groups is the transformation of the amines into isocyanates [27], e.g. with the aid of carbonyldiimidazole [322] and subsequent addition of the alcohol:

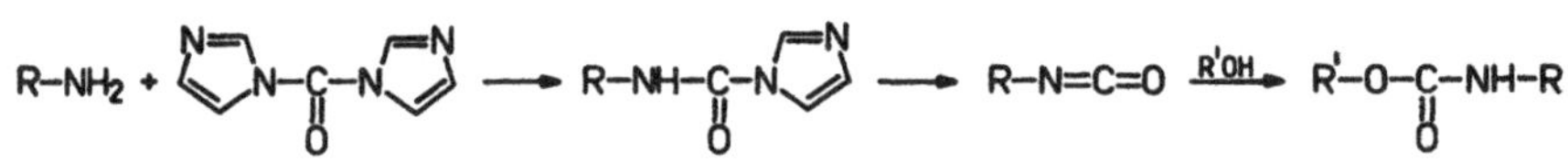

This approach was useful, for instance, in the introduction of the diphenylmethyloxycarbonyl (benzhydryloxycarbonyl) group [89] or the cholesteryloxycarbonyl group [295].

A certainly incomplete summary of reactive forms of alkyl carbonates proposed for the introduction of urethane-type protecting groups is shown in Table 8.

The methods which can be applied for the formation of urethanes are essentially the same as the ones which are used for the formation of the peptide bond. In fact, most acylating agents can be applied for both purposes. The convenience, however, to have a relatively stable reagent in hand, which can be stored (and sold) and is available when needed, favors the use of

Table 8. Introduction of Urethane-type Protecting Groups

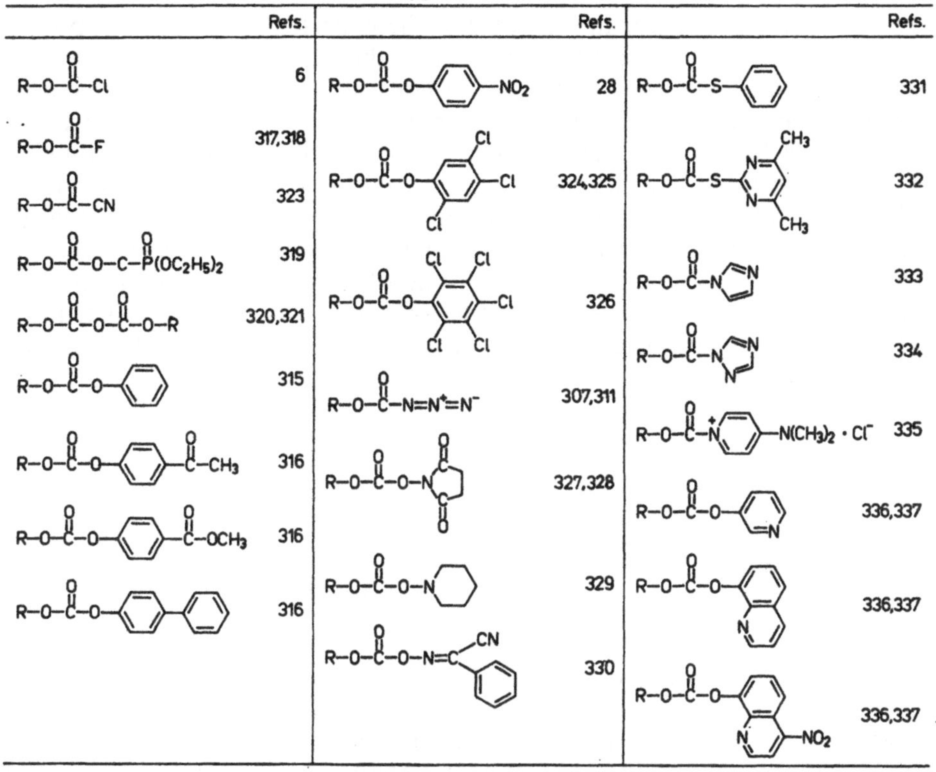

mixed carbonates. Thus, *p*-nitrophenyl carbonate was proposed [338] recently for the incorporation of the trimethylsilylethyloxycarbonyl [44]

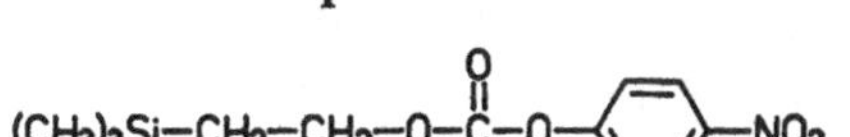

group. Some mixed carbonates might offer special advantages. For instance esters of 2-mercapto-3,5-dimethylpyrimidine [332] eliminate, on acylation, a pyrimidinethiol which is soluble in aqueous acids and thus readily removed from the reaction mixture. Even an incomplete Table 8 demonstrates that there is an ample choice of methods for urethane formation.

As a concluding comment on the preparation of mixed carbonates, we should mention that it is not always necessary to start from phosgene and the alcohols and phenols respectively. Transesterification of di-p-nitrophenyl carbonate with an alcohol in pyridine can provide the needed reagent [339], e.g.

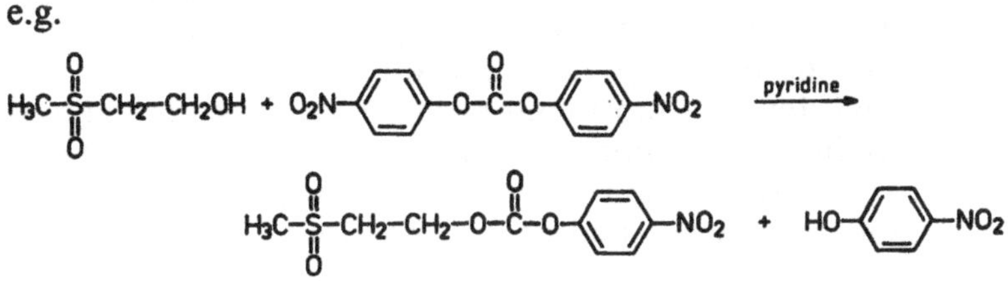

c. Removal of Urethane-type Protecting Groups

The *reductive* methods for the removal of the benzyloxycarbonyl group were briefly reviewed in the introductory part of this chapter. From the various reducing agents mentioned there, only a few are commonly used by peptide chemists. Catalytic hydrogenation [6, 7] remains the method of choice and it is applicable for most of the urethanes listed in Table 5 and for some in Table 6. The separately mentioned 2-ethynyl-isopropyloxycarbonyl group [291, 292] is also cleaved by hydrogenolysis. It is probably worthwhile to try to remove protecting groups by catalytic hydrogenation even when the structure of the blocking group provides no obvious clue for the outcome of the reaction. This is the case with the 9-fluoromethyloxycarbonyl (Fmoc) [34, 35] group which is quantitatively reduced [340, 341] to 9-methylfluorene and the carbamoic acid:

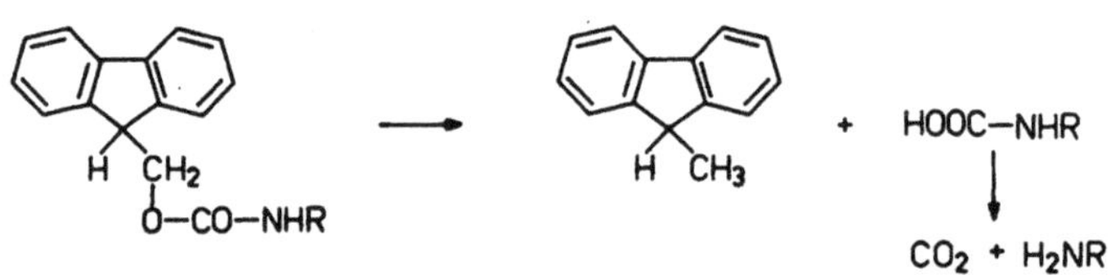

The execution of deblocking by hydrogenolysis remains unchanged in most laboratories: it is usually carried out at atmospheric pressure and room temperature in alcohol, or aqueous acetic acid, less frequently in dimethylformamide, with palladium on charcoal or palladium black catalysts. A word of caution should be added here: active palladium catalysts are pyrophoric and can ignite the solvent, particularly methanol.

The recommendation of Medzihradszky and Medzihradszky-Schweiger [342, 343] to perform catalytic reduction in the presence of organic bases has been rarely followed [344, 345], although this permits the removal of benzyloxycarbonyl groups from methionine containing peptides and leaves benzyl ethers unaffected [346]. The related, and still very original, catalytic reduction in liquid ammonia [12] might gain more application in the future. Transfer hydrogenation with various hydrogen donors [13—16] remains to be tested in major syntheses. In an interesting contribution [347], a new catalyst $K_3[Co(CN)_5]$ was proposed for the removal of benzyloxycarbonyl groups and benzyl esters. It leaves benzyl ethers on serine and threonine residues unchanged. The catalyst is readily prepared *in situ* from $CoCl_2$ and KCN.

The most important alternative, reduction with sodium in liquid ammonia [11], has been less frequently applied in recent years, yet, it can save a synthesis when problems arise in the final deprotection [348].

Acidolysis is, beyond doubt, a most convenient process for the removal of urethane-type protecting groups. For instance, in the case of the tert. butyloxycarbonyl group, a mere dissolution of the protected peptide in trifluoroacetic acid is sufficient for cleavage and decarboxylation of the carbamoic acid. The amine is obtained as the trifluoroacetate salt simply by evaporation or by precipitation with ether or ethyl acetate.

$$(CH_3)_3C{-}O{-}CO{-}NHR + CF_3COOH \longrightarrow (CH_3)_3C{-}O{-}COCF_3 + HOOC{-}NHR$$

$$HOOC{-}NHR + CF_3COOH \longrightarrow CO_2 + CF_3COO^- \cdot H_3N^+R$$

The intermediate carbocation and/or the tert. butyl trifluoroacetate [349] formed can alkylate some sensitive sites in amino acid side chains. This is an obvious shortcoming of acidolysis and must be corrected by the addition of scavengers [350]. Trifluoroacetic acid is often the reagent of choice because it is also an excellent solvent for the intermediates which are sometimes poorly soluble in other solvents. Thus, it was reasonable to attempt to use it for the cleavage of less acid sensitive protecting groups, such as the benzyloxycarbonyl group as well. This, however, required higher temperatures [350] or prolonged periods of time if deprotection was carried out at room temperature [351]. More recently, Kiso and his associates [352] observed that thioanisole accelerates acidolysis, probably through a push-pull mechanism. The same effect aids the cleavage of benzyl esters and

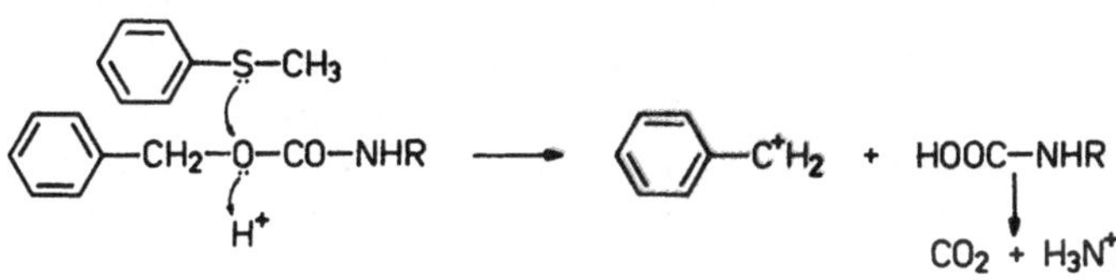

benzyl ethers as well and thioanisole enhanced the reaction rate also when acids other than trifluoroacetic acid were used for deprotection.

It would be difficult to list all the various acidic reagents which were proposed for the removal of blocking groups. Such a list would include HCl in acetic acid [76], in ethyl acetate [353], dioxane [354], diethyl phosphite [28], or water [355], HBr in acetic acid [24, 27], in liquid SO_2 [356], and as neat liquid [357]. Several sulfonic acids such as p-toluenesulfonic acid [358] or trifluoromethanesulfonic acid [359] were recommended, as was formic acid [360]. While strong acids induce, particularly in the presence of air, some decomposition of the tryptophan side chain, a specially developed reagent, β-mercaptoethanesulfonic acid [361], can be safely used in the deblocking of tryptophan containing peptides. Lewis acids, for instance boron trifluoride ethereate [362], boron tribromide [363], and boron tri-trifluoroacetate [364], were also proposed for the removal of acid sensitive masking groups. The most general and hence least selective reagent, liquid hydrogen fluoride [365] and the related pyridine polyhydrogenfluoride [366, 367], might cause side reactions not frequently encountered with less powerful acids. To mention only one, N $\rightarrow$ O acyl migration [368] is likely to occur if serine containing peptides are dissolved in liquid HF:

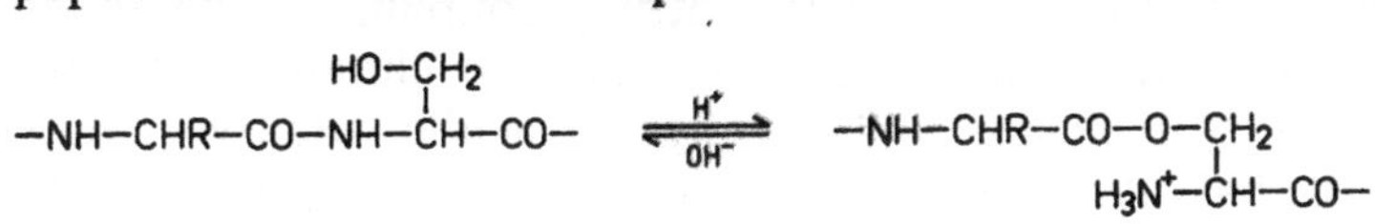

It is understandable, therefore, that highly acid-labile groups and weak acids that can be applied for their removal receive increased attention. Some early studies on the kinetics and mechanism of the cleavage of benzyloxycarbonyl groups [369, 370] provided a better insight in the details of the process and the exceptionally productive research of Sieber and Iselin [29] on the relative stability of a whole series of urethanes resulted in the addition of the biphenylisopropyloxycarbonyl (Bpoc) group to the armament of peptide chemists. This very sensitive group is selectively removed in the presence of tert.butyloxycarbonyl groups, which are hardly affected if the acidity of the system, a solution in 90% aqueous trifluoroethanol, is carefully maintained at pH 2, controlled with a glass electrode. In some cases the Bpoc group seemed to be too delicate: it was sensitive even to the slightly acidic additive 1-hydroxybenzotriazole [371]. Recent studies in the laboratory of the author revealed that the Bpoc group can be quantitatively removed by 1-hydroxybenzotriazole and the resulting amine salt acylated in the subsequent coupling step without the addition of a tertiary amine.

The embarassing richness of reagents recommended for acidolysis might contribute to the further development of peptide synthesis, but only if valid comparisons are made between various approaches. For such comparisons analytical procedures are needed, such as the determination of tert.butyloxycarbonyl groups [372], etc. Also, fine details in mechanisms, like those revealed by the conspicuous difference in the acid sensitivities of the bornyloxycarbonyl [373] and isobornyloxycarbonyl groups [270, 271], must be studied for a

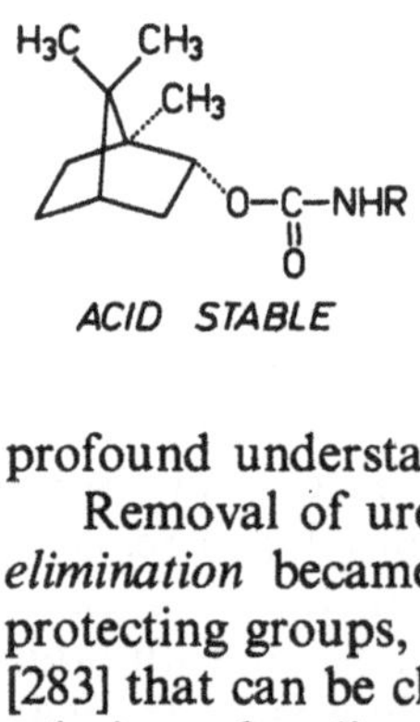

ACID STABLE

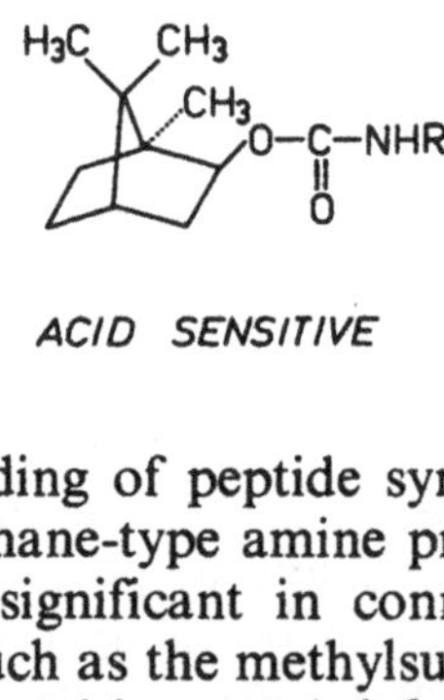

ACID SENSITIVE

profound understanding of peptide synthesis.

Removal of urethane-type amine protecting groups by base induced *β-elimination* became significant in connection with highly alkali sensitive protecting groups, such as the methylsulfonylethyloxycarbonyl (MSc) group [283] that can be cleaved by very brief exposure to a dilute aqueous-organic solution of sodium hydroxide at or below room temperature. Still, the β-elimination approach reached general acceptance only when the 9-fluorenylmethyloxycarbonyl (Fmoc) group [34, 35] was proposed for the protection of the amino function. The sensitivity of the Fmoc group, particularly to secondary amines, allows deblocking with dilute solutions of piperidine or diethylamine in dimethylformamide under mild conditions. The risk of premature removal [374] of the Fmoc group during coupling by the amino component in the reaction mixture can practically be eliminated by the addition of 1-hydroxybenzotriazole. This slightly acidic additive simultaneously reduces the basicity of the system and the time of exposure of the

Fmoc amino acid or peptide to adverse conditions. To face up to the problems created by the formation of a tertiary amine from the dibenzofulvene produced and the secondary amine used for deprotection,

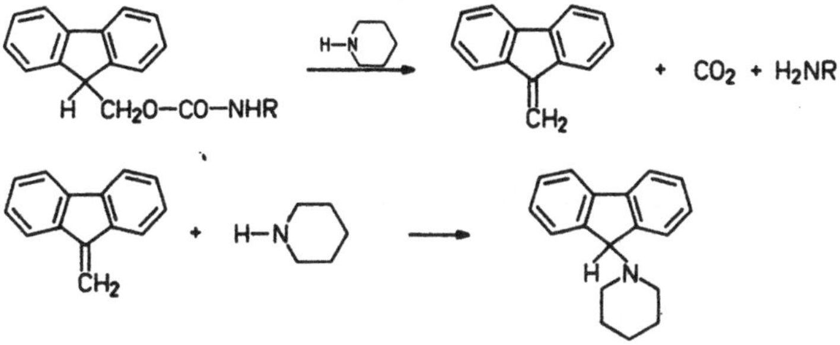

Carpino and Williams [375] developed an insoluble secondary amine by reacting piperazine with the Merrifield resin [76]. This polymeric reagent swells in dichloromethane, cleaves the Fmoc group and removes the by-product, dibenzofulvene, from the reaction mixture:

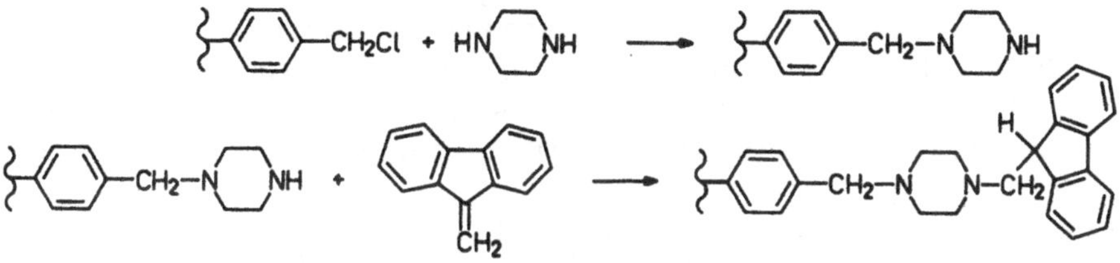

It seems to be worthwhile to point out some less conventional elimination reactions, e.g. the removal of the 5-benzisoxalylmethyloxycarbonyl group [285]

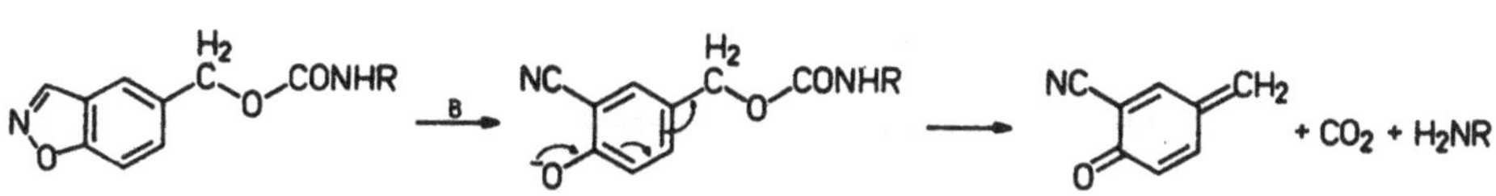

or that of the 4-acetoxybenzyloxycarbonyl group [286]

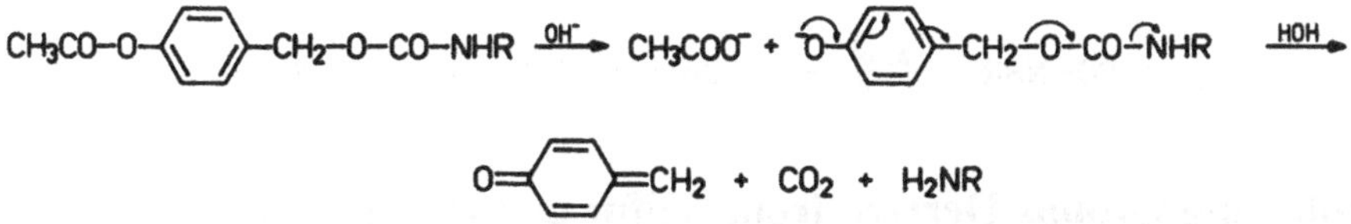

The dihydroxyboronatobenzyloxycarbonyl [296] protection could have been added to Table 5 since it can be cleaved, like other substituted benzyloxycarbonyl groups, by reductive methods or by acids, but it also could belong to Table 7 because, after oxidation with hydrogen peroxide, elimination of the group takes place in a manner which is analogous to the removal of the acetoxybenzyloxycarbonyl group.

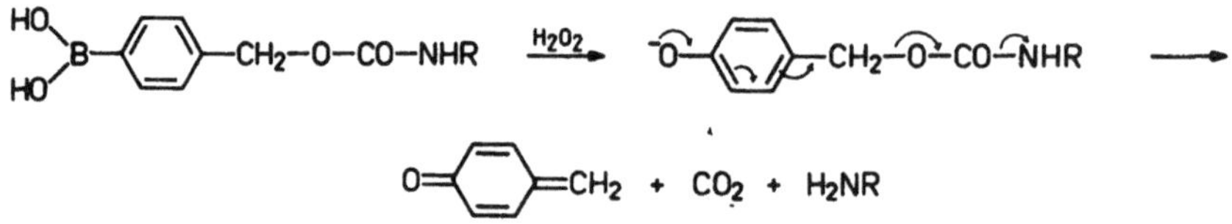

Also, the dihydroxyboronato moiety is lost on treatment of the protected peptide with water in the presence of silver diammine ions and benzyloxycarbonyl derivatives are formed.

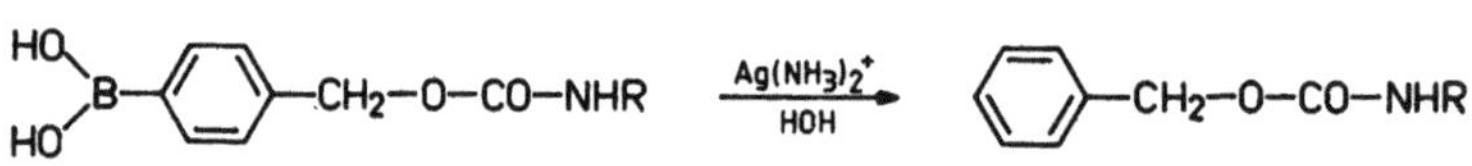

Multiple choices for deblocking can be an attractive feature of protecting groups. Thus, the 1-piperidyloxycarbonyl group [297], which is quite resistant to acids, is cleaved by zinc in acetic acid and also by electrolytic reduction. The cyano-tert. butyloxycarbonyl group [288] can be removed by treatment with aqueous solutions of potassium bicarbonate or triethylamine at p_H 10, but also by a prolonged exposure to trifluoroacetic acid at room temperature. The trimethylsilylethyloxycarbonyl group [44], removed with fluoride ions (tetraalkylammonium fluorides), is sensitive also to zinc chloride in trifluoroethanol [338].

It may not be the most practical method but certainly represents a most original thought to cleave alkyloxycarbonyl groups with β-halogen substituents such as the β-chloroethyloxycarbonyl group, or other urethanes derived from 2-bromoethanol-2,2,2-trichloroethanol or 2,2-dibromopropanol, with *supernucleophiles*, particularly with cobalt complexes: cobaloxime (bis-dimethylglyoximato cobalt) or cobalt(I) phthalocyanine [376, 377].

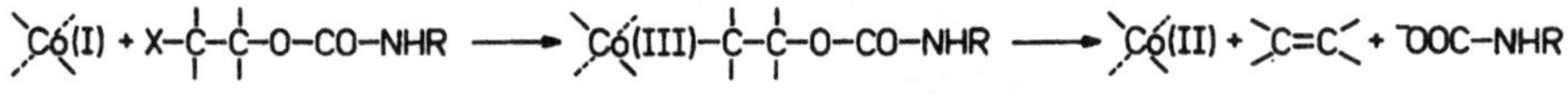

It must be noted, however, that β-halogenoalkyloxycarbonyl groups, for instance the 2-iodoethyloxycarbonyl group [378], are similarly split by zinc in acetic acid or methanol.

$$Zn + I{-}CH_2{-}CH_2{-}O{-}CO{-}NHR \xrightarrow[MeOH]{AcOH\ or} (ZnI)^+ + H_2C{=}CH_2 + {}^-OOC{-}NHR$$

4. Protecting Groups Derived from Sulfur and Phosphorus

The ready availability of *p*-toluenesulfonyl chloride (a by-product in the manufacturing of the sweetener saccharin) led to the early application [379] of the *p*-toluenesulfonyl (tosyl, Tos) group for the protection of the amino function. Yet, preparation of tosylamino acids is straightforward only if no complicating side chains are present;

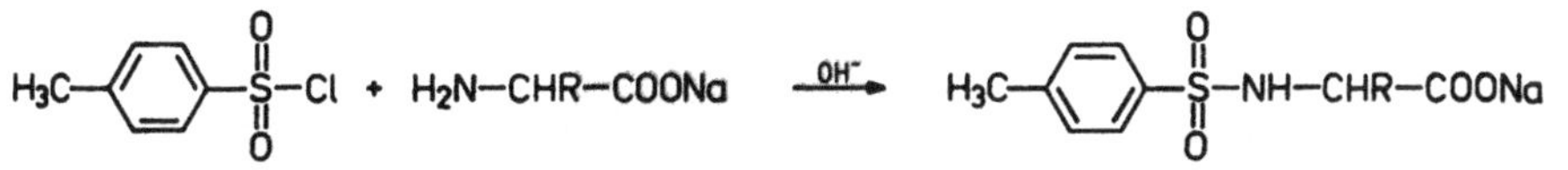

for instance tosylation of glutamic acid is accompanied by side reactions. Unless special precautions are taken [380], acylation of the amino group of glutamic acid with tosyl chloride results in ring closure to the pyrrolidone derivative tosyl-pyroglutamic acid.

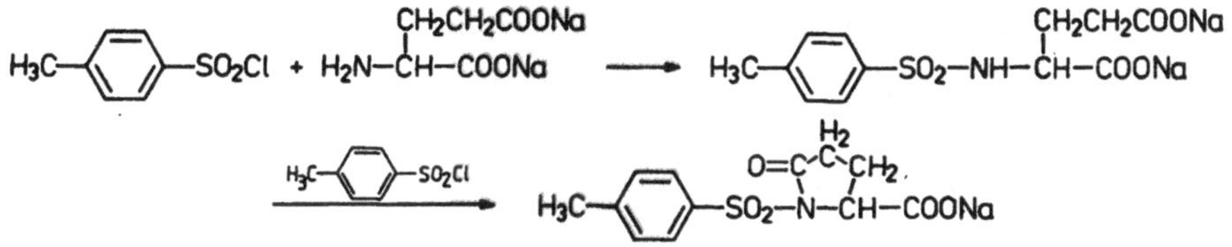

Similar problems arise in the tosylation of ornithine. Further details on the chemistry of tosylation can be found in an excellent review by Rudinger [380].

Removal of the tosyl group is not as easy as it would be desirable for delicate peptides. Sulfonamides are quite resistant to acids and only extremely strong acids such as hydrogen fluoride [381] or trifluoromethanesulfonic acid [359] can cleave them. Hydrobromic acid in acetic acid leaves the tosyl group unaffected as long as mild conditions prevail. For removal at room temperature, the presence of phenol and long reaction times are required [382, 383]. Reductive methods are more effective. The rather unattractive hydrogen iodide-phosphonium iodide reagent [8] was displaced by reduction with sodium in liquid ammonia [10], which remains the method of choice for deprotection of tosyl peptides. The execution of the reaction is simple, but the mechanism is not yet fully understood [384]. More recent recommendations, e.g. electrolytic reduction [385, 386], photolysis [45] or photohydrolysis in the presence of electron donors [387] have not been followed thus far in the synthesis of complex peptides.

A further disadvantage of the tosyl protection is that the extreme electron withdrawing effect of the blocking group renders the remaining hydrogen atom of primary amines sufficiently acidic to be abstracted by relatively mild bases.

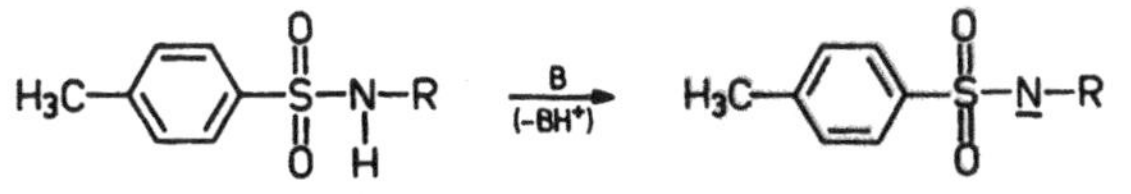

The negatively charged nitrogen as an intramolecular nucleophile might cause cyclization as shown in the case of pyroglutamic acid formation. It can compete with the amino component for the acylating intermediate in several methods of coupling. Also, when the carboxyl group is converted to a highly reactive derivative, the presence of an anionic and a potential anionic center within the same molecule leads to its disintegration [388–391]:

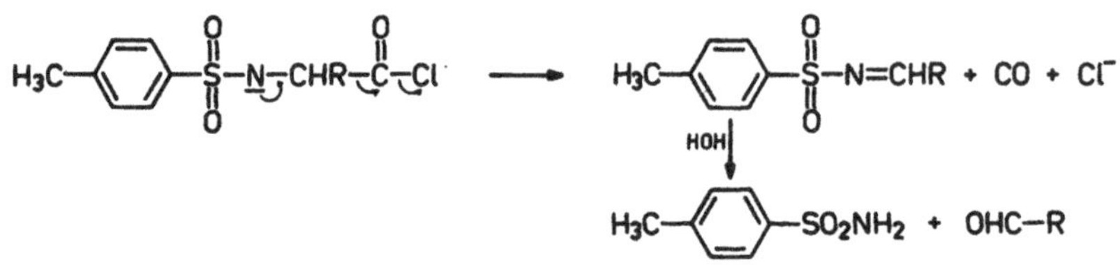

Protection of the amino function by acylation with *aliphatic sulfonic acids* has also been suggested. The benzylsulfonamides [392] obtained from amino acids and benzylsulfonyl chloride in aqueous dioxane

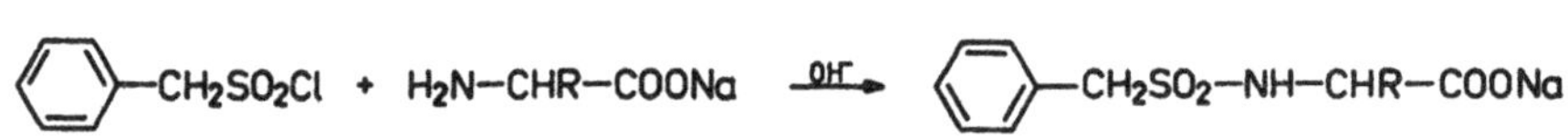

are cleaved by reduction with sodium in liquid ammonia and also by hydrogenation in the presence of a Raney-nickel catalyst. No practical application of the benzylsulfonyl group could be found in the literature, but the related *p*-tolylmethylsulfonyl or 4-methylbenzylsulfonyl group [393] has been used, albeit only for the protection of lysine side chains, in the syntheses of two biologically active peptides. The sulfonamide bond was cleaved with liquid hydrogen fluoride. Less drastic conditions are needed for the removal of the 2-phenyl-2-keto-ethylsulfonyl group [394]

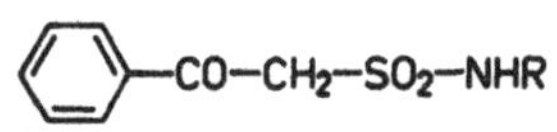

which like the phenacyl group, is sensitive to reduction with zinc in acetic acid.

It is interesting to note that from the derivatives of 6-covalent sulfur only the *p*-toluenesulfonyl group gained major significance in peptide synthesis. No blocking group based on 4-covalent sulfur plays a similar role, but protection of amines in the form of amides of sulfenic acids, derivatives of divalent sulfur, turned out to be quite practical. Several sulfenyl groups were recommended for masking the amino function: the triphenylmethylsulfenyl (or tritylsulfenyl) group [37], the relatively acid stable 2,4-dinitrophenylsulfenyl [41, 395], and pentachlorophenylsulfenyl [41] groups or the very acid sensitive 2-nitro-4-methoxyphenylsulfenyl group [396].

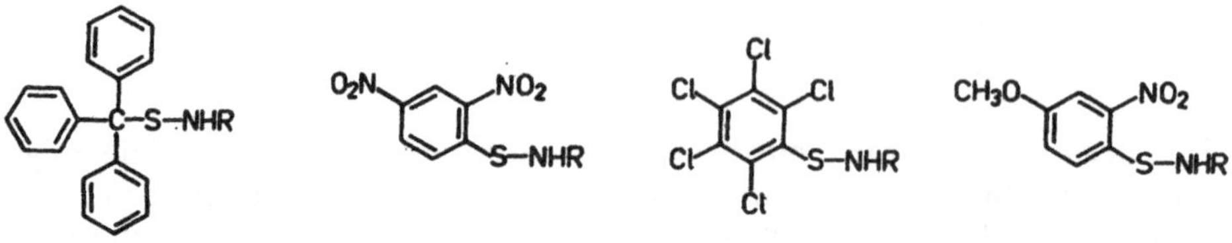

Only the *o*-nitrophenylsulfenyl (Nps) group [36, 37]

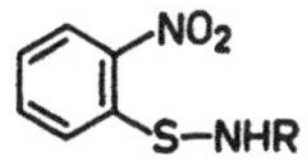

became an important amine protecting group frequently used in the synthesis of complex peptides.

Introduction of the Nps protection is fairly simple with the commercially available acid chloride

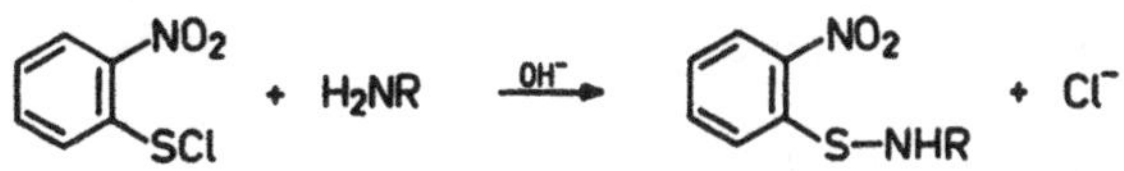

or with the stable *o*-nitrophenylsulfenyl thiocyanate in the presence of silver nitrate [397]

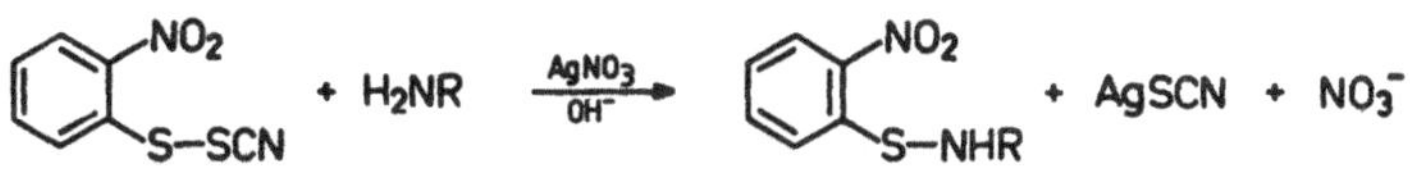

The *p*-nitrophenyl ester of *o*-nitrophenylsulfenic acid is also suitable as acylating agent [397].

Because of the acid-sensitivity of the Nps group, the blocked amino acids are usually isolated and stored in the form of their dicyclohexylammonium salts. Prior to activiation of the carboxyl groups, these salts are usually reconverted to the free acid by careful treatment with weak acids such as citric acid or with an aqueous solution of potassium hydrogen sulfate [398] used in moderate excess.

Initially [36, 37] the Nps group was looked upon as a highly acid labile masking group, removable with strong acids, but rapidly and under mild conditions. Dilute solutions of HCl in ether, ethyl acetate, acetone or other organic solvents caused almost instantaneous deblocking. Completion of the reversible reaction

could be achieved in methanol [399, 400], in which the acid chloride yields the methyl ester of o-nitrophenylsulfenic acid, or by the addition of appropriate scavengers. Various other acids, such as p-toluenesulfonic acid, perchloric acid or cation exchange resins in hydrogen cycle were also applied for deprotection as was benzenesulfonic acid imide [399], an acid that forms crystalline salts

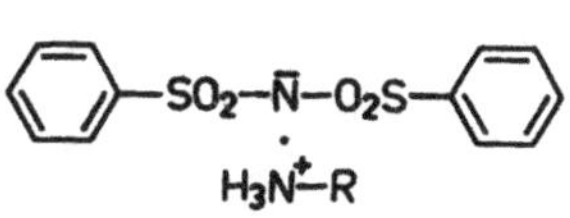

with many of the amines produced in deprotection. The Nps group became even more popular, however, when several alternatives to acidolysis were found for its removal. Reductive desulfurization with Raney-nickel [401] was an interesting approach but limited in scope, since it cannot be applied to peptides which contain cysteine or methionine residues. Deblocking with nucleophiles, on the other hand, started a new line of further developments. A whole gamut of reagents was found [39, 41] to be applicable for this purpose: sulfites, hydrogen sulfide, rhodanides [402], hydrogen cyanide, hydrogen iodide, hydrazoic acid, thiourea, thioacetamide, thiourethane, thioglycolic acid, mercaptoethanol and other mercaptanes. The presence of some acetic acid seems to be necessary for smooth reaction; perhaps it serves as a proton source for the neutralization of the liberated amines. Thiophenols, e.g. o-nitrothiophenol [40], have the advantage of forming inosoluble disulfides with the o-nitrophenylsulfenyl moiety [41–43]

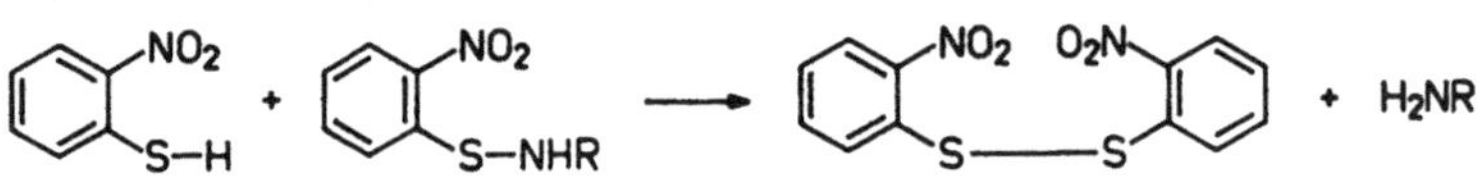

and this thought was further enhanced by the use of polymeric forms [403] of 4-mercapto-2-nitrobenzoic acid [42] and 2-mercaptopyridine [43]. To indicate the possible extension of useful reagents, we mention thiosulfate and dithionate [38, 404] and dithioerythritol [405]. The earlier noted [250] sensitivity of the Nps group to the additive 1-hydroxybenzotriazole [406] could be exploited for deprotection which is especially smooth in trifluoroethanol. Yet, it is not clear at this time whether the deblocking agent acts as a nucleophile or as an acid. The experiments show that one mole of the 1-hydroxybenzotriazole, used in excess, was reduced to benzotriazole in the process [407].

The possibility of removing the Nps group with thiophenol and its derivatives opened up a new avenue in synthesis, simultaneous deprotection and coupling. The new approach is based on acylating agents which release a thiophenol during acylation.

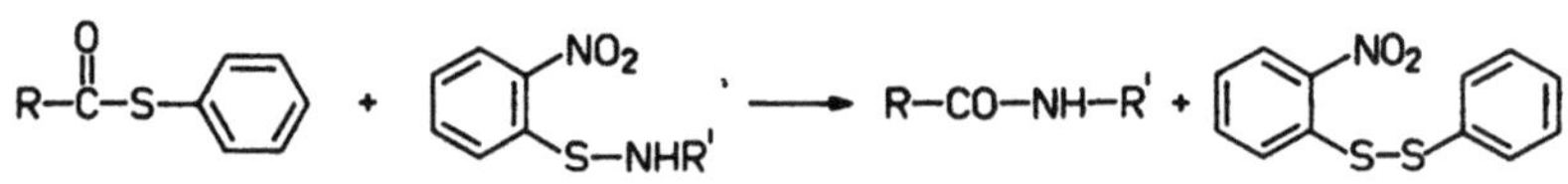

Demonstration of this principle [408, 409] required experiments under rather harsh conditions, but the idea itself is valuable and could be further persued.

There are some limitations in the use of Nps protection. The smooth addition of o-nitrophenylsulfenyl chloride to the indole nucleus of tryptophane [410] is cause for concern, both in the introduction of the Nps group for the protection of this amino acid and in the step of deprotection if the latter is carried out with HCl:

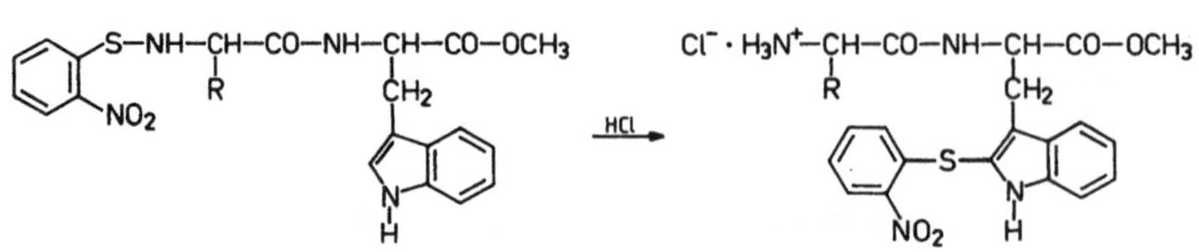

Similarly disturbing is the transfer of the Nps group from N to S resulting in disulfides from peptides with unprotected cysteine side chains [411].

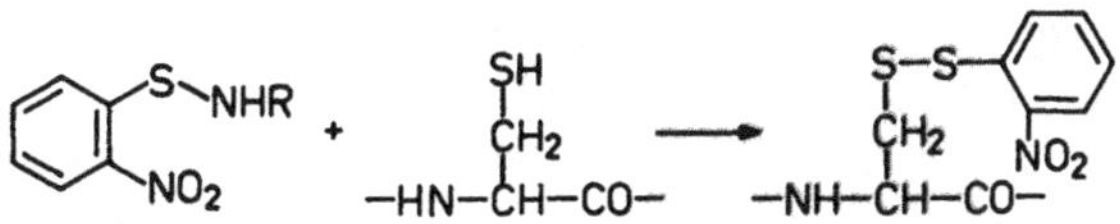

Nevertheless, the advantages of protection with the Nps group, not lastly the numerous options available for its removal, probably outweigh the disadvantages. On reading the literature, one gains the impression that the use of Nps-protection has not diminished in recent years, in fact it may even be increasing.

Only a few protecting groups are derived from *phosphorus*. Among these the dibenzylphosphoryl group and its ring-substituted derivatives [412] are too bulky to allow the coupling of the protected amino acids.

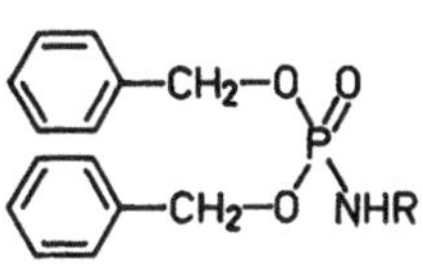

In contrast, the diphenylphosphine group [31] is quite promising. It provides protection which remains intact on hydrogenation, hydrazinolysis or treatment with alkali. Also, it prevents racemization and does not interfere with commonly used coupling procedures. Most importantly, the process of cleavage, with moderately strong acids, does not involve carbocations or other alkylating agents:

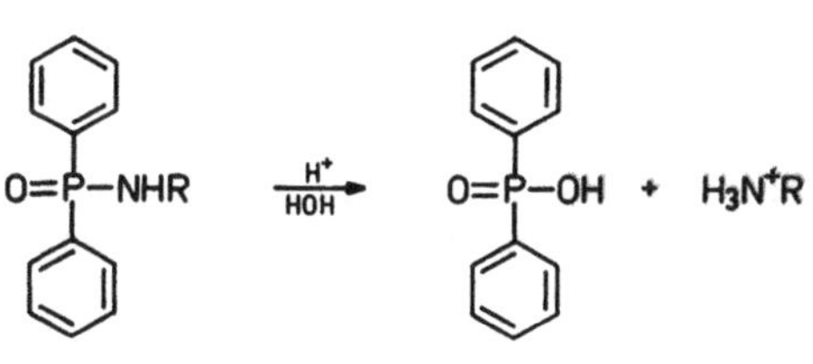

References of Chapter III

1. Photaki, K., Yiotakis, A. E.: J. Chem. Soc. (Perkin I), 259 (1976)
2. Bodanszky, M., Fink, M. L., Klausner, Y. S., Natarajan, S., Tatemoto, K., Yiotakis, A. E., Bodanszky, A.: J. Org. Chem. *42*, 149 (1977)
3. Fischer, E.: Ber. dtsch. Chem. Ges. *35*, 1095 (1902)
4. Wessely, F., Kemm, E.: Hoppe-Seilers Z. Physiol. Chem. *174*, 306 (1928); Wessely, F., Kemm, E., Mayer, J.: *ibid.*, *180*, 64 (1929)
5. MacLaren, J. A.: Aust. J. Chem. *11*, 360 (1958)
6. Bergmann, M., Zervas, L.: Ber. dtsch. Chem. Ges. *65*, 1192 (1932)
7. Rosenmund, K. W., Zetzsche, F., Heise, F.: Ber. dtsch. Chem. Ges. *54*, 2038 (1921)

8. Fischer, E.: Ber. dtsch. Chem. Ges. *48*, 93 (1915)
9. du Vigneaud, V., Audrieth, L. F., Loring, H. S.: J. Amer. Chem. Soc. *52*, 4500 (1930)
10. du Vigneaud, V., Behrens, O. K.: J. Biol. Chem. *117*, 27 (1937)
11. Sifferd, R. H., du Vigneaud, V.: J. Biol. Chem. *108*, 753 (1935)
12. Kuromizu, K., Meienhofer, J.: J. Amer. Chem. Soc. *96*, 4978 (1974); cf. also Felix, A. M., Jimenez, M. H., Mowles, T., Meienhofer, J.: Org. Synth. *59*, 159 (1980)
13. Anantaramaiah, G. M., Sivanandaiah, K. M.: J. Chem. Soc. (Perkin I), 490 (1977); cf. also Jackson, A. E., Johnstone, A. W.: Synthesis, 685, (1977)
14. Felix, A. M., Heimer, E. P., Lambros, T. J., Tzougraki, C., Meienhofer, J.: J. Org. Chem. *43*, 4194 (1978)
15. Anwer, M. K., Khan, S. A., Sivanandaiah, K. M.: Synthesis, 751 (1978)
16. Sivanandaiah, K. M., Gurusiddappa, S.: J. Chem. Res. (S), 108 (1979)
17. Pless, J., Guttmann, S.: In "Peptides 1966" (Beyermann, H. C., Van de Linde, A., Maassen van den Brink, W., eds.) p. 50, Amsterdam: North Holland Publ. 1967
18. Hayakawa, T., Fujiwara, Y., Noguchi, J.: Bull. Chem. Soc. Jpn. *40*, 1205 (1967)
19. Freidinger, R. M., Hirschmann, R., Veber, D. F.: J. Org. Chem. *43*, 4800 (1978)
20. Clubb, M. E., Scopes, P. M., Young, G. T.: Chimia *14*, 373 (1960); cf. also Scopes, P. M., Walshaw, K. B., Welford, M., Young, G. T.: J. Chem. Soc., 782 (1965)
21. Kollonitsch, J., Gabor, V., Hajos, A.: Nature *177*, 841 (1956; Kollonitsch, J., Hajos, A., Gabor, V.: Chem. Ber. *89*, 2288 (1956)
22. Kamber, B., Rittel, W.: Helv. Chim. Acta *51*, 2061 (1968)
23. Veber, D. F., Milkowski, J. D., Varga, S., Denkewalter, R. G., Hirschmann, R.: J. Amer. Chem. Soc. *94*, 5456 (1972)
24. Ben-Ishai, D., Berger, A.: J. Org. Chem. *17*, 1564 (1952)
25. Weygand, F., Hunger, K.: Chem. Ber. *95*, 1 (1962); cf. also ref. 27
26. Carpino, L. A.: J. Amer. Chem. Soc. *79*, 98 (1957)
27. McKay, F. C., Albertson, N. F.: J. Amer. Chem. Soc. *79*, 4686 (1957)
28. Anderson, G. W., McGregor, A. C.: J. Amer. Chem. Soc. *79*, 6180 (1957)
29. Sieber, P., Iselin, B.: Helv. Chim. Acta *51*, 614, 622 (1968)
30. Yajima, H., Fujii, N., Ogawa, H., Irie, H., Shinagawa, S., Fujino, M.: In "Peptide Chemistry 1976" (Nakajima, T., ed.), p. 57, Protein Res. Foundation, Osaka, 1977
31. Kenner, G. W., Moore, G. A., Ramage, R.: Tetrahedron Lett., 3623 (1976)
32. Crane, C. W., Rydon, H. N.: J. Chem. Soc., 766 (1947)
33. Mamalis, P., Rydon, H. N.: J. Chem. Soc., 1049 (1955)
34. Carpino, L. A., Han, G. Y.: J. Amer. Chem. Soc. *92*, 5748 (1970)
35. Carpino, L. A., Han, G. Y.: J. Org. Chem. *37*, 3404 (1972); *ibid. 38*, 4218 (1973)
36. Goerdeler, J., Holst, A.: Angew. Chem. *71*, 775 (1959)
37. Zervas, L., Borovas, D., Gazis, E.: J. Amer. Chem. Soc. *85*, 3660 (1963)
38. Ekström, M. B., Sjöberg, B.: Acta Chem. Scand. *19*, 1245 (1965)
39. Fontana, A., Marchiori, F., Moroder, L., Scoffone, E.: Tetrahedron Lett., 2985 (1966)
40. Brandenburg, D.: Tetrahedron Lett. 6201 (1966)
41. Kessler, W., Iselin, B.: Helv. Chim. Acta *49*, 1330 (1966)
42. Juillerat, M., Bargetzi, J. P.: Helv. Chim. Acta *59*, 855 (1976)
43. Tun-Kyi, A.: Helv. Chim. Acta *61*, 1086 (1978)

44. Carpino, L. A., Tsao, J. H., Ringsdorf, H., Fell, E., Hettisch, G.: J. Chem. Soc. Chem. Commun., 358 (1978)
45. D'Souza, L., Day, R. A.: Science *160*, 882 (1968)
46. Chamberlin, J. W.: J. Org. Chem. *31*, 1658 (1966)
47. Wieland, T., Birr, C.: In Peptides, Proc. 8th Eur. Peptide Symp., Nordwijk, 1966 (Beyerman, H. C., Van de Linde, A., Maassen van den Brink, eds.) p. 103, Amsterdam: North Holland Publ. 1966
48. Patchornik, A., Amit, B., Woodward, R. B.: J. Amer. Chem. Soc. *92*, 6333 (1970)
49. Sheehan, J. C., Umezawa, K.: J. Org. Chem. *38*, 3771 (1973)
50. Sheehan, J. C., Wilson, R. M., Oxford, A. W.: J. Amer. Chem. Soc. *93*, 7222 (1971)
51. Rich, D. H., Gurwara, S. K.: J. Amer. Chem. Soc. *97*, 1575 (1975)
52. Jering, H., Schorp, G., Tschesche, H.: Hoppe-Seyler's Z. Physiol. Chem. *355*, 1129 (1974)
53. Kloss, G., Schröder, E.: Hoppe-Seyler's Z. Physiol. Chem. *336*, 248 (1964)
54. Royer, G. P., Anantharamaiah, G. M.: J. Amer. Chem. Soc. *101*, 3394 (1979)
55. Barton, M. A., Lemieux, R. U., Savoie, J. Y.: J. Amer. Chem. Soc. *95*, 4501 (1973)
56. Prestidge, R. L., Harding, D. R. K., Battersby, J. E., Hancock, W. S.: J. Org. Chem. *40*, 3287 (1975)
57. Gross, E., Witkop, B.: J. Biol. Chem. *237*, 1856 (1962)
58. Sjöberg, B. O. H., Ekström, B. A.: Belg. Pat. 620519 (1962; Chem. Abstr. *59*, P 11502a (1963)
59. Bayer, E., Schmidt, K.: Tetrahedron Lett., 2051 (1973)
60. Bodanszky, M., Natarajan, S.: J. Org. Chem. *40*, 2495 (1975); cf. also Natarajan, S., Bodanszky, M.: *ibid. 41*. 1269 (1976)
61. Sakakibara, S.: In Peptides, Proc. Fifth Amer. Pept. Symp. (Goodman, M., Meienhofer, J., eds.) p. 436, New York: Wiley 1977
62. Haslam, E.: Tetrahedron *36*, 2409 (1980)
63. Roeske, R. W.: In "The Peptides" vol. 3, p. 102 (Gross, E., Meienhofer, J., eds.), New York: Academic Press 1981
64. Schwartz, H., Arakawa, K.: J. Amer. Chem. Soc. *81*, 5691 (1959)
65. Schwyzer, R., Sieber, P.: Helv. Chim. Acta *42*, 972 (1959)
66. Kisfaludy, L., Löw, M.: Acta Chim. Hung. *44*, 33 (1965)
67. Guttmann, S., Pless, J.: Angew . Chem. *77*, 53 (1965)
68. Stewart, F. H. C.: Austral. J. Chem. *18*, 1877 (1965)
69. Stewart, F. H. C.: Austral. J. Chem. *19*, 1067 (1966); *ibid. 20*, 2243 (1967)
70. Stewart, F. H. C.: Austral. J. Chem. *24*, 2193 (1971)
71. Stelakatos, G. C., Solomos-Aravidis, C., Karayannakis, P., Kolovos, M. G., Photaki, I.: In Peptides 1980 (Brunfeldt, K., ed.), p. 133, Copenhagen: Scriptor 1981
72. Wieland, T., Racky, W.: Chimia *22*, 375 (1968)
73. Stewart, F. H. C.: Austral. J. Chem. *18*, 1699 (1965)
74. Camble, R., Garner, R., Young, G. T.: J. Chem. Soc., 1911 (1969)
75. Burton, J., Fletcher, G. A., Young, G. T.: Chem. Commun., 1057 (1971)
76. Merrifield, R. B.: J. Amer. Chem. Soc. *85*, 2149 (1963)
77. Bethell, M., Bigley, D. B., Kenner, G. W.: Chem. Ind. (London), 653 (1963); cf. also Hardegger, E., El Heweihi, Z., Robinet, F. G.: Helv. Chim. Acta *31*, 439 (1948)
78. Aboderin, A. A., Delpierre, G. R., Fruton, J. S.: J. Amer. Chem. Soc. *87*, 5469 (1965)

79. Abderhalden, E., Suzuki, S.: Hoppe-Seyler's Z. Physiol. Chem. *176*, 601 (1928)
80. Miller, H. K., Waelsch, H.: J. Amer. Chem. Soc. *74*, 1092 (1952)
81. Cipera, J. D., Nicholls, R. V. V.: Chem. Ind., 16 (1955)
82. Kollonitsch, J., Vita, J.: Nature *178*, 1307 (1956)
83. Büchi, H., Steen, K., Eschenmoser, A.: Angew. Chem. *3*, 62 (1964)
84. Gisin, B. F.: Helv. Chim. Acta *56*, 1476 (1973)
85. Stewart, F. C. H.: Austral. J. Chem. *21*, 2543 (1968)
86. Wang, S. S.: J. Amer. Chem. Soc. *95*, 1328 (1973); cf. also Bodanszky, M., Fagan, D. T.: Int. J. Peptide Protein Res. *10*, 375 (1977)
87. Bywood, R., Gallagher, G., Sharma, G. K., Walker, D.: J. Chem. Soc. Perkin I, 2019 (1975); cf. also Horner, L., Fernekes, H.: Chem. Ber. *94*, 712 (1961)
88. Stelakatos, G. C., Paganou, A., Zervas, L.: J. Chem. Soc. (C), 1191 (1966)
89. Hiskey, R. G., Adams, J. B.: J. Amer. Chem. Soc. *87*, 3969 (1965)
90. Ho, T. L., Olah, G. A.: Angew. Chem. Int. Ed. *15*, 774 (1976)
91. Jung, M. E., Lyster, M. A.: J. Amer. Chem. Soc. *99*, 968 (1977)
92. Kornblum, N., Scott, A.: J. Amer. Chem. Soc. *96*, 590 (1974)
93. Curtius, T., Goebel, F.: J. prakt. Chem. [2] *37*, 150 (1888)
94. Fischer, E.: Ber. dtsch. Chem. Ges. *39*, 2893 (1906)
95. Brenner, M., Huber, W.: Helv. Chim. Acta *36*, 1109 (1953)
96. Hanby, W. E., Waley, S. G., Watson, J.: J. Chem. Soc., 3239 (1950)
97. Rachele, J. R.: J. Org. Chem. *28*, 2898 (1963)
98. Taschner, E., Wasiliewski, C.: Liebigs Ann. Chem. *640*, 142 (1961)
99. Guttmann, S., Boissonnas, R. A.: Helv. Chim. Acta *41*, 1852 (1958)
100. McLaren, J. A.: Austral. J. Chem. *11*, 360 (1958)
101. Kenner, G. W., Seely, J. H.: J. Amer. Chem. Soc. *94*, 3259 (1972)
102. McDermott, J. R., Benoiton, N. L.: Can. J. Chem. *51*, 1915 (1973)
103. Bartlett, P. A., Johnson, W. S., Tetrahedron Lett., 4459 (1970)
104. Sheehan, J. C., Daves, G. D., Jr.: J. Org. Chem. *29*, 2006 (1964)
105. Nefkens, G. H. L., Tesser, G. I., Nivard, R. J. F.: Rec. Trav. Chim. Pays-Bas *82*, 941 (1963)
106. Taylor-Papadimitriou, J., Yovanidis, C., Paganou, A., Zervas, L.: J. Chem. Soc. C, 1830 (1967)
107. Tam, J. P., Cunningham-Rundles, W. F., Erickson, B. W., Merrifield, R. B.: Tetrahedron Lett., 4001 (1977)
108. Bodanszky, M., Martinez, J.: J. Org. Chem. *43*, 3071 (1978)
109. Ledger, R., Stewart, F. H. C.: Austral. J. Chem. *20*, 787 (1967)
110. Hendrickson, J. B., Kandall, C.: Tetrahedron Lett., 343 (1970)
111. Eckstein, F.: Angew. Chem. Int. Ed. *4*, 876 (1965)
112. Woodward, R. B.: Angew. Chem. *78*, 557 (1966)
113. Bates, G. S., Diakur, J., Masamune, S.: Tetrahedron Letters, 4423 (1976)
114. Trzupek, L. S., Go, A., Kopple, K. D.: J. Org. Chem. *44*, 4577 (1979)
115. Ueda, K.: Bull. Chem. Soc. Jpn. *52*, 1879 (1979)
116. Taschner, E., Liberek, B.: Bull. Acad. Pol. Sci. (Ser. Sci. Chim. Geol. Geogr.) *7*, 877 (1959)
117. Ohno, M., Anfinsen, C. B.: J. Amer. Chem. Soc. *92*, 4098 (1970)
118. Amaral, M., Barrett, G. C., Rydon, H. N., Willett, J. E.: J. Chem. Soc. C, 807 (1966)
119. Miller, A. W., Stirling, C. J. M.: J. Chem. Soc. C, 2612 (1968)
120. Amaral, M. J. S. A.: J. Chem. Soc. C, 2495 (1969)
121. Diaz, J., Guegan, R., Beaumont, M., Benoit, J., Clement, J., Fauchard, C., Galtier, D., Millan, J., Muneaux, C., Maneaux, Y., Vedel, M., Schwyzer, R.: Bioorg. Chem. *8*, 429 (1979)

122. Sieber, P.: Helv. Chim. Acta *60*, 2711 (1977)
123. Riniker, B., Eisler, K., Kamber, B., Müller, H., Rittel, W., Sieber, P.: In Peptides 1978 (Siemion, I. Z., Kupryszewski, G., eds.) p. 631, Wroclaw: Wroclaw Univ. Press 1979
124. Roeske, R. W.: J. Org. Chem. *28*, 1251 (1963)
125. Roeske, R.: Chem. Ind. (London), 1121 (1959)
126. Taschner, E., Chimiak, A., Bator, B., Sokolowska, T.: Liebigs. Ann. Chem. *646*, 134 (1961)
127. Losse, G., Zeidler, D., Grieshaber, T.: Liebigs Ann. Chem. *715*, 196 (1968)
128. Birkofer, L., Ritter, A.: Chem. Ber. *93*, 424 (1960); cf. also Rühlmann, K., Hills, J.: Liebigs Ann. Chem. *683*, 211 (1965)
129. Schnabel, E., Oberdorf, A.: in "Peptides 1968" (Bricas, E., ed.), p. 261, Amsterdam: North Holland Publ. 1968
130. Blotny, G., Taschner, E.: Bull. Acad. Polon. Sci. Ser. Sci. Chim. *14*, 615 (1966)
131. Gordon, M., Miller, J. G., Day, A. R.: J. Amer. Chem. Soc. *70*, 1946 (1948)
132. Kenner, G. W.: Angew. Chem. *71*, 741 (1959)
133. Galpin, I. J., Handa, B. K., Hudson, D., Jackson, A. G., Kenner, G. W., Ohlsen, S. R., Ramage, R., Singh, B., Tyson, R. G.: Peptides 1976 (Loffet, A., ed.), p. 246 Edition de l'Université de Bruxelles (1976)
134. Chaudhury, A. M., Kenner, G. W., Moore, S., Ramage, S. R., Richards, P. M., Thorpe, W. D.: In "Peptides 1976" (Loffet, A., ed.), p. 257. Editions de l'Université de Bruxelles, Belgium, 1976
135. Taschner, E., Blotny, G., Bator, B., Wasielewski, C.: Bull. Acad. Polon. Sci. Ser. Sci. Chem. *12*, 755 (1964)
136. Barth, A.: Liebigs Ann. Chem. *683*, 216 (1965)
137. Ito, H.: Synthesis, 465 (1979)
138. Bodanszky, M.: Nature *175*, 685 (1955); cf. also Acta Chim. Hung. *10*, 335 (1957)
139. Wieland, T., Heinke, B.: Liebigs Ann. Chem. *615*, 184 (1958)
140. Goodman, M., Steuben, K. C.: J. Amer. Chem. Soc. *81*, 3980 (1959); *84*, 1279 (1962)
141. Iselin, B., Schwyzer, R.: Helv. Chim. Acta *43*, 1760 (1960)
142. Jones, J. H., Young, G. T.: J. Chem. Soc. C, 437 (1968)
143. Johnson, B. J., Jacobs, P. M.: Chem. Commun., 73 (1968)
144. Johnson, B. J., Trask, E. G.: J. Org. Chem. *33*, 4521 (1968)
145. Cheung, H. T., Blout, E. R.: J. Org. Chem. *30*, 315 (1965)
146. Bennett, E. L., Niemann, C.: J. Amer. Chem. Soc. *70*, 2610 (1948)
147. Milne, H. B., Stevens, C. M.: J. Amer. Chem. Soc. *72*, 1742 (1950)
148. Waldschmidt-Leitz, E., Kühn, K.: Chem. Ber. *84*, 381 (1951)
149. Milne, H. B., Peng, C. H.: J. Amer. Chem. Soc. *79*, 645 (1957)
150. Milne, H. B., Most, C. F., Jr.: J. Org. Chem. *33*, 169 (1968)
151. Lewalter, J., Birr, C.: Liebigs Ann. Chem. *740*, 48 (1970)
152. Wolman, Y., Gallop, P. M., Patchornik, A.: J. Amer. Chem. Soc. *83*, 1263 (1961); Wolman, Y., Gallop, P. M., Patchornik, A., Berger, A.: *ibid. 84*, 1889 (1962)
153. Kelly, R. B.: J. Org. Chem. *28*, 453 (1963)
154. Milne, H. B., Kilday, W.: J. Org. Chem. *30*, 64 (1965); Milne, H. B., Carpenter, F. H.: *ibid.*, *33*, 4476 (1968)
155. Wieland, T., Lewalter, J., Birr, C.: Liebigs Ann. Chem. *740*, 31 (1970)
156. Barton, D. H. R., Girijavallabhan, M., Sammes, P. G.: J. Chem. Soc. Perkin I, 929 (1972)

157. Hofmann, K., Magee, M. Z., Lindenmann, A.: J. Amer. Chem. Soc. *72*, 2814 (1950)
158. Schwyzer, R.: Angew. Chem. *71*, 742 (1959)
159. Preston, J., Weinstein, B.: Experientia *24*, 265 (1967)
160. Wang, S. S.: J. Amer. Chem. Soc. *95*, 1328 (1973)
161. Macrae, R., Young, G. T.: J. Chem. Soc. Chem. Commun., 446 (1974)
162. Bierzycka, D., Biernat, J. F.: Roczniki Chem. *44*, 2477 (1970)
163. Yajima, H., Kiso, Y.: Chem. Pharm. Bull. *19*, 420 (1971)
164. Velluz, L., Amiard, G., Heymes, R.: Bull. Soc. Chim. France, 1012 (1954)
165. Velluz, L., Anatol, J., Amiard, G.: Bull. Soc. Chim. France, 1449 (1954)
166. Liwschitz, Y., Zilkha, A.: J. Amer. Chem. Soc. *76*, 3698 (1954)
167. Liwschitz, Y., Zilkha, A., Amiel, Y.: J. Amer. Chem. Soc. *78*, 3067 (1956)
168. Quitt, P., Hellerbach, J., Vogler, K.: Helv. Chim. Acta *46*, 327 (1963); cf. also Scheibler, H., Baumgarten, P.: Ber. dtsch. Chem. Ges. *55*, 1358 (1922)
169. Hanson, R. W., Law, H. D.: J. Chem. Soc., 7285 (1965)
170. Helferich, B., Moog, L., Jünger, A.: Ber. dtsch. Chem. Ges. *58*, 872 (1925)
171. Hillmann-Elies, A., Hillmann, G., Jatzkewitz, J.: Z. Naturforsch. *8b*, 445 (1953)
172. Amiard, G., Heymes, R., Velluz, L.: Bull. Soc. Chim. France *191*, 1464 (1955)
173. Zervas, L., Theodoropoulos, D. M.: J. Amer. Chem. Soc. *78*, 1359 (1956)
174. Schnabel, E.: Liebigs Ann. Chem. *673*, 171 (1964)
175. Bezas, B., Zervas, L.: J. Amer. Chem. Soc. *83*, 719 (1961)
176. Guttmann, S.: Helv. Chim. Acta *46*, 1975 (1963)
177. Hiskey, R. G., Beacham, L. M., Matl, V. G.: J. Org. Chem. *37*, 2472, 2478 (1972)
178. Halstrøm, J., Brunfeldt, K.: Hoppe-Seyler's Z. Physiol. Chem. *353*, 1204 (1972)
179. Riniker, B., Kamber, B., Sieber, P.: Helv. Chim. Acta *57*, 831 (1974)
180. Bergmann, M., Ennslin, H., Zervas, L.: Ber. dtsch. Chem. Ges. *58*, 1034 (1925)
181. Bergmann, M., Zervas, L.: Hoppe-Seyler's Z. Physiol. Chem. *152*, 282 (1926)
182. Wieland, T., Schäfer, W.: Liebigs Ann. Chem. *576*, 104 (1952)
183. McIntire, F. C.: J. Amer. Chem. Soc. *69*, 1377 (1947); cf. also Zervas, L., Konstas, S.: Chem. Ber. *93*, 435 (1960)
184. Sheehan, J. C., Grenda, V. J.: J. Amer. Chem. Soc. *84*, 2417 (1962)
185. Hope, A. P., Halpern, B.: Tetrahedron Lett., 2261 (1972); Austral. J. Chem. *29*, 1591 (1976)
186. Abdipranoto, A., Hope, A. P., Halpern, B.: Austral. J. Chem. *30*, 2711 (1977)
187. Halpern, B., James, L. B.: Austral. J. Chem. *17*, 1282 (1964); Halpern, B.: *ibid.* *18*, 417 (1965); Halpern, B., Cross, A. B.: Chem. Ind., 1183 (1965)
188. Deer, A., Fried, J. H., Halpern, B.: Austral. J. Chem. *20*, 797 (1967)
189. Dane, E., Drees, F., Konrad, P., Dockner, T.: Angew. Chem. *74*, 873 (1962)
190. Dane, E., Dockner, T.: Chem. Ber. *98*, 780 (1965)
191. Dane, E., Dockner, T.: Angew. Chem. Int. Ed. *3*, 439 (1964)
192. Balog, A., Breazu, D., Vargha, E., Gönczy, F., Beu, L.: Rev. Roumaine Chem. *15*, 1375 (1970)
193. Yamashiro, D., Aaning, H. L., du Vigneaud, V.: Proc. Natl. Acad. Sci. USA *54*, 166 (1968)
194. Yamashiro, D., du Vigneaud, V.: J. Amer. Chem. Soc. *90*, 487 (1968)
195. Hardy, P. M., Samworth, D. J.: J. Chem. Soc. Perkin I, 1954 (1977)
196. Sheehan, J. C., Guziec, F. S.: J. Amer. Chem. Soc. *94*, 6561 (1972); J. Org. Chem. *38*, 3034 (1973)
197. Fischer, E., Warburg, O.: Ber. dtsch. Chem. Ges. *38*, 3997 (1905); cf. also Curtius, T.: *ibid.* *16*, 753 (1883)

198. du Vigneaud, V., Dorfmann, R., Loring, H.: J. Biol. Chem. *98*, 577 (1932)
199. Stoll, A., Petrzilka, T.: Helv. Chim. Acta *35*, 589 (1952)
200. Hofmann, K., Stutz, E., Spühler, G., Yajima, A., Schwartz, E. T.: J. Amer. Chem. Soc. *82*, 3727 (1960)
201. Okawa, K., Hase, S.: Bull. Chem. Soc. Jpn. *36*, 754 (1963)
202. Muramatsu, I., Itoi, M., Tsuji, M., Hagitani, A.: Bull. Chem. Soc. Jpn. *37*, 756 (1964)
203. Thomas, J. O.: Tetrahedron Lett., 335 (1967)
204. Sheehan, J. C., Yang, D. D. H.: J. Amer. Chem. Soc. *80*, 1154, 1158 (1958)
205. Losse, G., Zönnchen, W.: Liebigs Ann. Chem. *636*, 140 (1960)
206. Geiger, R., Siedel, W.: Chem. Ber. *101*, 3386 (1968)
207. Geiger, R., Siedel, W.: Chem. Ber. *102*, 2487 (1969)
208. Losse, G., Nadolski, D.: J. Pract. Chem. *24*, 118 (1964)
209. Lefrancier, P., Bricas, E.: Bull. Soc. Chim. France, 3668 (1965)
210. Sigmund, F., Wessely, F.: Hoppe-Seyler's Z. Physiol. Chem. *147*, 91 (1925)
211. Bartlett, P. D., Jones, R. H.: J. Amer. Chem. Soc. *79*, 2153 (1957)
212. Bartlett, P. D., Dittmer, D. C.: J. Amer. Chem. Soc. *79*, 2159 (1957)
213. Denkewalter, R. G., Schwam, H., Strachan, R. G., Beesley, T. E., Veber, D. F., Schoenewaldt, E. F., Barkemeyer, H., Paleveda, W. J., Jr., Jacob, T. A., Hirschmann, R.: J. Amer. Chem. Soc. *88*, 3163 (1966)
214. Holley, R. W., Holley, A. D.: J. Amer. Chem. Soc. *74*, 3069 (1952)
215. Masaki, M., Kitahara, T., Kurita, H., Oata, M.: J. Amer. Chem. Soc. *90*, 4508 (1968)
216. Fontana, A., Scoffone, E.: Gazetta Chim. Ital. *98*, 1261 (1968)
217. Steglich, W., Batz, H. G.: Angew. Chem. Int. Ed. *10*, 75 (1971)
218. Glass, J. D., Pelzig, M., Pande, C. S.: Int. J. Peptide Protein Res. *13*, 28 (1979)
219. Undheim, K., Fjelstad, E.: J. Chem. Soc. Perkin I, 829 (1973)
220. Panetta, C. A.: J. Org. Chem. *34*, 2773 (1969)
221. Cuiban, F.: Rev. Romaine Chem. *17*, 897 (1972); *18*, 449 (1973)
222. Wieland, T.: Acta Chim. Acad. Sci. Hung. *44*, 5 (1965)
223. D'Angeli, F., Filiva, F., Scoffone, E.: Tetrahedron Lett., 605 (1965)
224. Di Bello, F., Giormani, V., D'Angeli, F.: J. Chem. Soc. (C), 350 (1969)
225. Di Bello, C., Filiva, F., D'Angeli, F.: J. Org. Chem. *36*, 1818 (1971)
226. Weygand, F., Csendes, E.: Angew. Chem. *64*, 136 (1952)
227. Weygand, F., Geiger, R.: Chem. Ber. *89*, 647 (1956)
228. Schallenberg, E. E., Calvin, M.: J. Amer. Chem. Soc. *77*, 2779 (1955); cf. also Hauptschein, M., Stokes, C. S., Nodiff, E. A.: ibid. *74*, 4005 (1952)
229. Weygand, F., Geiger, R.: Chem. Ber. *92*, 2099 (1959)
230. Weygand, F., Röpsch, A.: Chem. Ber. *92*, 2095 (1959)
231. Benoiton, L., Rydon, H. N., Willett, J. E.: Chem. Ind. (London), 1060 (1960)
232. Panetta, C. A., Casanova, T. G.: J. Org. Chem. *35*, 4275 (1970)
233. Goldberger, R. F., Anfinsen, C. B.: Biochemistry *1*, 401 (1902)
234. Weygand, F., Frauendorfer, E.: Chem. Ber. *103*, 2437 (1970)
235. Hubbuch, A., Danho, W., Zahn, H.: Peptides (Goodman, M., Meienhofer, J., eds.), p. 550, New York: Wiley 1977
236. Meyers, C., Glass, J. D.: Proc. Natl. Acad. Sci. USA *72*, 2193 (1975)
237. Busse, W. D., Hansen, S. R., Carpenter, F. H.: J. Amer. Chem. Soc. *96*, 5949 (1974)
238. Lawson, W. B., Gross, E., Folk, C. M., Witkop, B.: J. Amer. Chem. Soc. *84*, 1715 (1962)
239. Kidd, D. A. A., King, F. E.: Nature *162*, 776 (1948); cf. also King, F. E., Kidd, D. A. A.: J. Chem. Soc., 3315 (1949)

240. Sheehan, J. C., Frank, V. S.: J. Amer. Chem. Soc. *71*, 1856 (1949)
241. Fling, M., Minard, F. N., Fox, S. W.: J. Amer. Chem. Soc. *69*, 2466 (1947)
242. Billman, J. H., Harting, W. F.: J. Amer. Chem. Soc. *70*, 1473 (1948)
243. Nefkens, G. H. L.: Nature *185*, 309 (1960)
244. Nefkens, G. H. L., Tesser, G. I., Nivard, R. J. F.: Rec. Trav. Chim. Pays Bas *79*, 688 (1960)
245. Ing, H. R., Manske, R. H. F.: J. Chem. Soc., 2348 (1926)
246. Schumann, I., Boissonnas, R. A.: Helv. Chim. Acta *35*, 2235 (1952)
247. Keller, O., Rudinger, J.: Helv. Chim. Acta *58*, 531 (1975)
248. Barany, G., Merrifield, R. B.: J. Amer. Chem. Soc. *99*, 7363 (1977)
249. Barany, G., Merrifield, R. B.: J. Amer. Chem. Soc. *102*, 3084 (1980)
250. Geiger, R., König, W.: in "The Peptides" (Gross, E., Meienhofer, J., eds.), Vol. 3, p. 3, New York: Academic Press 1981
251. Wünsch, E.: Methoden der Org. Chem. (Houben-Weil), Vol. 15/1 (Wünsch, E., ed.), p. 46, Stuttgart: Thieme 1974
252. Boissonnas, R. A., Preitner, G.: Helv. Chim. Acta *36*, 875 (1953)
253. Kisfaludy, L., Dualszky, S.: Acta Chim. Acad. Sci. Hung. *24*, 301 (1960)
254. Bláha, K., Rudinger, J.: Coll. Czech. Chem. Commun. *30*, 585 (1965)
255. Meienhofer, J.: Peptides. Proc. 6th Eur. Peptide Symp., Athens, 1963 (Zervas, L., ed.), p. 55, Oxford: Pergamon Press 1966
256. Erickson, B. W., Merrifield, R. B.: J. Amer. Chem. Soc. *95*, 3757 (1973)
257. Channing, D. M., Turner, P. B., Young, G. T.: Nature *167*, 457 (1951)
258. Yamashiro, D., Li, C. H.: J. Amer. Chem. Soc. *95*, 1310 (1973)
259. Carpenter, F. H., Gish, D. T.: J. Amer. Chem. Soc. *74*, 3818 (1952)
260. Gish, D. T., Carpenter, F. H.: J. Amer. Chem. Soc. *75*, 950 (1953)
261. Amit, B., Zehavi, U., Patchornik, A.: Israel J. Chem. *12*, 103 (1974)
262. Noda, K., Terada, S., Izumiya, N.: Bull. Chem. Soc. Jpn. *43*, 1883 (1970)
263. Letsinger, R. L., Kornet, M. J.: J. Amer. Chem. Soc. *85*, 3045 (1963)
264. Schwyzer, R., Sieber, P., Zatsko, K.: Helv. Chim. Acta *41*, 491 (1958)
265. Losse, G., Jeschkeit, H., Willenberg, E.: Angew. Chem. Int. Ed. *3*, 307 (1974)
266. Jeschkeit, H., Losse, G., Neubert, K.: Chem. Ber. *99*, 2803 (1966)
267. Veber, D. F., Paleveda, W. J., Jr., Lee, Y. C., Hirschmann, R.: J. Org. Chem. *42*, 3286 (1977)
268. Coyle, S., Keller, O., Young, G. T.: J. Chem. Soc. Chem. Commun., 939 (1975)
269. Sakakibara, S., Fukuda, T., Kishida, Y., Honda, I.: Bull. Chem. Soc. Jpn. *43*, 3322 (1970)
270. Jäger, G., Geiger, R.: Liebigs Ann. Chem., 1535 (1973)
271. Fujino, M., Shinagawa, S., Nishimura, O., Fukuda, T.: Chem. Pharm. Bull. *20*, 1017 (1972)
272. Hiskey, R. G., Smithwick, E. L.: J. Amer. Chem. Soc. *89*, 437 (1967)
273. Sakakibara, S., Shin, M., Fujino, M., Shimonishi, Y., Inouye, S., Inukai, N.: Bull. Chem. Soc. Jpn. *38*, 1522 (1965)
274. Sakakibara, S., Itoh, M.: Bull. Chem. Soc. Jpn. *40*, 646 (1967)
275. Birr, C., Lochinger, W., Stahnke, G., Lang, P.: Liebigs Ann. Chem. *763*, 162 (1972)
276. Coyle, S., Keller, O., Young, G. T.: J. Chem. Soc. Perkin I, 1459 (1979)
277. Kalbacher, H., Voelter, W.: Angew. Chem. Int. Ed. *17*, 944 (1978)
278. Tun-Kyi, A., Schwyzer, R.: Helv. Chim. Acta *59*, 1642 (1976)
279. Brady, S. F., Hirschmann, R., Veber, D. F.: J. Org. Chem. *42*, 143 (1977)
280. Bailey, W. J., Griffith, J. R.: Polym. Prepr. *5*, 266 (1964)
281. Haas, W. L., Krumkalns, E. V., Gerzon, K.: J. Amer. Chem. Soc. *88*, 1988 (1966)

282. Kader, A. T., Stirling, C. J. M.: J. Chem. Soc., 258 (1964)
283. Tesser, G. I., Balvert-Geers, I. C.: Int. J. Peptide Protein Res. *7*, 295 (1975)
284. Kunz, H.: Chem. Ber. *109*, 2670, 3693 (1976)
285. Kemp, D. S., Hoyng, C. F.: Tetrahedron Lett., 4625 (1975)
286. Wakselman, M., Guibé-Jampel, E.: Chem. Commun., 593 (1973)
287. Woodward, R. B., Heusler, K., Gosteli, J., Naegeli, P., Oppolzer, W., Ramage, R., Ranganathan, S., Vorbrüggen, H.: J. Amer. Chem. Soc. *88*, 852 (1966)
288. Wünsch, E., Spangenberg, R.: Chem. Ber. *104*, 2427 (1971)
289. Stevens, C. M., Watanabe, R.: J. Amer. Chem. Soc. *72*, 725 (1950)
290. Olofson, R. A., Yamamoto, Y. S., Wancowicz, D. J.: Tetrahedron Lett., 1563 (1977)
291. Carpino, L. A., Parameswaran, K. N., Kirkley, R. K., Spiewak, J. W., Schmitz, E.: J. Org. Chem. *35*, 3291 (1970)
292. Southard, G. L., Zaborowski, B. R., Pettee, J. M.: J. Amer. Chem. Soc. *93*, 3302 (1971)
293. Cassady, D. R., Easton, N. R.: J. Org. Chem. *29*, 2032 (1964)
294. Brechbühler, H., Büchi, H., Hatz, E., Schreiber, J., Eschenmoser, A.: Helv. Chim. Acta *48*, 1746 (1965)
295. Schellenberg, P., Schühle, H.: Angew. Chem. *73*, 770 (1961)
296. Kemp, D. S., Roberts, D. C.: Tetrahedron Lett., 4629 (1975)
297. Stevenson, D., Young, G. T.: Chem. Commun., 900 (1967)
298. Stevenson, D., Young, G. T.: J. Chem. Soc. (C), 2389 (1969)
299. Ehrensvärd, G. C. H.: Nature *159*, 500 (1947)
300. Lindenmann, A., Khan, N. H., Hofmann, K.: J. Amer. Chem. Soc. *74*, 476 (1952)
301. Barany, G.: Int. J. Peptide Protein Res. *19*, 321 (1982)
302. Wünsch, E., Moroder, L., Nyfeler, R., Jaeger, E.: Hoppe-Seyler's Z. Physiol. Chem. *363*, 197 (1982)
303. Carter, H. E., Frank, R. L., Johnston, H. W.: Organic Syntheses Coll., Vol. III, p. 167 (Horning, E. C., ed.), New York: Wiley 1955
304. Greenstein, J. P., Winitz, M.: The Chemistry of Amino Acids, p. 887, New York: Wiley 1961
305. Zahn, H., Falkenburg, H. R.: Liebigs Ann. Chem. *636*, 117 (1960); cf. also Zahn, H., Schmidt, F.: Makromol. Chem. *36*, 1 (1960)
306. Sofuku, S., Mizimura, M., Hagitani, A.: Bull. Chem. Soc. Jpn. *42*, 278 (1969)
307. Sofuku, S., Mizimura, M., Hagitani, A.: Bull. Chem. Soc. Jpn. *43*, 177 (1970)
308. Sakakibara, S., Honda, I., Naruse, M., Kanaoka, M.: Experientia *25*, 576 (1969)
309. Carpino, L. A.: J. Amer. Chem. Soc. *79*, 4427 (1957)
310. Carpino, L. A., Giza, C. A., Carpino, B. A.: J. Amer. Chem. Soc. *81*, 995 (1959)
311. Schnabel, E.: Liebigs Ann. Chem. *702*, 188 (1967)
312. Polzhofer, K. P.: Tetrahedron Lett., 2305 (1969)
313. Grzonka, Z., Lammek, B.: Synthesis *9*, 661 (1974)
314. Khan, S. A., Sivanandaiah, K. M.: Indian J. Chem. *15B*, 80 (1977)
315. Sieber, P., Iselin, B.: Helv. Chim. Acta *52*, 1525 (1969)
316. Schnabel, E., Schmidt, G., Klauke, E.: Liebigs Ann. Chem. *743*, 69 (1971)
317. Schnabel, E., Herzog, H., Hoffmann, P., Klauke, E., Ugi, I.: Experientia *7*, 380 (1968); Liebigs Ann. Chem. *716*, 175 (1968)
318. Moroder, L., Wackerle, L., Wünsch, E.: Hoppe-Seyler's Z. Physiol. Chem. *357*, 1647 (1976)
319. Tarbell, D. S., Insalaco, M. A.: Proc. Natl. Acad. Sci. USA *57*, 233 (1968)

320. Tarbell, D. S., Uamamoto, Y., Pope, B. M.: Proc. Natl. Acad. Sci. USA *69*, 730 (1972)
321. Moroder, L., Hallett, A., Wünsch, E., Keller, O., Wersin, G.: Hoppe-Seyler's Z. Physiol. Chem. *357*, 1651 (1976)
322. Staab, H. A., Benz, W.: Angew. Chem. *73*, 66 (1961)
323. Carpino, L. A.: J. Org. Chem. *29*, 2820 (1964)
324. Broadbent, W., Morley, J., Stone, J. E.: J. Chem. Soc. C, 2632 (1967)
325. Klieger, E.: Liebigs Ann. Chem. *724*, 204 (1969)
326. Fujino, M., Hatanaka, C.: Chem. Pharm. Bull. *15*, 2015 (1967)
327. Frankel, M., Ladkany, D., Gilon, C., Wolman, Y.: Tetrahedron Lett., 4765 (1964)
328. Gross, H., Bilk, L.: Liebigs Ann. Chem. *716*, 212 (1969)
329. Jones, J. H., Young, G. T.: Chem. Ind., 1722 (1966)
330. Itoh, M., Hagiwara, D., Kamiya, T.: Tetrahedron Lett., 4393 (1975)
331. Duggan, A. J., Roberts, F. E.: Tetrahedron Lett., 595 (1979)
332. Nagasawa, T., Kuriowa, K., Narita, K., Isowa, Y.: Bull. Chem. Soc. Jpn. *46*, 1269 (1973)
333. Klee, W., Brenner, M.: Helv. Chem. Acta *44*, 2151 (1961)
334. Bram, G.: Tetrahedron Lett., 469 (1973)
335. Guibe'-Jampel, E., Wakselman, M.: Chem. Commun., 267 (1971)
336. Rzeszotarska, B., Palka, G.: Bull. Acad. Polonaise des Sciences Ser. Sci. Chim. *16*, 23 (1968)
337. Rzeszotarska, B., Wiejak, S.: Angew. Chem. *80*, 364 (1968)
338. Wünsch, E., Moroder, L., Keller, O.: Hoppe-Seyler's Z. Physiol. Chem. *362*, 1789 (1981)
339. Bodanszky, M., Tolle, J. C.: unpublished
340. Atherton, E., Bury, C., Sheppard, R. C., Williams, B. J.: Tetrahedron Lett., 3041 (1979)
341. Martinez, J., Tolle, J. C., Bodanszky, M.: J. Org. Chem. *44*, 3596 (1979)
342. Medzihradszky-Schweiger, H., Medzihradszky, K.: Acta Chim. Hung. *50*, 339 (1966)
343. Medzihradszky-Schweiger, H.: Acta Chim. Hung. *76*, 437 (1973)
344. Bodanszky, M., Martinez, J., Priestly, G. P., Gardner, J. D., Mutt, V.: J. Med. Chem. *21*, 1030 (1978)
345. Bodanszky, M., Bodanszky, A., Deshmane, S. S., Martinez, J., Said, S. I.: Bioorg. Chem. *8*, 399 (1979)
346. Medzihradszky-Schweiger, H.: Ann. Univ. Sci. Budapestiensis de Rolando Eötvös Nominatae, Sectio Chimica *13*, 35 (1972)
347. Losse, G., Stiehl, H. U.: Z. Chem. *21*, 188 (1981)
348. Tzougraki, C., Makofske, R. C., Gabriel, T. F., Wang, S. S., Kutny, R., Meienhofer, J.: J. Amer. Chem. Soc. *100*, 6248 (1978)
349. Lundt, B. F., Johansen, N. L., Vølund, A., Markussen, J.: Int. J. Pept. Protein Res. *12*, 258 (1978)
350. Weygand, F., Steglich, W.: Z. Naturforsch. *14b*, 472 (1959)
351. Bodanszky, M., Klausner, Y. S., Lin, C. Y., Mutt, V., Said, S. I.: J. Amer. Chem. Soc. *96*, 4973 (1974)
352. Kiso, Y., Ukawa, K., Akita, T.: J. Chem. Soc. Chem. Commun., 101 (1980)
353. Schwyzer, R., Kappeler, H., Iselin, B., Rittel, W., Zuber, H.: Helv. Chim. Acta *42*, 1702 (1959)
354. Stewart, J. M., Woolley, D. W.: Nature *206*, 619 (1960)
355. White, J.: J. Biol. Chem. *106*, 141 (1934)
356. Idelson, M., Blout, E. R.: J. Amer. Chem. Soc. *80*, 4631 (1958)

357. Brenner, M., Curtius, H. C.: Helv. Chim. Acta *46*, 2126 (1963)
358. Chauvette, R. R., Pennington, P. A., Ryan, C. W., Cooper, R. D. E., Jose, L., Wright, I. G., Van Heyningen, E. M., Huffmann, G. W.: J. Org. Chem. *36*, 1259 (1971)
359. Yajima, H., Fujii, N., Ogawa, H., Kawatani, H.: J. Chem. Soc. Chem. Commun., 407 (1974)
360. Halpern, B., Nitecki, D. E.: Tetrahedron Lett., 3031 (1967)
361. Loffet, A., Dremier, C.: Experientia *27*, 1003 (1971)
362. Hiskey, R. G., Beacham, L. M. III, Matl, V. G., Smith, J. N., Williams, E. B., Jr., Thomas, A. M., Wolters, E. T.: J. Org. Chem. *36*, 488 (1971)
363. Felix, A. M.: J. Org. Chem. *39*, 1427 (1974)
364. Pless, J., Bauer, W.: Angew. Chem. *85*, 142 (1973)
365. Sakakibara, S., Shimonishi, Y.: Bull. Chem. Soc. Jpn. *38*, 1412 (1965)
366. Olah, G. A., Nojima, M., Kerekes, I.: Synthesis, 779 (1973)
367. Matsuura, S., Niu, C. H., Cohen, J. S.: J. Chem. Soc. Chem. Commun., 451 (1976)
368. Sakakibara, S., Shin, K. H., Hess, G. P.: J. Amer. Chem. Soc. *84*, 4921 (1962)
369. Homer, R. B., Moodie, R. B., Rydon, H. N.: J. Chem. Soc., 4403 (1965)
370. Bláha, K., Rudinger, J.: Collect. Czechoslov. Shem. Commun. *30*, 599 (1965)
371. König, W., Geiger, R.: in Chemistry and Biology of Peptides (Meienhofer, J. ed.), p. 343. Ann Arbor, Michigan: Ann Arbor Sci. Publ. 1972
372. Kutassy, S., Kajtár, J., Medzihradszky-Schweiger, H.: Acta Chim. Hung. *57*, 123 (1968)
373. Carpino, L. A.: Chem. Commun., 858 (1966)
374. Bodanszky, M., Deshmane, S. S., Martinez, J.: J. Org. Chem. *44*, 1622 (1979)
375. Carpino, L. A., Williams, J. R.: J. Chem. Soc. Chem. Commun., 450 (1978)
376. Eckert, H., Schrauzer, G. N., Ugi, I.: Tetrahedron *31*, 1399 (1975)
377. Eckert, H., Ugi, I.: Angew. Chem. *88*, 717 (1976; Angew. Chem. Int. Ed. *15*, 681 (1976)
378. Grimshaw, J.: J. Chem. Soc., 7136 (1965)
379. Schönheimer, R.: Hoppe-Seyler's Z. Physiol. Chem. *154*, 203 (1926)
380. Rudinger, J.: in "The Chemistry of Polypeptides" (Katsoyannis, P. G., ed.), p. 87, New York: Plenum Press 1973
381. Mazur, R. H., Plume, G.: Experientia *24*, 661 (1968)
382. Weisblat, D. I., Magerlein, B. J., Meyers, D. R.: J. Amer. Chem. Soc. *75*, 3630 (1953)
383. Poduska, K., Rudinger, J., Šorm, F.: Collect. Czech. Chem. Commun. *20*, 1174 (1955)
384. Rudinger, J., Maassen van den Brink-Zimmermannova, H.: Helv. Chim. Acta *56*, 2216 (1973)
385. Horner, L., Neumann, H.: Chem. Ber. *98*, 3462 (1965)
386. Okumura, K., Iwasaki, T., Matsuoka, M., Matsumoto, K.: Chem. Ind., 929 (1971)
387. Hamada, T., Nishida, A., Matsumoto, Y., Yonemitsu, O.: J. Amer. Chem. Soc. *102*, 3978 (1980)
388. Wiley, R. H., Davis, H. L., Gensheimer, D. E., Smith, N. R.: J. Amer. Chem. Soc. *74*, 936 (1952)
389. Wiley, R. H., Davis, R. P.: J. Amer. Chem. Soc. *76*, 3496 (1954)
390. Beecham, A. F.: Chem. Ind., 1120 (1955)
391. Beecham, A. F.: J. Amer. Chem. Soc. *79*, 3257, 3262 (1957)
392. Milne, H. B., Peng, C. H.: J. Amer. Chem. Soc. *79*, 639 (1957)

393. Fukuda, T., Kitada, C., Fujino, M.: J. Chem. Soc. Chem. Commun., 220 (1978)
394. Hendrickson, J. B., Bergeron, R.: Tetrahedron Lett., 345 (1970)
395. Ito, H., Ogawa, K., Iida, T., Ichikizaki, I.: Chem. Pharm. Bull. *26*, 296 (1978)
396. Wolman, Y.: Israel J. Chem. *5*, 231 (1967)
397. Šavrda, J., Veyrat, D. H.: J. Chem. Soc. (C), 2180 (1970)
398. Spangenberg, R., Thamm, P., Wünsch, E.: Hoppe-Seyler's Z. Physiol. Chem. *352*, 655 (1971)
399. Poduska, K., Maassen van den Brink-Zimmermannová, H.: Coll. Czech. Chem. Commun. *33*, 3769 (1968)
400. Poduska, K.: Coll. Czech. Chem. Commun. *33*, 3779 (1968)
401. Meienhofer, J.: Nature *205*, 73 (1965)
402. Wünsch, E., Spangenberg, R.: Chem. Ber. *105*, 740 (1972)
403. Stern, M., Warshawsky, A., Fridkin, M.: Int. J. Peptide Protein Res. *17*, 531 (1981)
404. Foss, O.: Acta Chem. Scand. *1*, 307 (1947)
405. Polzhoffer, K. P., Nay, K. H.: Tetrahedron *27*, 1997 (1971)
406. König, W., Geiger, R.: Chem. Ber. *103*, 788 (1970)
407. Bodanszky, M., Bednarek, M. A., Bodanszky, A.: Int. J. Peptide Protein. Res. *20*, 387 (1982)
408. Mitin, Y. V., Vlasov, G. P.: Dokl. Akad. Nauk. S.S.S.R. *179*, 353 (1968)
409. Šavrda, J., Veyrat, D. H.: Tetrahedron Letters 6253 (1968)
410. Wünsch, E., Fontana, A., Drees, F.: Z. Naturforsch. *22b*, 607 (1967)
411. Phocas, I., Yovanidis, C., Photaki, I., Zervas, L.: J. Chem. Soc. (C), 1506 (1967)
412. Zervas, L., Dilaris, L.: J. Amer. Chem. Soc. *77*, 5354 (1955)

IV. Semipermanent Protection of Side Chain Functions

A. Carboxyl Groups of Aspartyl and Glutamyl Residues

Protection of the side chain carboxyl groups of aspartyl and glutamyl residues is not always necessary. For instance, carboxyl groups in the side chains of the carboxyl component can be left unmasked if the α-carboxyl of the C-terminal residue is activated via the hydrazide and azide:

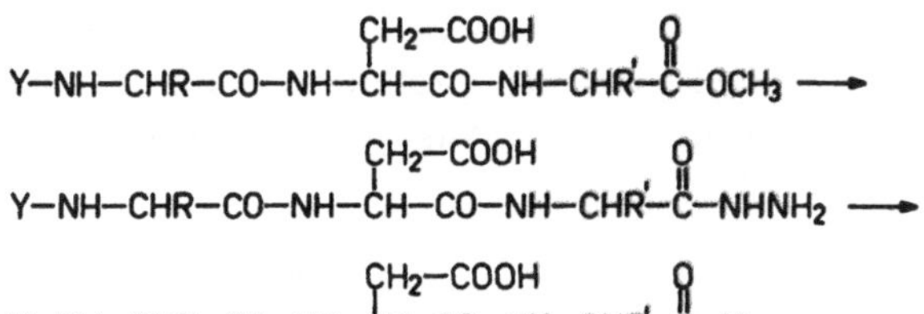

The minor complication caused by the formation of hydrazinium salts in the conversion of the ester to the hydrazide can be remedied by extraction with weak acids. Protection of the side chain carboxyls is even less mandatory in the amino component:

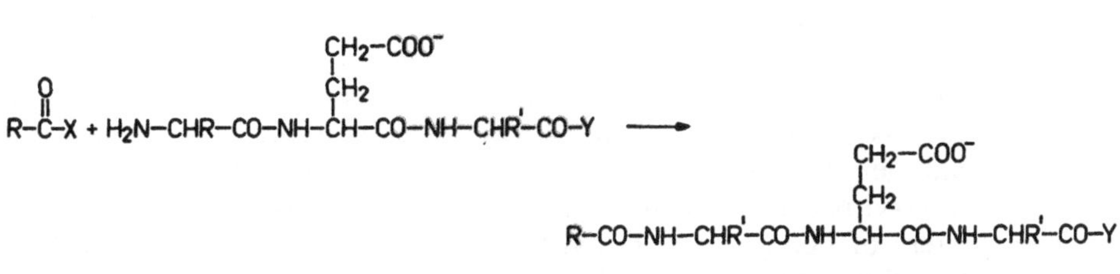

In fact, unprotected carboxyls might be preferred when global protection leads to intermediates which are insoluble in the available organic solvents. Still, protection of the side chain carboxyls, generally in the form of esters, is considered by many investigators as the better approach to complex peptides because no concern has to be felt about the counterions, such as hydrazinium or trialkylammonium ions associated with the carboxylate groups and perhaps more significantly, no interaction with the acylating agent has to be expected. On the other hand, free carboxyls or more exactly the carboxylates can react with the activated carboxyl component to form mixed anhydrides or active esters, which, in turn, can cause intramolecular side reactions resulting in cyclization [1] or intermolecular reactions causing the formation of branched peptide chains [2]:

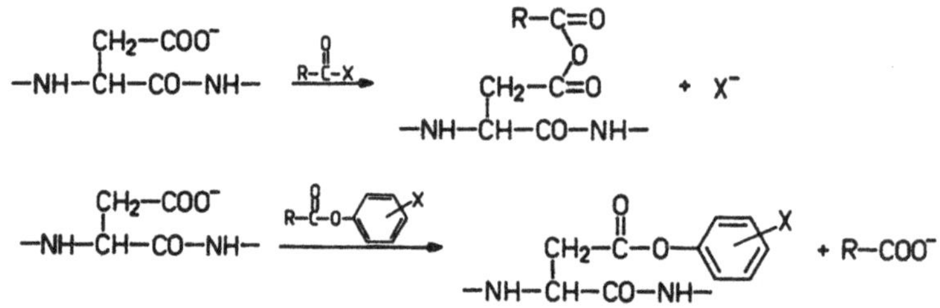

Blocking of the side chain carboxyls may require two different masking groups, one for the α, the other for the ω-carboxyl function. In any case, carboxyl protection in dicarboxylic acids necessitates the specific introduction of selectively removable ester groups [3]. A classical solution for this general problem was offered by the β and γ benzyl esters of aspartic and glutamic acid, respectively. β-Benzyl-L-aspartic acid can be obtained by partial saponification of aspartic acid dibenzyl ester [4], preferably with lithium hydroxide in acetone [5], while the controlled cleavage of the diester with hydroiodic acid [5] affords the α-ester:

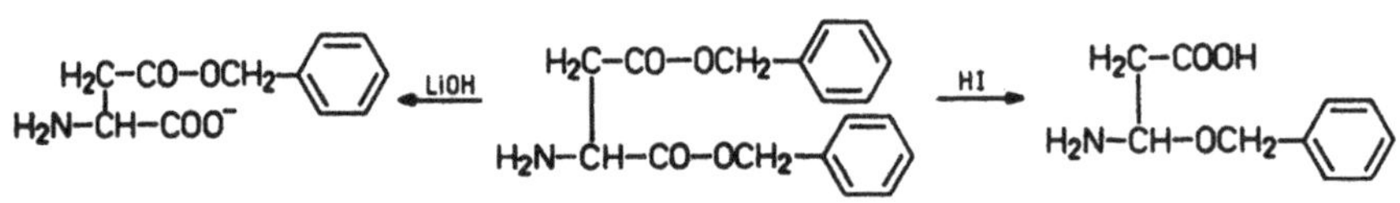

Transesterification of dicarboxylic acid with methyl or ethyl acetate in the presence of perchloric acid yields aspartic acid β-esters and glutamic acid γ-esters because the positive charge on the amino group interferes with protonation of the α-carboxyl group [6]. An alternative approach to selective esterification is offered by the copper(II) complexes of the dicarboxylic acids. In these complexes only the carboxyl group is engaged as a ligand while the side chain carboxyls can form alkali salts and are available for esterification with alkyl halogenides [7]. The reverse process, in which selective hydrolysis at the ω-carboxyl is catalyzed by Cu(II) ions [8] is somewhat more efficient.

The α-esters of benzyloxycarbonyl-L-aspartic acid could be obtained by esterification of benzyloxycarbonyl-L-asparagine followed by de-amidation of the carboxamide group with nitrous acid [9]. The preparation of mixed diesters was completed by the perchloric acid catalyzed transesterification of the newly formed carboxyl group with tert. butyl acetate. The same series of reactions was extended [9] to benzyloxycarbonyl-L-glutamine as well:

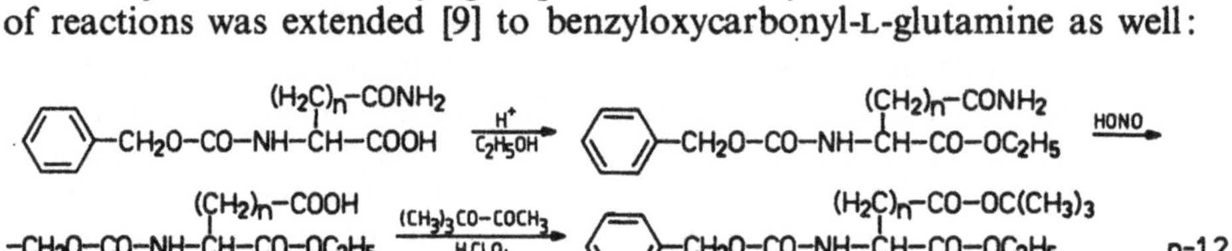

In an improved version of the de-amidation reaction nitrosonium hydrogen sulfate ($NO \cdot HSO_4$) in acetic acid was applied [10] and the α-methyl, ethyl, benzyl and *p*-nitrobenzyl esters of both benzyloxycarbonyl-L-aspartic acid and benzyloxycarbonyl-L-glutamic acid were prepared in good yield.

In the case of glutamic acid an additional pathway is open for selective esterification. It leads through the cyclic compounds benzyloxycarbonyl or *p*-toluenesulfonyl pyroglutamic acid which are converted to their tert. butyl esters and then opened up with alkali to yield pure α-esters of (N-protected) glutamic acid [11].

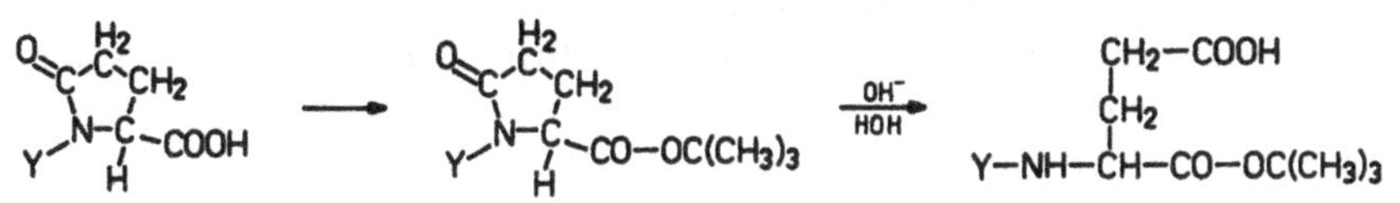

In a more conservative manner one can start with benzyloxycarbonyl-L-aspartic acid α-ethyl ester obtained through alcoholysis of the cyclic anhydride and separation of the α and β isomers. The α-ethyl ester is then transformed into the α-ethyl-β-tert. butyl ester and saponified with alkali to yield the pure β-tert. butyl ester or, after hydrogenation β-tert. butyl L-aspartate [12]:

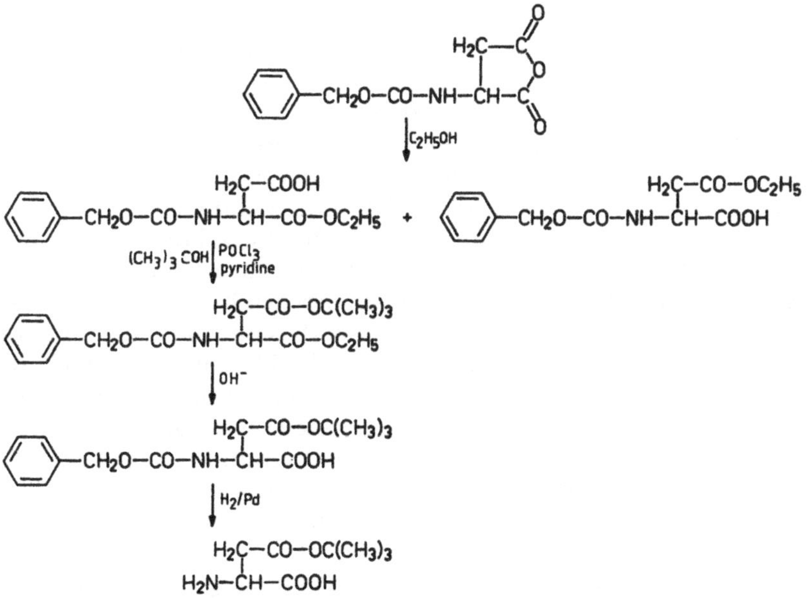

An analogous series of reactions is built on the treatment of benzyloxy-carbonyl-L-aspartic acid α-benzyl ester with isobutene in the presence of sulfuric acid. Hydrogenation then affords β-tert. butyl aspartate [13]:

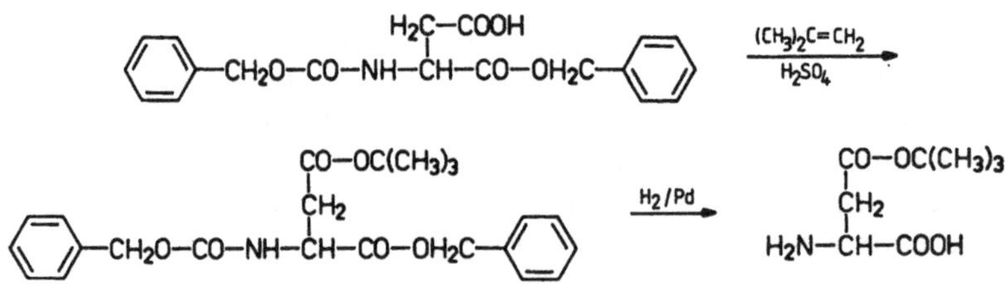

The methods of protection and the procedures for the removal of blocking groups discussed in connection with the carboxyl group in Chapter II are also applicable for the masking and unmasking of the side chain carboxyls of aspartyl and glutamyl residues. An essential feature of side chain protection should be pointed out here: the side chain functions are deblocked usually only at the end of the chain building process. Therefore β and γ benzyl esters are good in *combination* with acid labile groups used for the transient protection of the α-amino function. Similarly β and γ tert. butyl esters are excellent for the masking of carboxyl groups of aspartyl or glutamyl residues if hydrogenolysis is used for amine deprotection after each coupling step. The wide choice of amine and carboxyl protecting groups allows a large number of such logical and useful combinations [14].

B. Side Chain Amino Groups of Lysine and Ornithine

Of the twenty amino acids which are the constituents of proteins only lysine has an amino group in its side chain. Ornithine, frequently found in microbial peptides and also often built into hormone analogs and potentially useful as a precursor of arginine [15], is the next homolog of lysine and the two can be discussed together. In the incorporation of these residues masking of the side chain amino function is obviously mandatory. Otherwise two amino groups are simultaneously available for acylation in a subsequent coupling, and since their reactivities are not sufficiently different, a branching of the peptide chain will occur. Thus, if chain lengthening at α-amino groups is the objective, the ω-amino functions must remain blocked by a semipermanent masking group which is removed only at the end of the chain building process. The protecting groups which can be applied for the blocking of side chain amino groups are the same as the ones used for the masking of α-amino functions, but *two different groups* must be selected: a readily removable group for the α-amino groups and a semipermanent group for the side chain function. Problems of differential protection and selective deprotection were presented in considerable detail in a recent review by Fauchere and Schwyzer [14]. Here we must limit ourself to the essentials.

A simple combination for the protection of the amino groups in lysine is the use of the benzyloxycarbonyl group for blocking of the α-amino function and the *p*-toluenesulfonyl group for the masking of the side chain. This approach, illustrated with a synthesis of lysine vasopressin [16] (Scheme 1), permits partial deprotection after incorporation of the lysine residue and after each coupling step by hydrogenolysis or by a brief treatment with hydrobromic acid in acetic acid, since the tosyl group remains unaffected in these operations. It can be removed, however, by reduction with sodium in liquid ammonia at the conclusion of the synthesis, prior to ring closure.

Scheme 1. A Synthesis of Lysine Vasopressin [16]

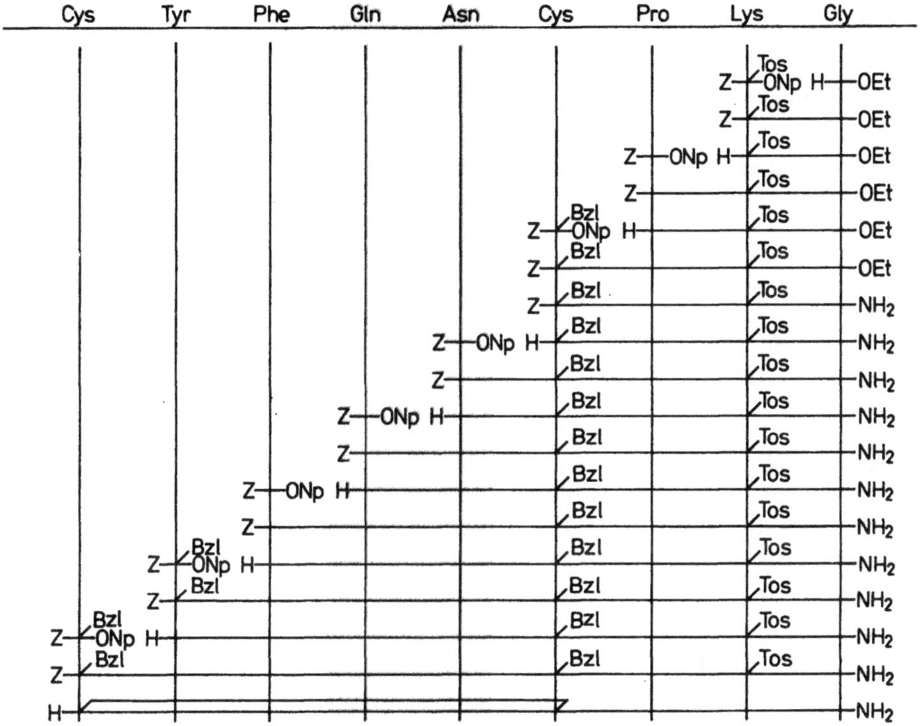

Such combinations of two different protecting groups necessitate the selective acylation of the individual amino functions. This can be done quite efficiently by the exploitation of the Cu(II) complex of lysine in which only the ε-amino group is available for acylation [17].

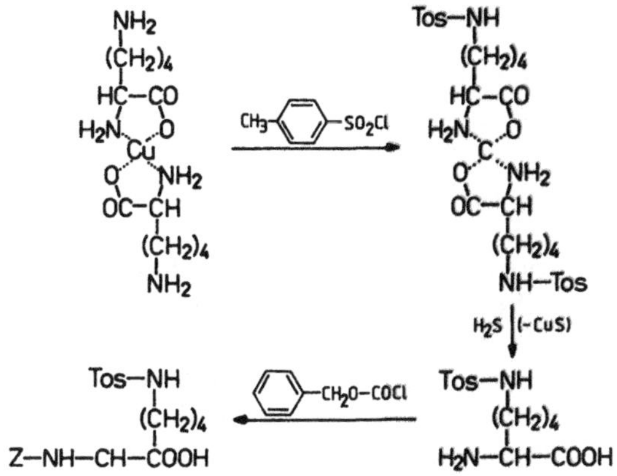

The same principle can be applied also for the selective derivatization of ornithine but not so unequivocally for α,γ-diaminobutyric acid or α,β-diaminopropionic acid.

An alternative pathway [18] for exclusively blocking the side chain amino group starts with $N^{\alpha},N^{\varepsilon}$-dibenzyloxycarbonyllysine (or the corresponding disubstituted derivative of ornithine) and proceeds through the acid chloride and N-carboxyanhydride:

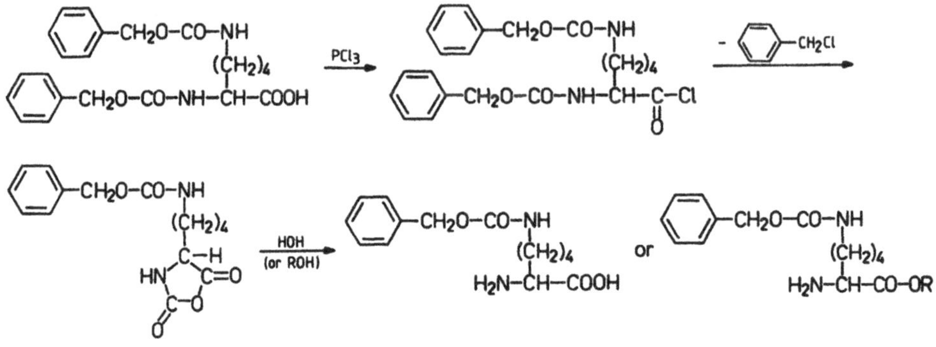

With moderately reactive acylating agents, such as trifluoroacetic acid thioethyl ester [19] it is possible to prepare N^{ε}-acyl derivatives of lysine that can be separated from a small amount of N^{α}-acyl derivative formed in the reaction, because the latter are more soluble in water.

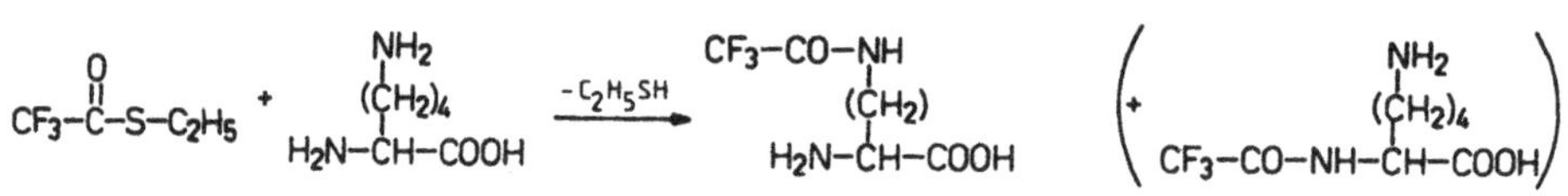

The literature abounds in combinations which allow the selective removal of the blocking at the α-amino functions without the loss of protection at the ε-amino function of lysine residues or the δ-amino groups of ornithine moieties. One of the most obvious pairs consists of the tert. butyloxycarbonyl and the benzyloxycarbonyl groups. The Boc group is unaffected by catalytic hydrogenation but is removed by acidolysis with moderately strong acids and the reverse is true for the Z group. Yet, there are certain limitations inherent in the Z-Boc combination. Catalytic reduction is impeded in sulfur containing peptides and, while hydrogenation in the presence of base [20] or in liquid ammonia [21] provides a remedy, the process can fail in some cases, e.g. when two or more methionine residues occur in the sequence [22]. The N^{α}-Z,N^{ε}-Boc approach could be applied successfully, e.g. in an outstanding synthesis of corticotropin [23], but probably cannot be extended for many other peptides. The alternative use of the same two protecting groups, Boc for α-protection and Z for the ε-amino groups suffers from the imperfect resistance of the Z group to trifluoroacetic acid, HCl in dioxane or other acidic reagents of moderate strength needed for the removal of the Boc group. Although usually only a very small fraction of the Z group is lost in a single operation, during the unmasking of several α-amino groups considerable damage occurs at the side chain amino groups and consequently a not negligible branching of the chain takes place at lysine residues. Many attempts were made to correct this situation; those of Schnabel and his associates [24] are particularly noteworthy. To increase selectivity, they proposed to use 70% aqueous trifluoroacetic acid rather than the neat acid. The side chain Z groups are fairly resistant to 70% trifluoroacetic acid and the losses are negligible. A possible hydrolysis of carboxamide groups by the aqueous acid can be avoided by replacing water with acetic acid in the mixture [25]. Several other acidic reagents, for instance ethanesulfonic acid [26], were recommended because they show enhanced selectivity. An additional and rather obvious solution to the problem lies in the modification of the Z group. Substitution on the aromatic ring with electron withdrawing substituents leads to an increase in the resistance of the group toward acids. Details of various implementations of this principle have already been discussed in Chapter III. A certain need for caution is indicated, however, in the dependence of the acid sensitivity of the Z group from various factors. The ready removal of the Z group by trifluoroacetic acid if thioanisole is added [27] to the reaction mixture suggests that the presence of other scavengers or perhaps of methionine residues in the sequence, could also alter the stability of blocking by the benzyloxycarbonyl group.

As an example of new combinations the proposed [28] application of the trichloroethyloxycarbonyl group [29] for side chain amine protection can be mentioned. This acid resistant blocking group allows the selective removal of Z groups by acidolysis, but not by hydrogenation, because, contrary to earlier observations [30] the trichloroethyloxycarbonyl group does not completely resist catalytic reduction.

Combinations in which the removal of individual protecting groups is based on different principles are called *orthogonal* [31]. A promising orthogonal combination consists of the base sensitive and acid stable 9-fluorenylmethyloxycarbonyl group [32] for α-amino groups and the acid sensitive tert. butyloxycarbonyl group for the side chain amino function of lysine residues. Full selectivity can be reached also by acetylation of the ε-amino groups [33] because the acetyl group remains unchanged under the commonly used deblocking conditions and is removed only by a specific acyl-lysine deacylase at an appropriate later stage. The method will gain practicality only if the specific enzyme becomes commercially available. A selection of useful combinations of protecting groups for the α and ε amino functions is shown in Table 9.

Finally, replacement of lysine residues by a precursor such as α-amino-ε-nitro caproic acid (or 2-amino-6-nitrohexanoic acid) [35–37]

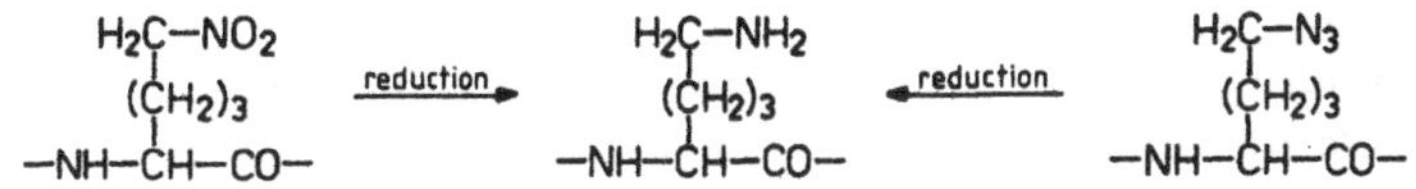

Table 9. Protecting Group Combinations for Lysine and Ornithine

N^{ε}-Protection	N^{α}-Protection
p-toluenesulfonyl	benzyloxycarbonyl tert. butyloxycarbonyl
benzyloxycarbonyl	*p*-methoxybenzyloxycarbonyl tert. butyloxycarbonyl adamantyloxycarbonyl biphenylisopropyloxycarbonyl isonicotinyloxycarbonyl *o*-nitrophenylsulfenyl formyl
tert. butyloxycarbonyl	benzyloxycarbonyl *p*-toluenesulfonyl *o*-nitrophenylsulfenyl biphenylisopropyloxycarbonyl 9-fluorenylmethyloxycarbonyl
acetyl,trifluoroacetyl,formyl, 2-chlorobenzyloxycarbonyl,4-chlorobenzyloxycarbonyl,4-nitrobenzyloxycarbonyl,phthalyl	tert. butyloxycarbonyl

* For references and additional combinations cf. ref. 34.

renders the application of two different protecting groups unnecessary. The incorporation of α-amino-ω-azido acids could also be considered. No differential protection is needed when a lysine residue is N-terminal, as in kallidin or in cholecystokinin.

C. Hydroxyl Groups in Serine, Threonine and Tyrosine

Protection of the side chain functions of the hydroxy-amino acids serine, threonine and tyrosine is discussed in the same section because the same methods of protection can be applied for all three of them. Some differences, however, in their need for protection must be noted. Unwanted acylation of the primary alcoholic hydroxyl group in serine occurs rather readily, while it is less easy to acylate the hindered secondary alcohol in the threonine side chain. In tyrosine, the reactivity of the side chain hydroxyl group very much depends upon the conditions which prevail in the reaction mixture during coupling. In the presence of bases deprotonation of the hydroxyl group occurs and a potent nucleophile is generated; the formerly inert phenol changes into a reactive phenolate. Because of these differences it is difficult to make general and definite recommendations for the choice between minimal and global protection. Often a compromise between the two extremes is the most reasonable approach with protected serine and unprotected threonine and tyrosine side chains. Alternatively, the hydroxyl groups of serine and tyrosine are protected and only the fairly unreactive threonine side chain is left without a blocking group. This can be done in schemes [38, 39] in which moderately reactive derivatives of the carboxyl component, e.g. azides or active esters, are used for coupling. Syntheses in which powerful acylating agents, such as mixed or symmetrical anhydrides, are applied generally require the masking of the side chain function of all three hydroxy-amino acids.

An obvious approach to hydroxyl protection could be the adaptation of the methods of carbohydrate chemistry that is the acylation of the hydroxyl group. Yet, while several attempts were made in this direction, neither acetylation nor benzoylation are satisfactory in peptide synthesis and the use of p-toluenesulfonyl and benzyloxycarbonyl groups for the masking of hydroxyl functions remains sporadic only. There are good reasons for this lack of success with O-acyl derivatives. Phenyl esters are sensitive to nucleophiles and might migrate from O to N, thus blocking a terminal amino group:

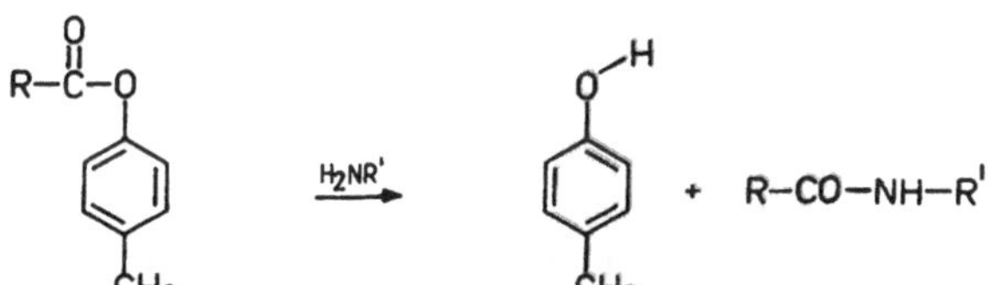

Therefore, one has to feel some concern about protection of tyrosine side chains by acetylation [40] benzoylation [41], or tosylation [42]. Carbonic acid derivatives, such as O-benzyloxycarbonyl-tyrosine [43], O-2-bromobenzyloxycarbonyl-tyrosine [44], or O-tert. butyloxycarbonyl-tyrosine [45] are less likely to cause similar complications and yet are seldom used. Attempts to revive the application of O-alkyloxycarbonyl derivatives of tyrosine were not entirely successful, although, e.g. O-ethoxycarbonyl-tyrosine [46] is easily prepared with ethyl chlorocarbonate and the masking is readily removed by hydrazinolysis. Similarly, the O-carbamoyl and the O-phenylcarbamoyl groups [47], which are cleaved by nucleophiles, found only limited use in practical syntheses.

O-Acyl-derivatives of serine and threonine are less potent acylating agents, but one still has to expect intramolecular transfer of acyl groups from O to N when, for instance, an N-terminal O-acetyl-serine residue is present in a coupling mixture:

Also, base catalyzed β-elimination is common in O-acyl derivatives of serine and threonine, particularly if the acid component of the ester provides a good leaving group. Thus, O-tosyl-serine and O-tosyl-threonine can produce dehydroalanine and dehydrobutyrine residues respectively:

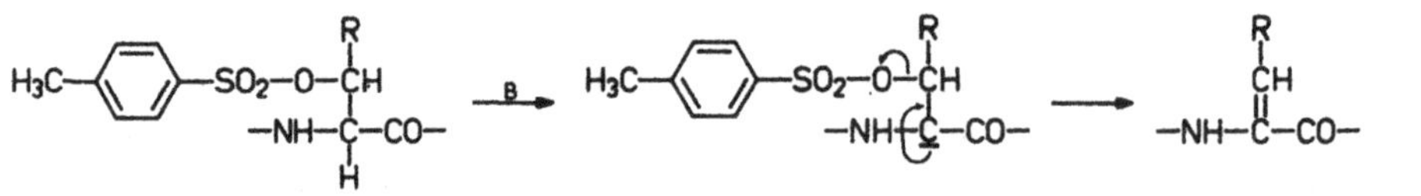

The 2,2,2-trifluoro-1-benzyloxycarbonylaminoethyl group [48], removable by catalytic

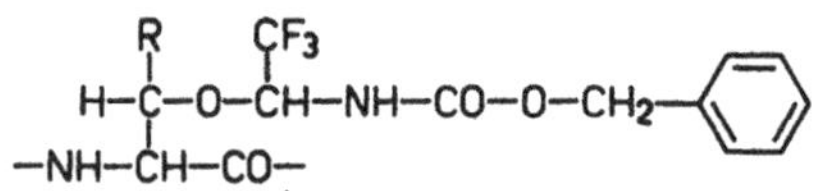

hydrogenation and also by acidolysis with HBr-acetic acid or with HF, remains a very interesting but perhaps not simple enough possibility for the blocking of serine and threonine side chains.

At this time the standard method for the blocking of hydroxyl groups is to convert them to their *benzyl or tert.butyl ethers*. Alkylation of the hydroxyls is somewhat more involved than their acylation and initially an indirect route had to be followed [49] for the preparation of O-benzyl serine. 2,3-Dibromopropionic acid esters, obtained from esters of acrylic acid, were treated with sodium benzylate. This lead to a selective displacement of the β-bromo substituent. Saponification of the ester and nucleophilic displacement of the α-substituent by ammonia gave O-benzyl-DL-serine which was then resolved by

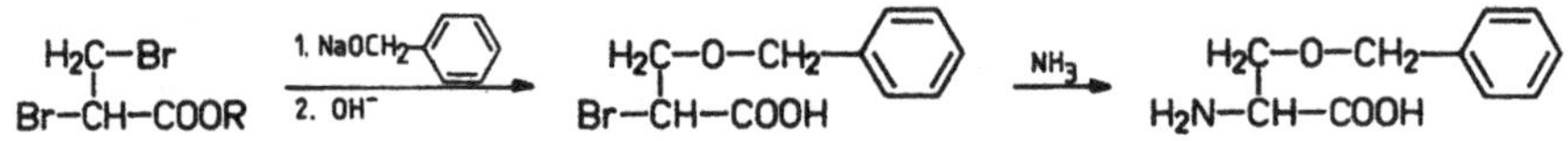

selective deacylation of the N-acetyl derivative with an enantiospecific enzyme [50]. This procedure is less suitable for the preparation of O-benzyl-L-threonine since it has two chiral centers and the separation of four isomers is necessary. More straightforward is the direct alkylation [51] of the hydroxyl group of N-acetyl-DL-threonine by treatment with sodium in liquid ammonia followed by the addition of benzyl bromide. An extension of this approach to tert.butyloxycarbonyl-L-serine is quite practical [52] but not for the synthesis of tert.butyloxycarbonyl-O-benzyl-L-threonine which was obtained in only very low yield. For threonine derivatives the acid catalyzed formation of the ether in a reaction between the amino acid, benzyl alcohol and benzenesulfonic acid [53] is more satisfactory. The resulting benzyl ether-benzyl ester is saponified under mild conditions.

Synthesis of tert. butyl ethers of serine and threonine is even less obvious. This can be achieved by the acid catalyzed addition of isobutene [54, 55] to esters of the amino acids and subsequent cleavage of the ester group, e.g. by catalytic reduction:

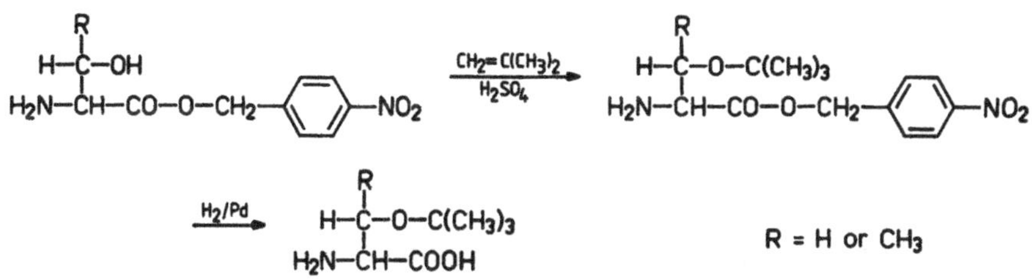

Benzylation of the phenolic hydroxyl of tyrosine proceeds smoothly if the copper(II) complex of the amino acid is treated with benzyl bromide in an alkaline solution [56]. The tert. butyl ether is formed in an acid catalyzed reaction of tyrosine esters with isobutene [54].

Benzyl ethers are cleaved by catalytic hydrogenation or by acidolysis, e.g. with hydrobromic acid in acetic acid. It still should be noted, that while hydrobromic acid smoothly cleaves benzyl ethers, it also catalyzes the esterification of the unmasked hydroxyl group of serine with acetic acid and O-acetyl serine derivatives form in significant amounts. Threonine is less easily esterified. This side reaction can be avoided by carrying out the acidolysis in trifluoroacetic acid rather than in acetic acid, but in that case a new problem arises: migration of the benzyl group from the phenolic hydroxyl of tyrosine to the carbon atom ortho to the hydroxyl group. The rearrangement [57] yields derivatives of 3-benzyl tyrosine [58]. The side reaction does not occur,

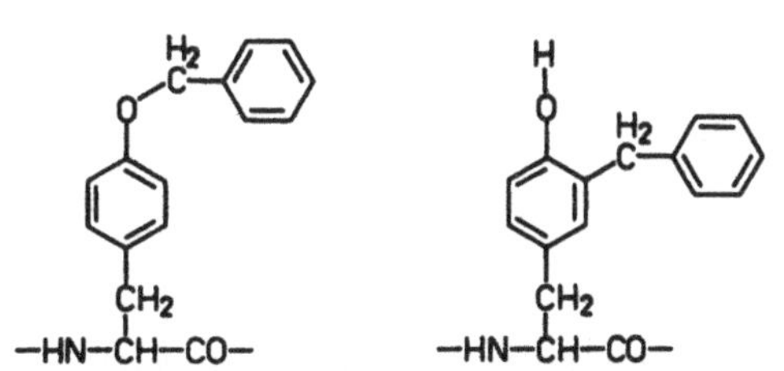

or at least not to a noticeable extent, in HBr-acetic acid. This problem and the means proposed for its solution will be discussed in more detail in the chapter on side reactions. Here we mention only a recent finding of Kiso and his associates [59] who could deprotect O-benzyltyrosine derivatives with a thioanisole-trifluoroacetic acid mixture without O → C rearrangement. A second interesting observation worth mentioning is the selective hydrogenolysis of O-benzylserine derivatives. In the presence of bases, such as cyclohexylamine, benzyloxycarbonyl groups can be cleaved by hydrogenation while the O-benzyl group remains intact [60].

In connection with the removal of O-alkyl groups we should point to derivatives of O-4-picolyl-tyrosine [61]. This acid resistant blocking group

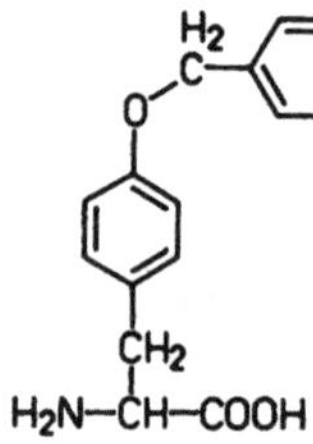

is cleaved by electrolytic reduction in dilute sulfuric acid.

O-Benzyl derivatives of hydroxyamino acids are very useful [62] but not necessarily optimal intermediates in peptide synthesis. Difficulties in their preparation might be overcome by improvements such as the ones implemented in the synthesis of N-tert. butyloxycarbonyl-O-methyl-L-serine and N-tert. butyloxycarbonyl-O-methyl-L-threonine which were obtained in good yield from the N-protected amino acids by treatment with sodium alkoxides and methyl iodide [63]. Yet, one has always to be concerned about alkylation during deprotection. Tert.butyl ethers cause less alkylation because of steric hindrance in the tert. butyl cations or in the tert. butyl esters which are the alkylating intermediates. The same steric hindrance however can unfavorably affect coupling rates when O-tert. butyl derivatives of hydroxyamino acids are activated. Undesirable steric effects are present also in hydroxy-amino acids protected in the form of diphenylmethyl [64] or triphenylmethyl [65] ethers. Thus, further developments are desirable in this area. Some proposed alternatives, e.g. protection of hydroxyl groups by incorporating them into acetals [66, 67],

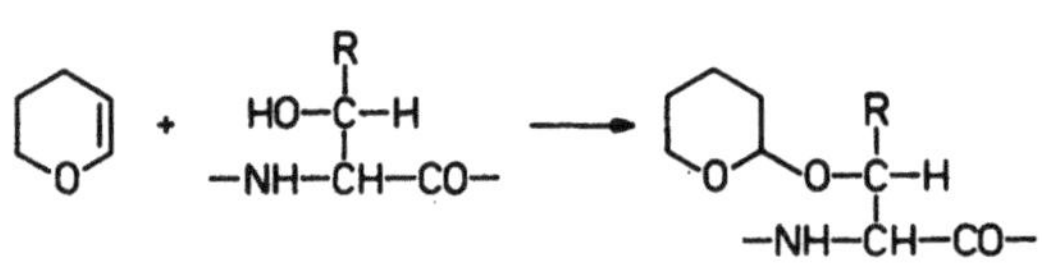

R = H or CH_3

remains to be evaluated.

In closing this section we refer to a comprehensive review by Stewart [68] on the protection of the hydroxyl group in peptide synthesis.

D. The Sulfhydryl Group of Cysteine

Blocking of the sulfhydryl group of cysteine residues is mandatory. Mercaptanes are quite reactive. Unlike alcohols they are potent nucleophiles which effectively compete with amino groups for acylating agents to form thiol esters. They can also be oxidized, even by air, to disulfides.

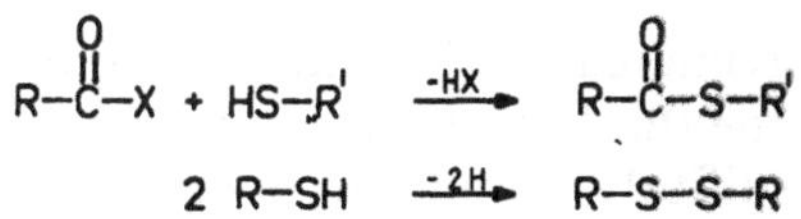

The latter reaction provides, however, an alternative to the masking of the thiol function. It is possible to incorporate the disulfide cystine, instead of cysteine, in peptides and to build two identical chains simultaneously [69]:

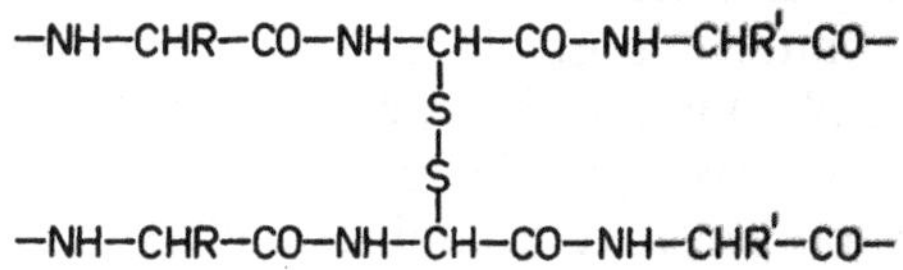

An obstacle in the execution of such schemes might be the insufficient solubility of some high molecular weight intermediates in the solvents which can be used in peptide synthesis.

Treatment of disulfides with alkali hydrogen sulfites yields *S-sulfonates* [70], so-called Bunte salts:

$$\text{R–S–S–R} + \text{NaSO}_3 \rightleftharpoons \text{R–S–SO}_3\text{Na} + \text{HS–R}$$

The regenerated sulfhydro compound can be oxidized to the disulfide and in this way the entire disulfide converted to S-sulfonate. The Bunte salts are important intermediates in the synthesis of complex peptides with disulfide bridges, such as insulin, but while they can also be applied for the protection of the SH function [71] their sensitivity to acids and thiols (cf. the above reversible reaction) is too great to render them generally useful for semipermanent protection.

Blocking in the form of *mixed disulfides* can also be subject to objections. The tendency of disulfides to disproportionate to symmetrical disulfides

$$2\ \text{R–S–S–R}' \longrightarrow \text{R–S–S–R} + \text{R}'\text{–S–S–R}'$$

is an obvious cause for concern. Yet this tendency, which is quite pronounced in some disulfides, e.g. in cystine, is almost absent in others, for instance in S-ethylmercapto derivatives [72]. The careful examination of

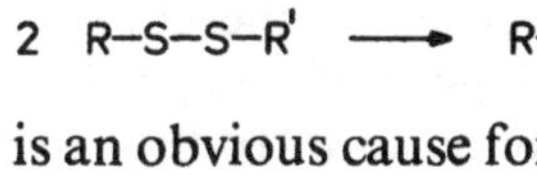

a whole series of disulfides from the point of view of disproportionation revealed [73, 74] that the S-isopropylmercapto and S-tert. butylmercapto derivatives of cysteine are particularly inert in this respect. The reason for this inertia is not merely steric hindrance

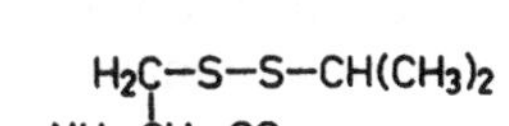

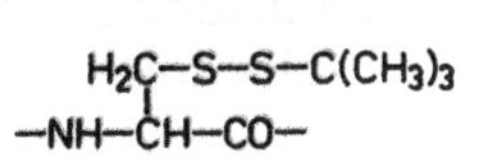

since other bulky substituents, e.g. the *p*-tolylmercapto derivative are quite prone to disproportionation. Unmasking of the thiol function is fairly simple because the mixed disulfides are cleaved by sulfitolysis and also by reduction with thiols such as thioglycol or thioglycolic acid.

The objection mentioned in connection with the protection of hydroxyl groups by acyl moieties, namely that they can migrate from O to N, is valid for S-acyl derivatives as well. In fact, thiolesters are more reactive than their oxygen analogs and hence the stability and inertness of S-acyl protecting groups is at least questionable. Yet, in spite of possible S → N migration several acyl groups were proposed for the blocking of the cysteine side chain. Of these the S-acetyl and S-benzoyl derivatives [75, 76] gained no major significance, obviously because they are too sensitive to amines and even to methanol. The less reactive carbonic acid derivatives are more auspicious. Thus, S-benzyloxycarbonyl [4, 77–79] and the more acid sensitive S-p-methoxybenzloxycarbonyl [80] and S-tert.butyloxycarbonyl [81, 82] cysteine derivatives could be applied in actual syntheses. The

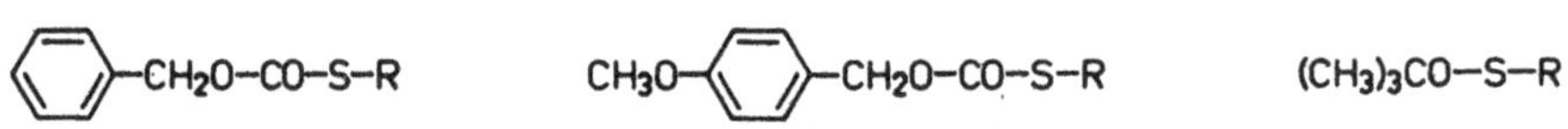

fairly stable S-ethylcarbamoyl group [83] was incorporated into peptides via an active aryl ester [84]:

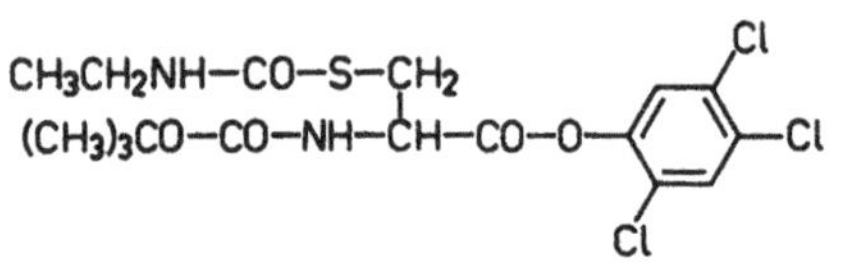

The ethylcarbamoyl group is removed by bases, such as sodium hydroxide, sodium alkoxides, liquid ammonia or hydrazine, as well as by mercury(II) or silver salts. A double protection was proposed by Jäger and Geiger [85] who used the reaction between an isocyanate and cysteine to introduce the new blocking group, a derivative of the ethylcarbamoyl group:

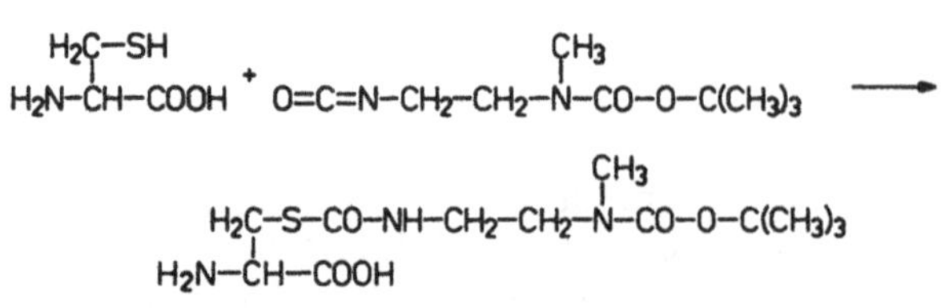

For the removal of the blocking first acidolytic cleavage is applied followed by neutralization which leads to cyclization and simultaneous unmasking of the thiol function:

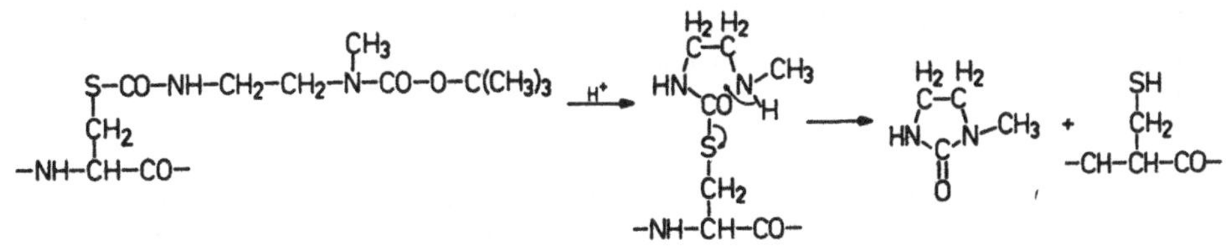

Instead of the N-tert. butyloxycarbonyl group the isocyanates can be provided with a benzyloxycarbonyl or adamantyloxycarbonyl group as well.

The most classical and probably also most popular approach to the protection of the sulfhydryl function is *alkylation*. The reactivity of thiols allows ready introduction of alkyl groups. For instance cysteine can be selectively S-benzylated in liquid ammonia [86] and also in aqueous alkali [87] with benzyl chloride or benzyl bromide respectively:

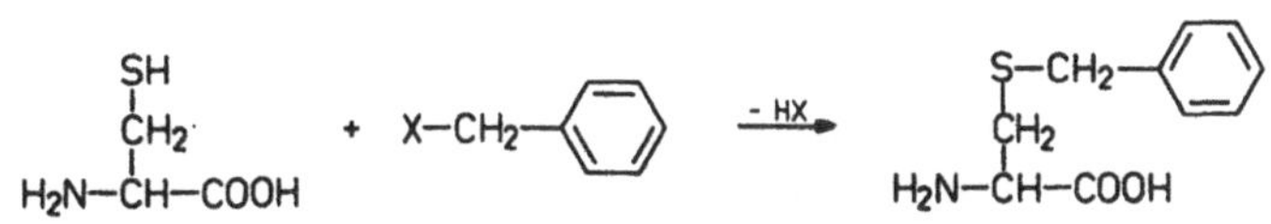

The resistance of the S-benzyl group toward acidolysis renders it suitable for semipermanent protection, since it can be used in combination with the benzyloxycarbonyl, the tert. butyloxycarbonyl and other acid sensitive groups. The S-benzyl group is cleaved only by extremely strong acids such as liquid hydrogen fluoride. The most commonly applied method of deprotection, however, is reduction with sodium in liquid ammonia [86]:

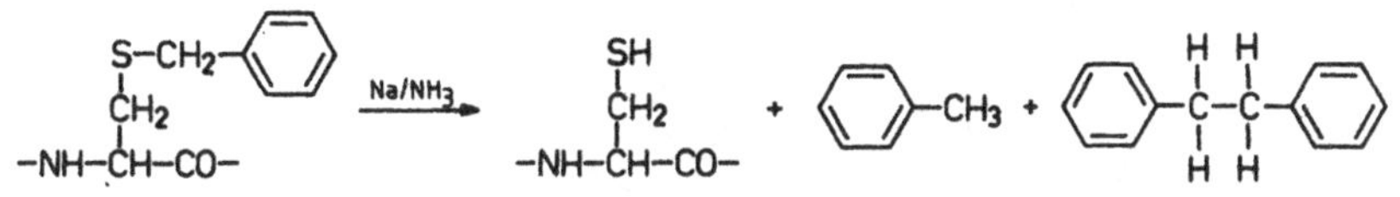

The presence of dibenzyl among the products of reduction indicates a free radical mechanism. While the reaction is quantitative and its execution elegant, there are peptides which are damaged during reduction with metals in liquid ammonia. For instance, if hydrogen donors are not excluded from the reaction mixture, the bond between proline and the preceding amino acid can suffer reductive cleavage:

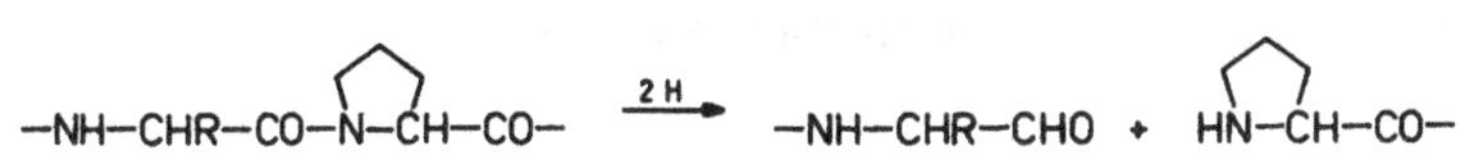

Therefore the search for S-alkyl groups which do not show the extreme acid resistance of the S-benzyl group was certainly justified. An alternative solution, an S-alkyl group that can be removed by catalytic hydrogenation was found in the S-*p*-nitrobenzyl group [88]. The contradictory observation that the *p*-nitrobenzyl group is merely reduced on hydrogenation to the *p*-aminobenzyl group but is not cleaved from the sulfur atom [89] was resolved by the finding [90] according to which the S-*p*-aminobenzyl group is removed by the action of mercury(II) salts used in the work-up process of the original study [88]:

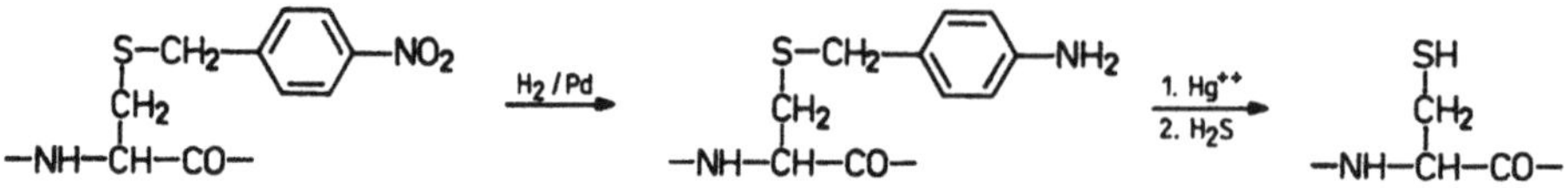

Since acidolysis of the benzyl group is based on the stability of the intermediate benzyl cation, it was reasonable to attempt enhancement of acid-sensitivity by the introduction of substituents which increase the stability of the cation. Substitution with one methoxy group did not achieve this goal: the 4-methoxybenzyl group [91] is cleaved from the sulfur atom only in boiling

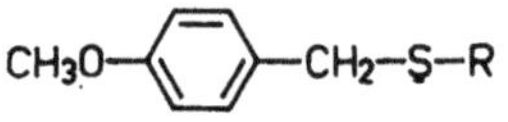

trifluoroacetic acid or in liquid HF and for its more gentle removal one has to resort to reduction with sodium in liquid ammonia. A whole series of modified benzyl groups were examined [92] as candidates for acid sensitive protection of the sulfhydryl function, but most of them, e.g. the 3,4-methylene-dioxybenzyl or the 9-(9-methylfluorenyl)-groups, were still too resistant to acids for practical application. A more promising line of research led to the S-diphenylmethyl [93, 94] and the S-4,4′-dimethoxydiphenyl [95] groups and to the widely accepted S-triphenylmethyl group [93–97].

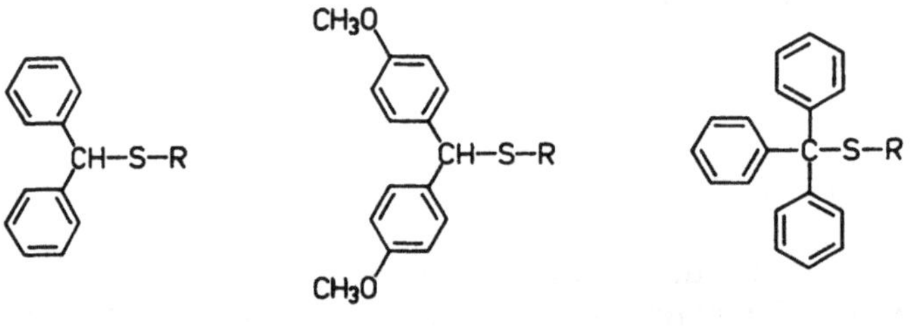

The reason behind the success of the S-trityl group lies in the diverse options for its removal. In addition to acidolysis, with HCl in chloroform [96], the sulfhydryl function can be unmasked also with salts of silver or mercury(II) and last but not least it can be changed, during the deblocking process, to the disulfide by treatment with iodine in methanol [98]:

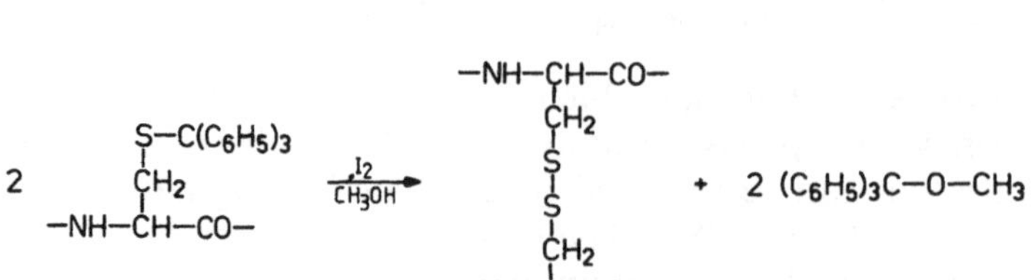

Through such oxidative deblocking mixed disulfides can also be prepared and the method is particularly suitable for cyclization via a disulfide bridge:

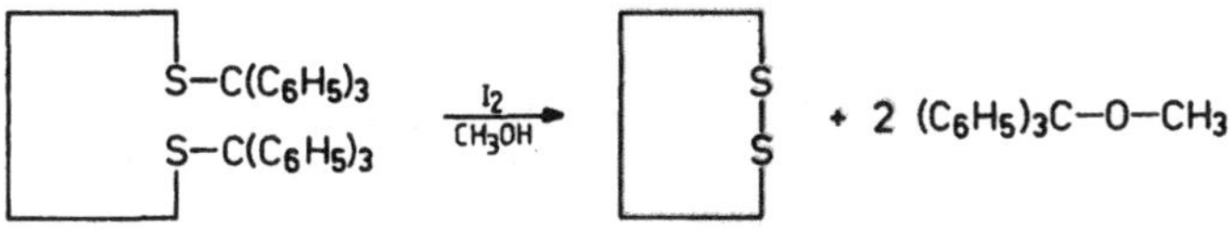

In spite of the ready oxidation of the thioether in methionine or substitution of the aromatic ring in tyrosine, even peptides which contain methionine or tyrosine residues can be treated, in methanol, with iodine. This interesting method of detritylation could therefore be applied in the synthesis of complex molecules [99] as well. A critical evaluation [100] of various methods of detritylation led to the removal of the S-trityl group with a 1:1 mixture of trifluoroacetic acid and ethanethiol (ethylmercaptane). The large excess of the mercaptane prevents the retritylation of the unmasked sulfhydryl group:

$$R{-}S{-}C(C_6H_5)_3 \xrightarrow{H^+} R{-}SH + C^+(C_6H_5)_3$$

$$(C_6H_5)_3C^+ + HS{-}CH_2CH_3 \longrightarrow (C_6H_5)_3C{-}S{-}CH_2CH_3 \quad (+H^+)$$

While this method of acidolysis might compete with oxidation by iodine it is not necessarily superior to it [101]. Several other acidic reagents were tried for the removal of the S-trityl group, e.g. hot trifluoroacetic acid, HBr in acetic acid or in trifluoroacetic acid and even HF. None of these reagents gained general acceptance for S-trityl cleavage, neither did silver or mercury(II) salts [102]. Sophisticated approaches to the conversion of S-trityl derivatives to disulfides through the application of dirhodane [103] or methoxycarbonylsulfenyl chloride ($CH_3O{-}CO{-}SCl$) [104, 105] remain to be tested in major syntheses.

The carbocation with stability comparable to that of the triphenylmethyl cation is obviously the tert. butyl ion. Yet, while S-tert. butyl derivatives of cysteine were known for a long time [55, 106] they were not applied in practical syntheses because too drastic conditions, e.g. liquid HF at room temperature [107], were needed for the removal of the blocking group. In more recent years several solutions were found for this tough problem. Among these the interesting approach of Pastuszak and Chimiak [107] circumvents the extreme stability of the S-tert. butyl group by displacing it with the *o*-nitrophenylsulfenyl group which in turn can be cleaved with mercaptoethanol (and also with sodium borohydride or with thioglycolic acid [109]):

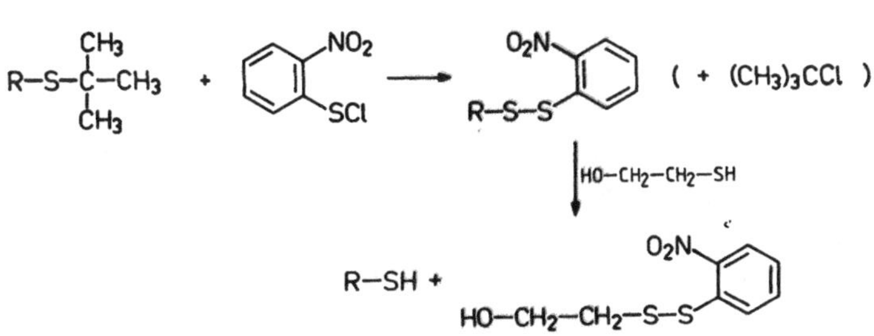

Two alternatives, deblocking with mercury(II) trifluoroacetate in trifluoroacetic acid [110] and treatment of S-tert. butyl derivatives with triphenylphosphine or tributylphosphine in trifluoroethanol [111] were already applied in the synthesis of biologically active peptides.

The versatile picolyl derivatives could also be adapted for the protection of the sulfhydryl function [112]. Blocking is reversed by electrolytic reduction on a mercury cathode. A further S-alkyl masking is afforded by the addition of thiols to methylenemalonic acid ester [113],

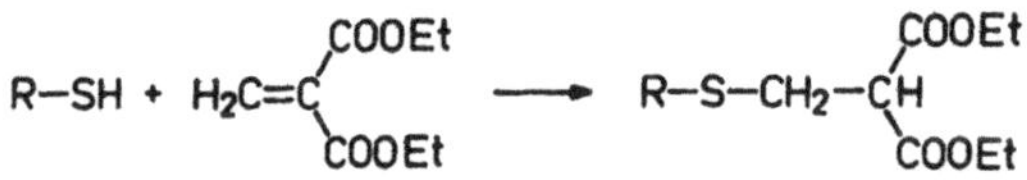

but cleavage of the protecting group requires treatment with alcoholic KOH.

A common principle can be discerned in a series of sulfhydryl blocking groups which are based on the acid sensitivity of acetals. Thioacetals are more resistant in this respect but the benzylthiomethyl group, introduced by Pimlot and Young [114], can be cleaved [115] with mercury(II) acetate in aqueous formic acid, although in order to prevent the formation of thiazolidines the reaction has to be followed by the addition of thioglycol and the introduction of H_2S.

Hemithioacetals, such as the addition products of dihydrofurane or dihydropyrane and thiols [116, 117], might introduce a minor complication: a new chiral center.

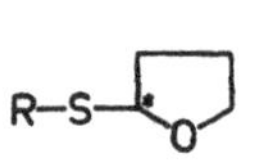

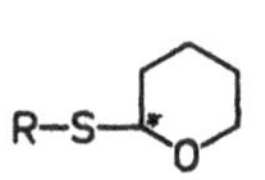

No new asymmetric carbon is present in the isobutyloxymethyl group [115]:

$$R{-}S{-}CH_2{-}O{-}CH_2{-}CH(CH_3)_2$$

The thiazolidine formed in the reaction of cysteine with acetone can also be looked upon as a carbonyl derivative or as a nitrogen analog of a thioketal. In the thiazolidine both the amino and the sulfhydro

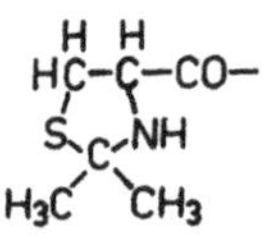

function of the amino acid are blocked [118, 119], but just because of this two-pronged protection the activated derivatives of the thiazolidine can be used only if the thus blocked residue is N-terminal. Also, the thiazolidine is not sufficiently sensitive to acids to allow unmasking of the two functional groups under mild conditions.

Related but more readily removable protecting groups are formed when the sulfur and nitrogen atoms are not members of a ring. An early proposal of Weygand and his associates [120], the 1-acylaminotrifluoroethyl groups,

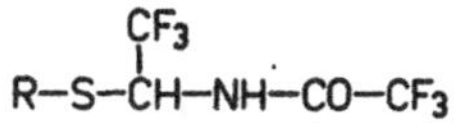

remains more or less unexplored, but the S-acetaminomethyl (Acm) group [121]

$R{-}S{-}CH_2{-}NH{-}CO{-}CH_3$

gained considerable popularity, probably because it can be easily cleaved [122] with mercury(II) ions at pH 4 in less than an hour, at room temperature and also with iodine, with concomitant oxidation to the disulfide.

An extensive article by Wünsch [123] includes a thorough treatment of the synthesis of symmetrical and unsymmetrical cystine peptides. A recent review by Hiskey [124] brings up to date an earlier article by the same author and his associates [125]. Two reviews by Photaki [126, 127] and one by Wolman [128] provide a comprehensive account of the literature of sulhydryl protection.

E. The Guanidino Group of Arginine

The three amino groups in guanidines form a single, monoacidic cation. Hence the basicity of the guanidino group in arginine is extreme and it remains protonated under the usual conditions of peptide synthesis. In

order to render guanidines nucleophilic, strong bases, such as sodium methoxide have to be applied. Deprotonated guanidines can be acylated with protected amino acids [129], but an unintended branching of the peptide chain at arginine residues is unlikely. Thus, protonation provides sufficient protection and salts (mostly hydrochlorides) of N^{α}-protected arginine derivatives were successfully incorporated time and again into peptide chains. Still, many investigators prefer, instead of protonation, to use arginine derivatives which have a suitable subtituent on the guanidine. One reason for this preference might be the unfavorable solubility properties of intermediates with guanidinium ions in arginine side chains; a second is the continued concern one has to feel about the counterions associated with the guanidinium cation. The protonated peptides behave like ion-exchange resins and can change anions during various operations. Furthermore, while it is simple to activate N^{α}-protected arginine derivatives with protonated side chains, the reactive intermediates seem to be not sufficiently stable to allow their isolation, purification and characterization and only recently was an active ester of protonated N$^{\alpha}$-benzyloxycarbonyl-L-arginine obtained [130] in crystalline form:

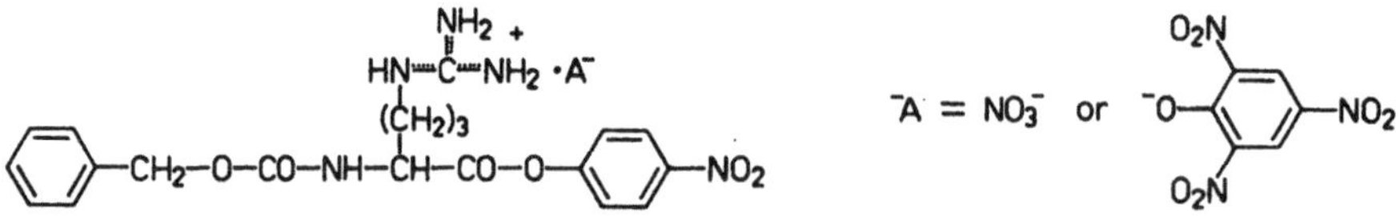

The earliest method of protection was stimulated by the availability of nitroarginine [131]. Yet, in spite of the ease by which this almost neutral amino acid can be acylated selectively on its α-amino group, e.g. to form the benzyloxycarbonyl derivatives [132],

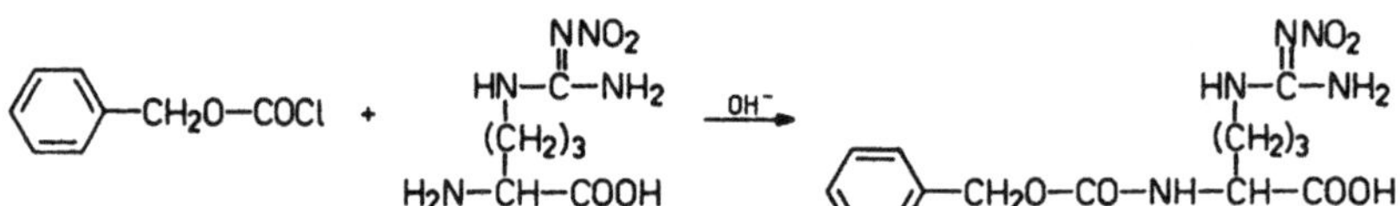

for many years activation and coupling of N-acyl-nitroarginine remained unsuccessful. Only with the advent of the mixed anhydride and carbodiimide procedures could arginyl peptides be obtained [133, 134] via nitroarginine derivatives. The reason for the initial difficulties became obvious in the attempted preparation of active esters of benzyloxycarbonyl-nitro-L-arginine [135, 136]. The nitroguanidine group has sufficient nucleophilic character to react with intramolecular electrophilic centers. Cyclization [137] of arginine derivatives results mostly in six membered rings, derivatives of 2-piperidone:

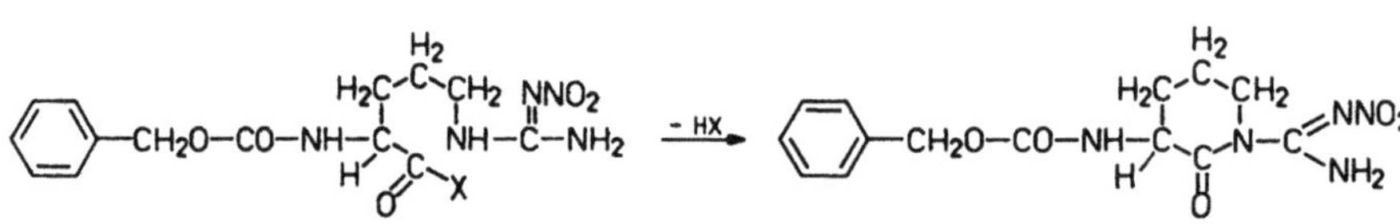

Once a nitroarginine residue is incorporated in a peptide chain, the nitro group performs the expected blocking of the guanidine function. A particular advantage, not shared by most other blocking groups used for this purpose, is the increase in solubility of the intermediates in organic solvents. There are several options for the removal of the nitro group. Reduction with metals or metal salts, e.g. zinc and hydrochloric acid [138], stannous chloride [139], or titanium(III) chloride [140], found limited application. The more generally used catalytic hydrogenation [132] is not always smooth and this prompted numerous attempts to develop improved versions such as transfer hydrogenation with cyclohexadiene as hydrogen donor [141], catalytic reduction with hydrogen in liquid ammonia [21] or hydrogenation in the presence of boron trifluoride ethereate [142]. Reduction with sodium in liquid ammonia is impractical [143] because too many byproducts are formed. There are, however, reports on side reactions also in connection with catalytic hydrogenation and with practically all methods used for the cleaving of the nitro group from guanidines. If catalytic reduction requires long periods of time, saturation of the aromatic nucleus in phenylalanine can occur [144]. The complex pathway in the electrolytic removal [145] of the blocking group is paralleled by the complexities of hydrogenation [146] and reduction with zinc in the presence of formic, acetic, trifluoroacetic or phosphoric acid [147]. Acidolysis of the nitro group requires acids as strong as hydrogen fluoride [107], boron tribromide [148], boron tri-trifluoroacetate [149], methanesulfonic acid [150], trifluoromethansulfonic acid [151] or pyridinium polyhydrogenfluoride [152]. When, however, such strong acids are applied under the conditions needed for acidolysis of the nitro groups, they usually cause some damage to sensitive sites of the peptide chain. A further shortcoming of protection by nitration is the absence of resonance stabilization which is so characteristic for the protonated guanidine group. Thus, nucleophiles like hydrazine or ammonia attack the nitroguanidine in arginine side chains and convert the arginine residues to ornithine moieties [15]:

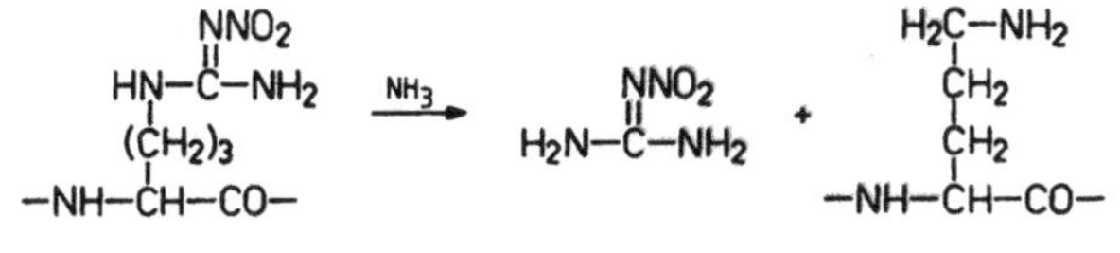

Substitution of the guanidine function with a single benzyloxycarbonyl group [153] does not prevent the formation of lactams (piperidones) during activation [154]:

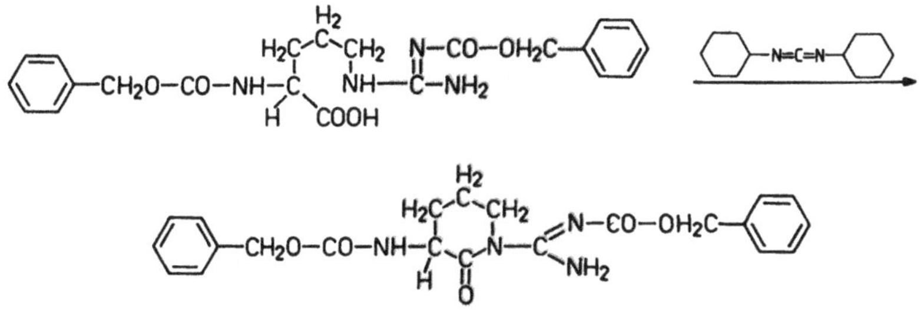

Also, when instead of the benzyloxycarbonyl group the tert. butyloxycarbonyl group was introduced for the protection of arginine side chains, a very useful intermediate (N^{α}-Z-N^{ε}-Boc-Arg) was obtained [155, 156] which allows selective removal of the masking from the α-amino group, yet, the possibility of lactam formation was not excluded. A series of potentially applicable acyl derivatives of arginine were examined [157] but even such a thorough search failed to yield a satisfactory solution for this serious problem. On the other hand *two* acyl groups hold more promise. Di-benzyloxycarbonyl guanidines [153] are not sufficiently stable [154] but both the adamantyloxycarbonyl [158] and the isobornyloxycarbonyl group [159] when used to block two nitrogens in the guanidino group provide the desired blocking of cyclization. Also, these masking groups resist catalytic hydrogenation and are removable by acidolysis with trifluoroacetic acid and thus they are probably the best approximations, so far, of the ideal protection of the guanidino function.

Substitution of the guanidino group with arylsulfonic acids is quite popular. The initially proposed benzenesulfonyl [160] and *p*-toluenesulfonyl groups [161] might represent some improvement over acyl groups because the substitution with the electron withdrawing arylsulfonyl groups renders the guanidine less sensitive toward nucleophiles like hydrazine or ammonia and, thus, does not enhance the production of ornithine derivatives. Lactam formation, however, is not impeded in tosyl-arginine. In fact bases abstract the rather acidic sulfonamide hydrogen and the remaining negatively charged nitrogen atom is readily substituted. Removal of the tosyl group from the guanidine requires reduction with sodium in liquid ammonia or acidolysis with very strong acids such as liquid hydrogen fluoride or trifluoromethanesulfonic acid. Both procedures involve certain risks. Since sulfonamides are proton donors, reductive cleavage of aminoacyl-proline bonds can occur in deprotection with sodium. In HF, formation of ornithine peptides was observed. Therefore, arylsulfonyl groups which are somewhat more sensitive to acids were designed in the expectation of improvements. The *p*-methoxybenzenesulfonyl group [162] is cleaved with the same strong acids but under less drastic conditions. Unfortunately another shortcoming of arylsulfonyl groups, their migration to the hydroxyl groups of tyrosine residues, could be observed in acidolysis [163]. The side reaction would be less serious if the arylsulfonyl group did not continue to migrate, i.e., from the phenolic hydroxyl to the adjacent (ortho) carbon atom.

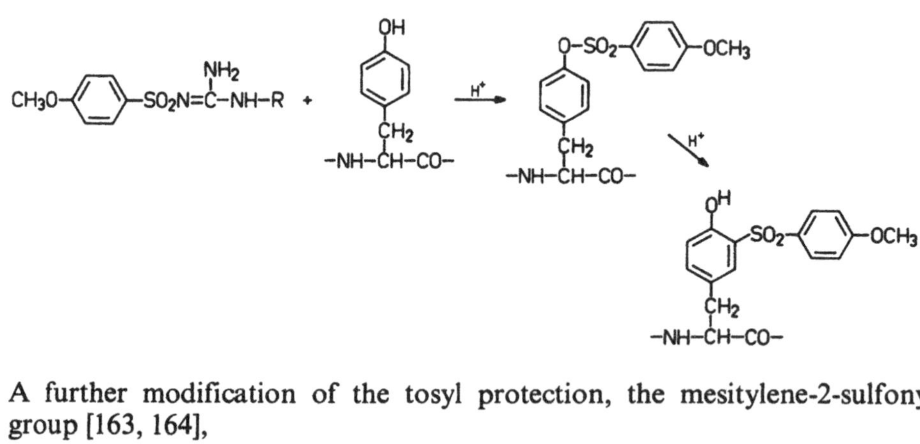

A further modification of the tosyl protection, the mesitylene-2-sulfonyl group [163, 164],

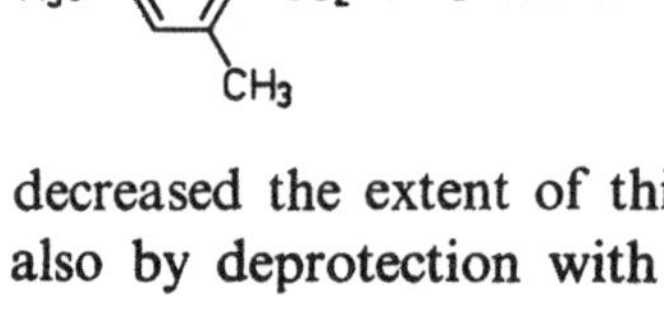

decreased the extent of this side reaction [165] which could be suppressed also by deprotection with trifluoromethanesulfonic acid in trifluoroacetic acid, in the presence of thioanisole [138, 166]. Clearly no perfect method is known at this time for the blocking of the arginine side chain. Many attempts were made toward improvements. For instance, hindered blocking groups such as the isopropyloxycarbonyltetrachlorobenzoyl group [157]

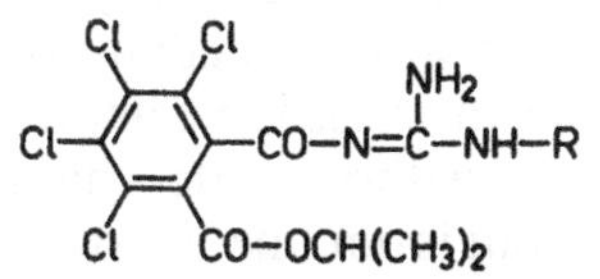

or the exhaustive alkylation of the guanidine with the bulky trityl group [167] were tried without practical consequences. The reaction of 1,3-diones with guanidines yields stable pyrimidine derivatives but 1,2-diones

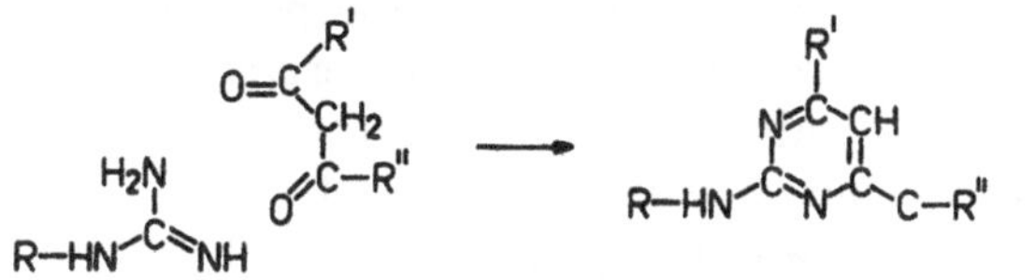

give condensation products from which the guanidine function can be regenerated. Several such 1,2-diones were applied for the reversible modification of proteins, e.g. phenylglyoxal [168] or camphor-quinone-10-sulfonic acid [169]. One of them, cyclohexane-1,2-dione [170, 171], holds certain promise for peptide synthesis because it can be removed with hydroxylamine under very mild conditions.

The difficulties inherent in the blocking of the guanidino group can be circumvented by the incorporation of ornithine rather than arginine residues into the peptide chain [172]. At a later stage of the synthesis the δ-amino groups are unmasked and guanylated, e.g. with 1-guanyl-3,5-dimethylpyrazole [173, 174]:

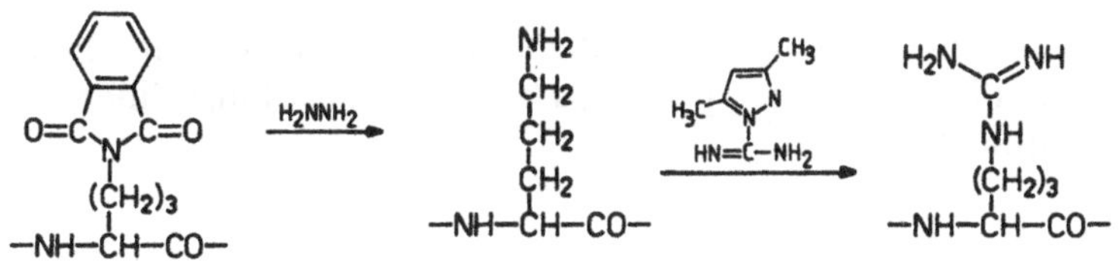

From the point of view of solubility of the intermediates it might be even more advantageous to build γ-cyanobutyrine-containing peptides and to convert these residues to ornithine and then to arginine residues at the conclusion of the synthesis:

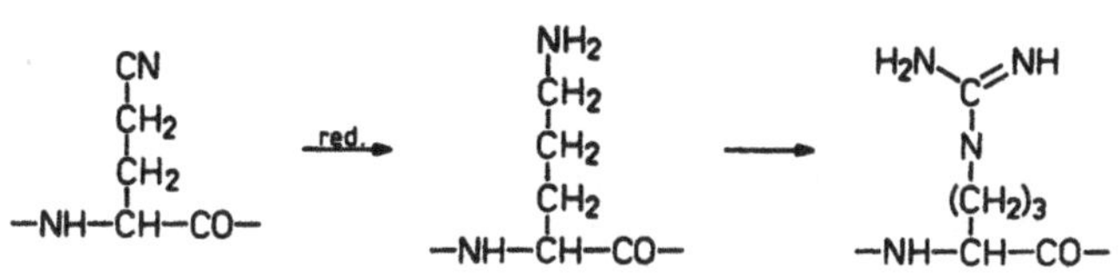

F. Imidazole in Histidine

The NH group in the imidazole ring of the histidine side chain suggests that it has to be blocked to avoid its unwandted acylation. The imidazole, however, is not readily acylated, or at least not permanently: acylimidazoles are good acylating agents. This is well demonstrated in the reactive intermediate of the coupling reactions mediated by carbonyldiimidazole, discussed in Chapter II. Thus histidine has often been built into peptides without side chain blocking. This approach has the advantage of simplicity but is not unequivocal. Imidazole can catalyze acyl transfers from N to O and lends a tentency to racemization of the activated residue, particularly if this is histidine itself. Also, intermediates which contain unprotected histidine residues can be soluble in aqueous acids, a sometimes inconvenient circumstance in the operations of purification or isolation. Last but not least the peptides with unblocked histidine are often obtained as a mixture of two materials, one with unprotonated imidazole, the other a salt, e.g. with trifluoroacetic acid. Such ambiguities stimulated considerable research toward protecting groups for the histidine side chain. This effort was complicated by the formation of two isomers on substitution of the imidazole ring of histidine:

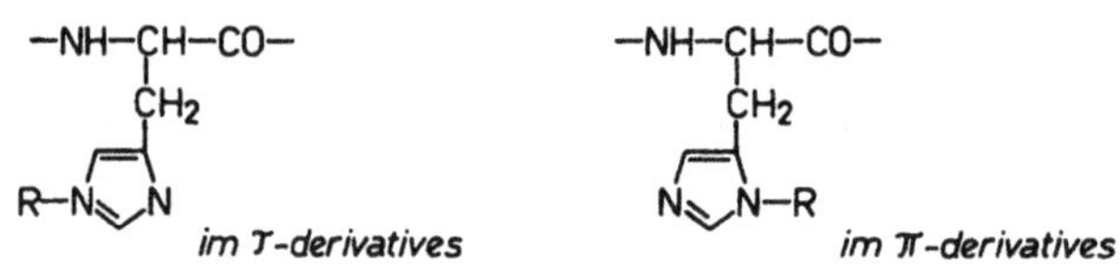

In the early literature on imidazole protecting groups the existence of two isomers was left without consideration. This might be the cause of conflicting findings on the stability of *im*-substituted histidine derivatives toward acids or hydrogenation. Only recent papers make clear distinction between τ and π substitution.

The first histidine derivative with a protecting group in its side chain, *im*-benzyl-L-histidine [175] was readily prepared by *alkylation* of histidine with benzyl chloride in liquid ammonia[9]

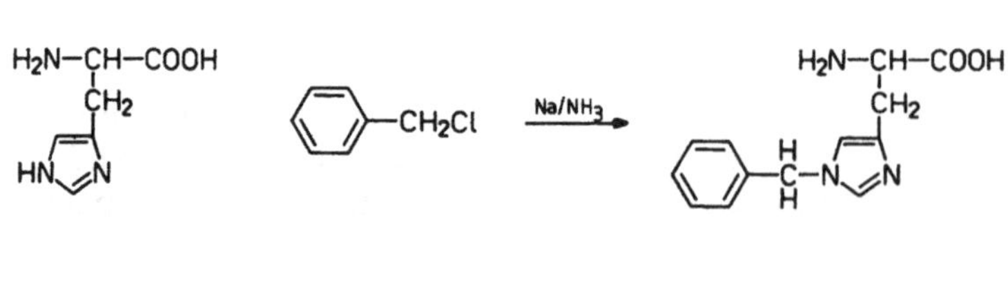

[9] Here and in the following structures the τ isomers are shown although there is mostly no evidence which of the two, or both, were obtained.

Alternatively, N^{α}-benzyloxycarbonyl histidine can be treated with two moles of benzyl bromide, in the presence of dicyclohexylamine, in dimethylformamide [176] followed by the saponification of the benzyl ester group:

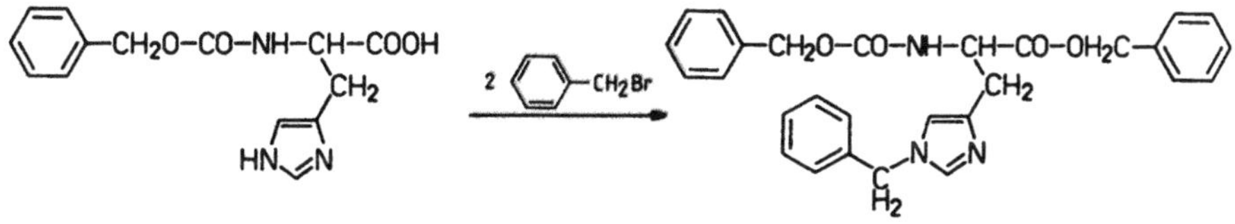

Removal of the *im*-benzyl group can be smoothly carried out by reduction with sodium in liquid ammonia. Catalytic hydrogenation [177] is sometimes too slow and might lead to the saturation of aromatic nuclei. Cleavage with acetic acid requires elevated temperatures [178]. The basic character of the alkylated product is a further imperfection of the benzyl protection. Thus probably several shortcomings inherent in the unprotected histidine side chain remain unchanged in *im*-benzyl histidine. The *im*-2-nitrobenzyl group, which can be removed from the histidine side chain by photolysis [179] and, albeit slowly, also by catalytic hydrogenation, might represent a major improvement.

Alkylation of histidine methyl ester with triphenylmethyl chloride yielded [180] N^{α}-,N^{im}-ditrityl-histidine methyl ester which was saponified and then treated with HCl in ethanol to afford the monosubstituted derivative [65, 181]:

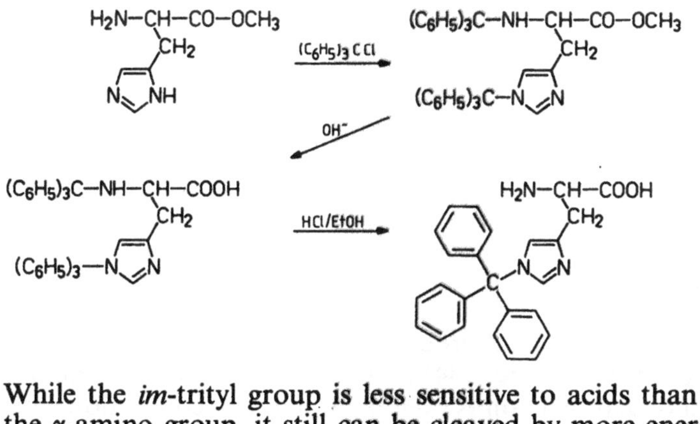

While the *im*-trityl group is less sensitive to acids than the trityl group on the α-amino group, it still can be cleaved by more energetic acidolysis, e.g. with HCl in ethyl acetate [182]. A very acid resistant modification of the *im*-trityl group was achieved through the introduction of the *im*-diphenyl-4-pyridylmethyl group [183], which requires reduction by electrolysis or catalytic

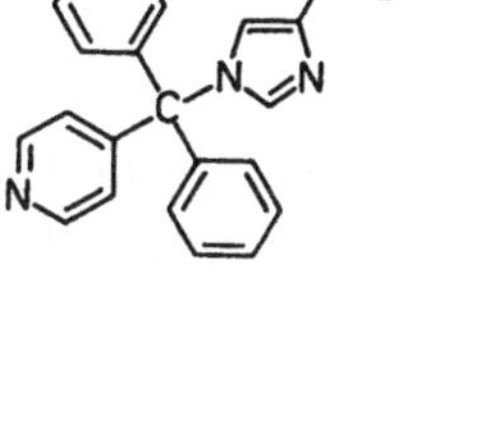

reduction or treatment with zinc in acetic acid. The considerable hindrance caused by these extremely bulky groups is somewhat reduced in the *im*-diphenylmethyl (or benzhydryl) group [184]:

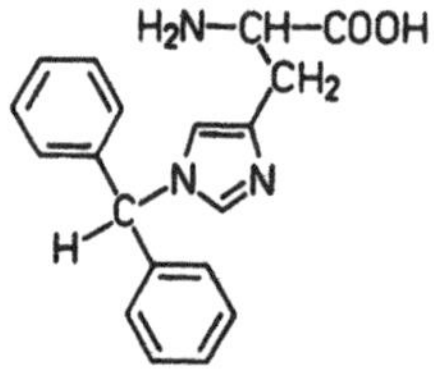

Arylation of the imidazole nitrogen with 2,4-dinitrofluorobenzene has been known for a long time from studies of amino acid sequences by the dinitrophenyl (Dnp) method, but adaptation of this substituent for the protection of the imidazole in histidine became possible only through the recognition by Shaltiel [185] that the Dnp group can be cleaved by thiolysis with thioglycolic acid, mercaptoethanol, dithiothreitol or thiophenol. The cleavage is easily monitored by spectrophotometry [186].

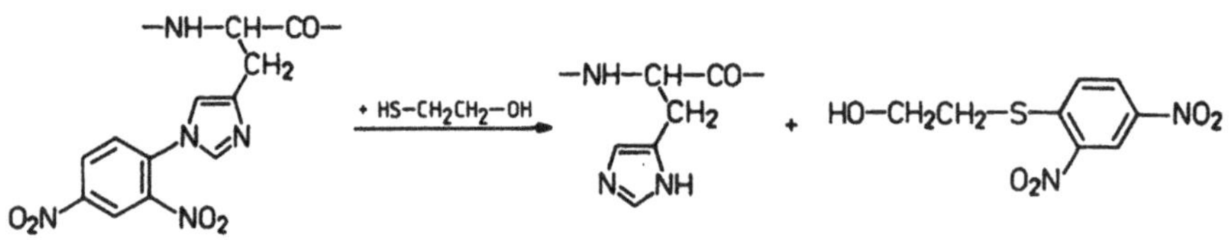

The position of the Dnp substituent could be established: it is on the τ-nitrogen of the imidazole [187].

Because of the strong electron withdrawing forces in the Dnp group the weak basic character of the imidazole is completely suppressed in *im*-Dnp histidine and therefore the side reactions which are related to basic character should not occur. Surprisingly, considerable racemization was noted in couplings where *im*-Dnp histidine was the C-terminal residue of the carboxyl component [188]. There are several other deficiencies in the Dnp protection of the histidine side chain. The Dnp group is cleaved also by nucleophiles other than thiols, e.g. by hydrazine [184] and potentially also by α-amino groups. Such premature deprotection is even more likely in connection with the 2,4,6-trinitrophenyl group [184] proposed for masking of the imidazole nitrogen.

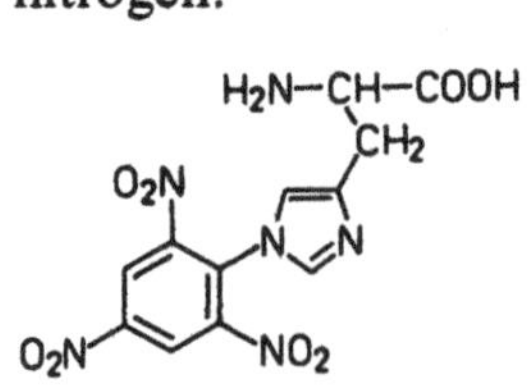

Acylation of the imidazole nitrogen is *a priori* suspect. Since acyl imidazoles, as mentioned before, are potent acylating agents it is not surprising that N$^{\alpha}$-,N^{im}-dibenzyloxycarbonyl histidine [189] is unstable [190]. The *im*-Z group is easily displaced by amines, including amino acid esters [191]. Thus, while *im*-Z derivatives have been used in actual syntheses [192] they might create more problems than they solve. Other urethanes, e.g. *im-p*-methoxybenzyloxycarbonyl [193], *im*-tert.butyloxycarbonyl [45], *im*-isobutyloxycarbonyl [194] and *im*-adamantyloxycarbonyl [195] derivatives were proposed and some of these found already application in actual syntheses. Yet, more promising with respect to stability and selectivity in removal seems to be the piperidinocarbonyl [196] group

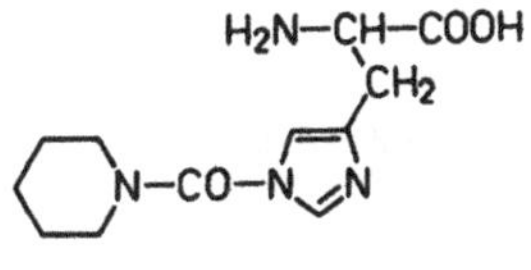

which is quite resistant to acids, not too sensitive to nucleophiles but is still removable by hydrazine, under mild conditions.

Arylsulfonyl groups created considerable interest. With the advent of liquid hydrogen fluoride as deprotecting reagent the protection of the imidazole nitrogen with the *p*-toluenesulfonyl group [197] became a practical method. Some migration of the *im*-tosyl group to α-amino groups cannot be completely excluded [198] but this side reaction was not serious enough to prevent the use of the *im*-tosyl group in the synthesis of biologically active peptides [199]. The displacement of the *im*-tosyl group by the additive 1-hydroxybenzotriazole was also noted [199]:

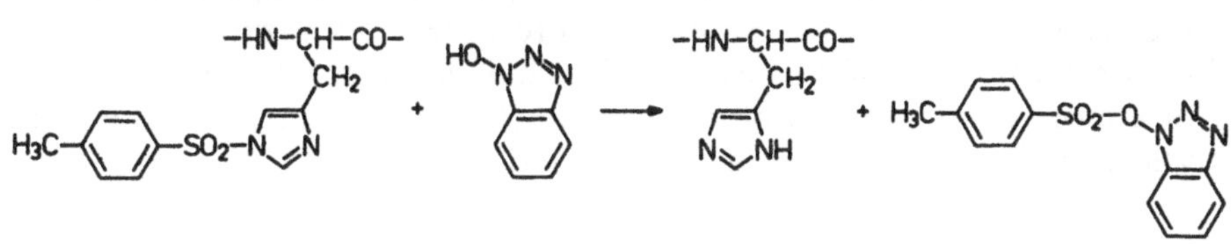

The *p*-methoxybenzenesulfonyl group [200] is cleaved with moderately strong acids, e.g. by trifluoroacetic acid in the presence of sulfur compounds of which dimethyl sulfide was found to be the most effective.

More recent efforts seek unequivocal substitution of the π-nitrogen of the imidazole ring. This kind of protection should preclude racemization via cyclic intermediates. Such a well characterized derivative, N$^{\alpha}$-benzyloxycarbonyl-N$^{\pi}$-phenacyl-L-histidine was prepared by Jones and Ramage [201]

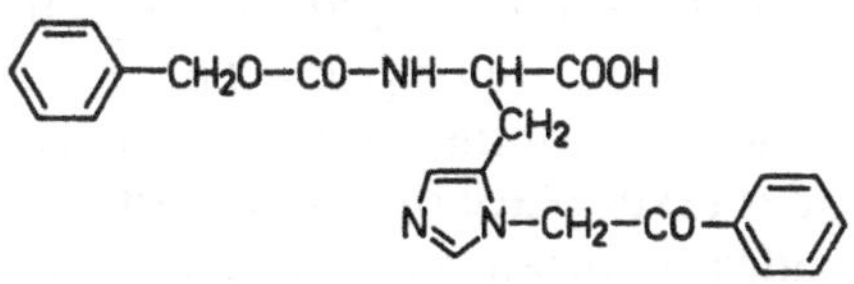

The N$^{\tau}$ isomer was also secured. The phenacyl group is resistant even to strong acids but is cleaved by zinc in aqueous acetic acid and also by photolysis.

The trifluoro-acylaminoethyl group [202] is probably the first example

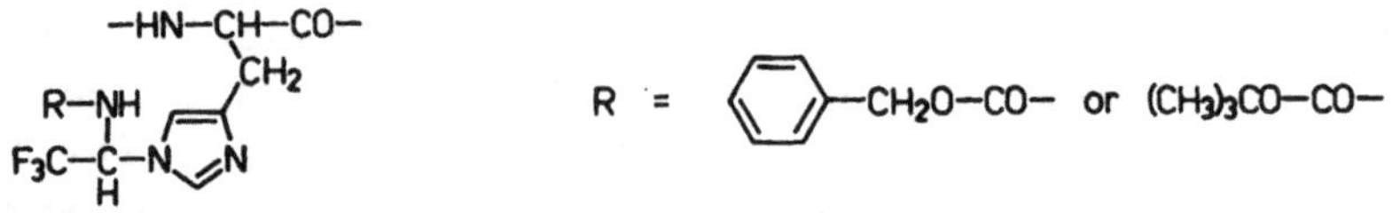

of imidazole protection in the form of an aldehyde derivative. Removal of the acyl group R leads to an unstable derivative of trifluoroacetaldehyde which is readily hydrolyzed by water:

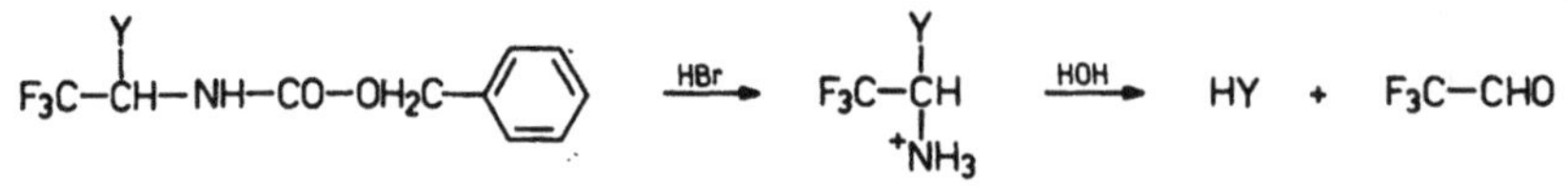

This sophisticated approach found little echo in the literature for a long time, but in recent years a derivative of formaldehyde was put to good use and the protected amino acid N^{α}-tert. butyloxycarbonyl-N^{π}-benzyloxymethyl-L-histidine [203] became commercially available. Since the side chain blocking

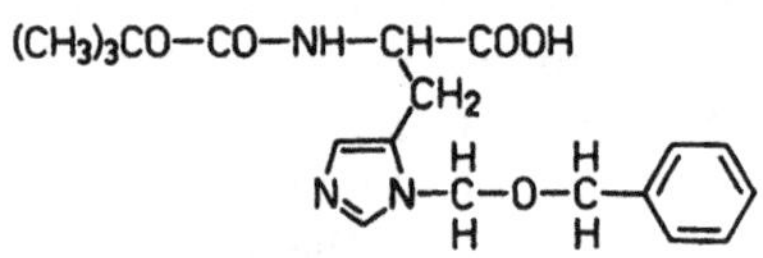

group can be cleaved, together with other benzyl masking groups, by acidolysis and the substitution at the π-nitrogen precludes racemization, the new method for the masking of the histidine side chain holds considerable promise.

For a more detailed account of the protection of the histidine side chain an acrticle by Geiger and König [204] should be consulted.

G. The Thioether in Methionine

In the synthesis of numerous methionine containing peptides the thioether function was left without protection. While indeed no acylation of the sulfur atom has to be expected, other problems are created by the presence of an unmasked thioether. Its poisoning effect on catalysts interferes with the removal of protecting groups by hydrogenolysis. This difficulty can be circumvented by carrying out the catalytic reduction in the presence of boron trifluoride etherate [142] or after the addition of bases [20] and also by transfer hydrogenation with formic acid as hydrogen donor [205]. There are, however, also side reactions which affect the methionine side chain itself. Thus, while oxidation of thioethers to sulfones takes place only under drastic conditions or requires metal catalysts, sulfoxides are produced from thioethers already on prolonged exposure to air and this readiness of thioethers to change into sulfoxides more or less excludes the oxidative removal of protecting groups from methionine containing peptides. Oxidation to sulfoxides (unlike oxidation to sulfones) is reversible. The harm caused by it is not necessarily fatal to peptides, but the presence of both unaltered and oxidized intermediates complicates their handling, analysis, etc. A further complexity is created by the formation of two diastereoisomers on oxidation which changes the thioether sulfur into a chiral center in the sulfoxides. A not less important side reaction is the alkylation of the thioether on acidolytic deprotections. Removal of benzyloxycarbonyl groups generates benzyl bromide [206] or benzyl trifluoroacetate [207] while tert. butyl trifluoroacetate [208] is formed in the cleavage of the tert. butyloxycarbonyl group or other masking groups based on tert. butyl cations with trifluoroacetic acid. All these by-products are good alkylating agents as are the intermediate carbocations themselves. Furthermore, methylation of the methionine side chain can take place when final deprotection of peptides is carried out with strong acids such as HF, methanesulfonic acid, or trifluoromethanesulfonic acid in the presence of anisole [209]. Through the initiative of Iselin [210] it became possible to prevent these side reactions by the (reversible) oxidation of the methionine thioether to the sulfoxide.

In the original process [210] oxidation with peracetic acid was used for blocking

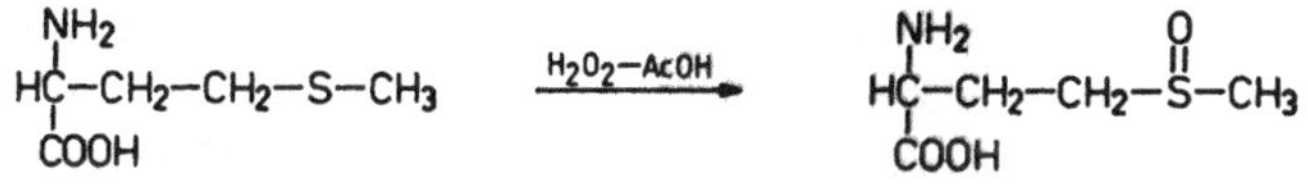

and reduction with mercaptoethanol or thioglycolic acid for the regeneration of the intact methionine side chain. Treatment with HBr in acetic acid yields the S-dibromo derivative which also can be reduced, e.g. by acetone:

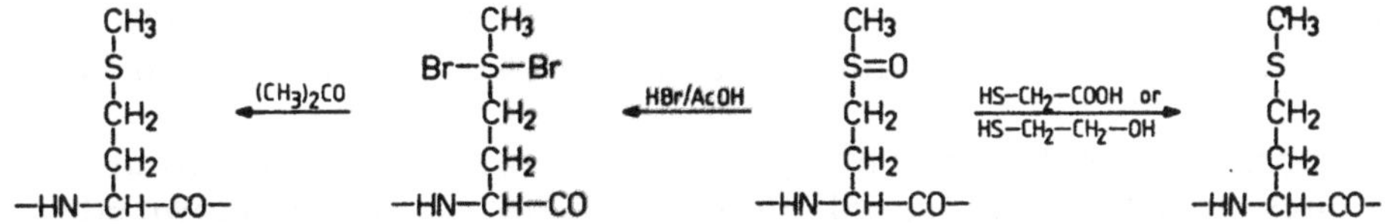

Later numerous modifications were suggested, such as oxidation of methionine with bromine water [211], with sodium metaperborate or periodate [212] and improved reducing agents, e.g. methyl mercaptoacetamide [213] or iodides in trifluoroacetic acid [214] where the iodine formed is reduced with thiols. Masking of the thioether seems to gain significance and there is only one disadvantage to be mentioned in connection with sulfoxides, namely that they tend to decrease the solubility of the intermediates in organic solvents.

The ease by which the sulfur of thioethers is alkylated [208, 209] prompted experiments for the utilization of this side reaction. Methylation of methionine derivatives seemed to be preferable, because other S-alkyl substituents could give rise to two different thioethers on dealkylation. A protected and activated derivative of the amino acid was treated with methyl *p*-toluenesulfonate [215] to yield the ternary sulfonium salt tert. butyloxycarbonyl-L-methionine *p*-nitrophenyl ester S-methyl *p*-toluenesulfonate in crystalline form. This intermediate could be applied

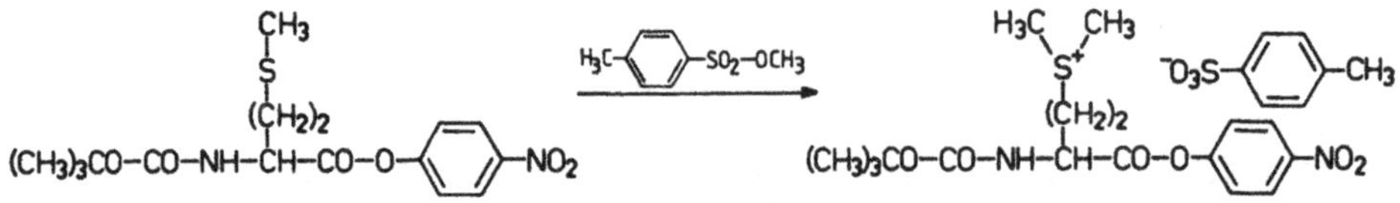

for the synthesis of peptide chains without reducing the solubility of the intermediary protected products. The sulfonium salt is reconverted, when the protection is no more needed, by thiolysis, e.g. with mercaptoethanol. Thus, methylation of methionine which is often used for the modification of proteins, is applicable also for the protection of the methionine side chain in peptide synthesis. The practical value of the method [216] remains to be established.

H. The Indole Nitrogen in Tryptophan

Indoles, in general, are sensitive to oxidation under acidic conditions. Therefore, in order to avoid such decomposition or at least to keep it at a minimum, acidolytic removal of protecting groups from tryptophan-containing peptides should be carried out at 0 °C and under argon or nitrogen. Also, acidolysis, if based on the generation of alkyl cations yielding alkylating agents such as benzyl bromide, benzyl trifluoroacetate, tert. butyl trifluoroacetate or methyl fluoride, etc., can cause alkylation of the indole nitrogen. This side reaction can extend then to several carbon atoms of the indole nucleus[10] (cf. Chapter V). Hence, protection of the tryptophan side chain seems to be fully justified although this point was mostly neglected in the past.

Substitution of the indole nitrogen with the benzyloxycarbonyl group [218] requires the presence of "naked" fluoride ions associated with crown ethers. The N^{in}-benzyloxycarbonyl group is stable toward moderately strong acids, although it is slowly cleaved by trifluoroacetic acid and faster by liquid HF. It can be removed by hydrolysis or by catalytic hydrogenation. The only indole protecting group, however, applied in practical peptide syntheses is the N^{in}-formyl group. Formylation was noted as a side reaction in the treatment of tryptophan-containing peptides with HCl in formic acid [219]. Subsequently this reaction was applied [220, 221] for the protection of the tryptophan side chain against oxidation and alkylation in peptide synthesis. The N^{in}-formyl group can be removed with dilute solutions of piperidine

[10] These side reactions can be suppressed with alkyl scavengers or by the use of mercaptoethanesulfonic acid [217] as the acidic reagent in deprotections.

in water [222] or with hydrazine [220, 222]. Under mild basic conditions or in liquid ammonia the formyl group can migrate from the indole to α-amino groups (cf. page 83 in ref. [204]). The formyl group prevents the alkylation of the indole nucleus [223] but it promotes its partial saturation by hydrogenation.

I. The Carboxamide Groups of Asparagine and Glutamine

In most schemes designed for the synthesis of peptides the carboxamide groups in the side chains of asparagine and glutamine residues are left without protection. The resistance of the carboxamide nitrogen to acylation or alkylation justifies this attitude, but certain side reactions related to asparagine and glutamine stimulated some search for masking groups of the side chain amide function. Such side reactions are the formation of cyano groups from carboxamides during activation

and also the conversion of glutamine residues to pyroglutamyl moieties in processes of deprotection or to glutarimides during coupling:

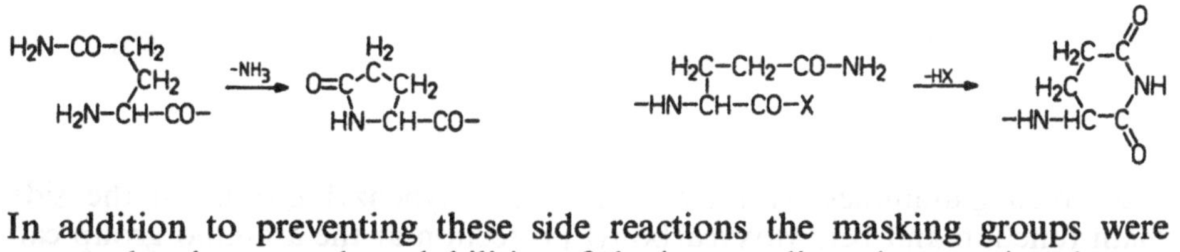

In addition to preventing these side reactions the masking groups were expected to improve the solubilities of the intermediates in organic solvents, since the carboxamide groups have an unfavorable effect in this respect.

Masking of the side chain carboxamides through condensation with xanthydrol was the first attempt [224, 225] in this direction. The bulky substituent, however, does not improve

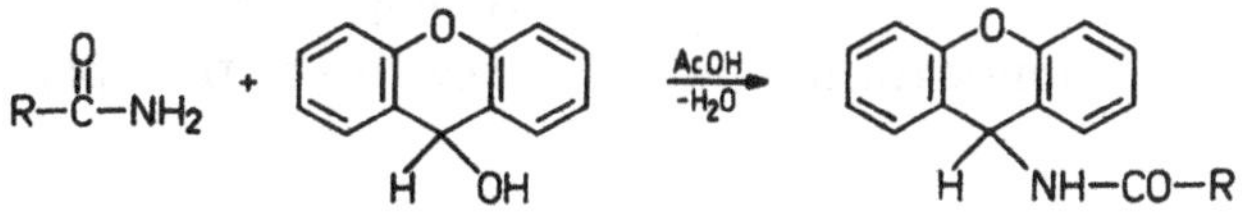

the solubility of the protected peptides, and regeneration of the carboxamide function requires treatment with strong acids with little hope for selectivity in deprotection. More sensitivity toward acids could be expected from the benzhydryl group [226] or from its substituted derivatives such as the 4,4′-dimethoxybenzhydryl group [227]:

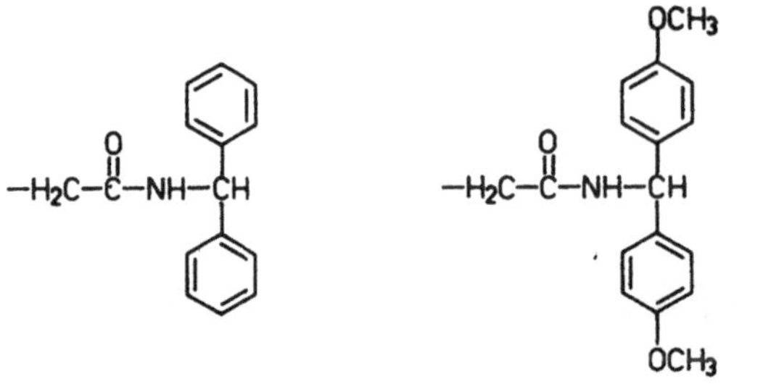

The conditions of acidolysis needed for the removal of these masking groups were not sufficiently different from those used for the removal of other blocking groups cleaved by acids. For instance on treatment of peptides with an N-terminal N^{α}-benzyloxycarbonyl-N^{γ}-(4,4′-dimethoxybenzhydryl) glutamine residue with a boiling mixture of trifluoroacetic acid and anisole, pyroglutamyl peptides could be obtained in good yield [228]. To some extent such difficulties prevail also in blocking of carboxamides with benzyl groups rendered acid sensitive by electron releasing substituents [226, 229]:

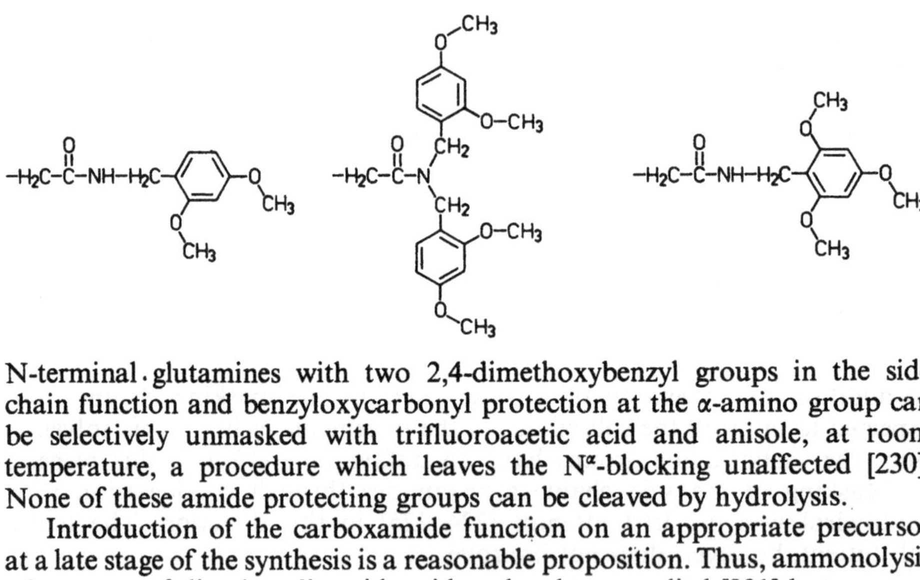

N-terminal glutamines with two 2,4-dimethoxybenzyl groups in the side chain function and benzyloxycarbonyl protection at the α-amino group can be selectively unmasked with trifluoroacetic acid and anisole, at room temperature, a procedure which leaves the N^{α}-blocking unaffected [230]. None of these amide protecting groups can be cleaved by hydrolysis.

Introduction of the carboxamide function on an appropriate precursor at a late stage of the synthesis is a reasonable proposition. Thus, ammonolysis of ω-esters of dicarboxylic acid residues has been applied [231] but was not followed in general practice. This process may be problematic because it should be accompanied by cyclic imide formation and hence also by transpeptidation (cf. Chapter V):

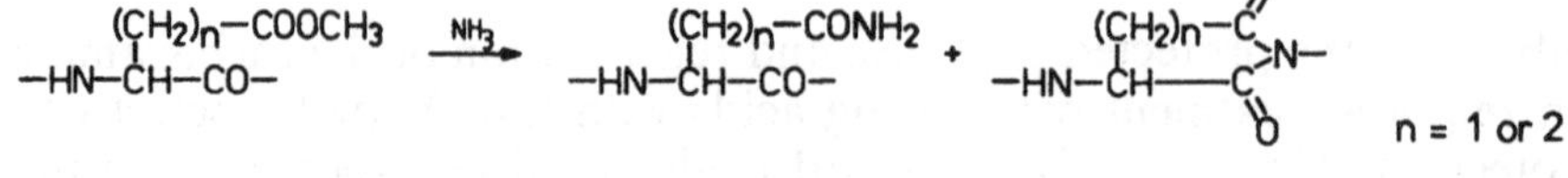

The possible addition of water to nitriles, e.g. to the side chain of β-cyano-alanine residues, has also been demonstrated [232, 233] but requires further evaluation

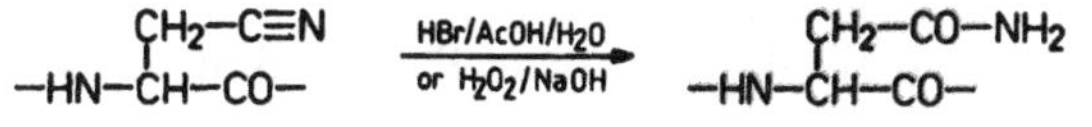

A certainly practical approach to glutaminyl peptides is the ammonolytic ring opening of pyroglutamyl residues [234, 235]:

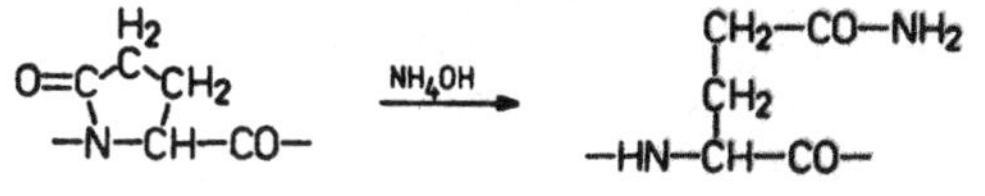

References of Chapter IV

1. Bodanszky, M., Natarajan, S.: J. Org. Chem. *40*, 2495 (1975)
2. Natarajan, S., Bodanszky, M.: J. Org. Chem. *41*, 1269 (1976)
3. Rudinger, J.: Collect. Czechoslov. Chem. Commun. *24*, 95 (1959)
4. Berger, A., Katchalski, E.: J. Amer. Chem. Soc. *73*, 4084 (1951)
5. Bryant, P. M., Moore, R. H., Pimlott, P. J., Young, G. T.: J. Chem. Soc., 3868 (1959)
6. Biernat, J. F., Rzeszotarska, B., Taschner, E.: Liebigs Ann. Chem. *646*, 125 (1961)
7. Ledger, R., Stewart, F. H. C.: Austral. J. Chem. *18*, 1477 (1965)
8. Prestidge, R. L., Harding, D. R. K., Battersby, J. E., Hancock, W. S.: J. Org. Chem. *40*, 3287 (1975)
9. Taschner, E., Chimiak, A., Biernat, J. F., Wasielewski, C., Sokolowska, T.: Liebigs Ann. Chem. *663*, 188 (1963)
10. Nefkens, G. H. L., Nivard, R. J. F.: Rec. Trav. Chim. Pay-Bas *84*, 1315 (1965)
11. Taschner, E., Wasielewski, C., Sokolowska, T., Biernat, J. F.: Justus Liebigs Ann. Chem. *646*, 127 (1961)
12. Wünsch, E., Zwick, A.: Hoppe-Seyler's Z. Physiol. Chem. *328*, 235 (1962)
13. Schwyzer, R., Dietrich, H.: Helv. Chim. Acta *44*, 2003 (1961)
14. Fauchère, J. L., Schwyzer, R.: in "The Peptides" (Gross, E., Meienhofer, J., eds.) Vol. 3, p. 203, New York: Academic Press 1981
15. Fruton, J. S.: in "Adv. Protein Chem.", Vol. 5, p. 64, New York: Academic Press 1949
16. Bodanszky, M., Meienhofer, J., du Vigneaud, V.: J. Amer. Chem. Soc. *82*, 3195 (1960)
17. Kurtz, A. C.: J. Biol. Chem. *140*, 705 (1941)
18. Bergmann, M., Zervas, L., Ross, W. F.: J. Biol. Chem. *111*, 245 (1935)
19. Schallenberg, E. F., Calvin, M.: J. Amer. Chem. Soc. *77*, 2279 (1955)
20. Medzihradszky-Schweiger, H., Medzihradszky, K.: Acta Chim. Hung. *50*, 339 (1966)
21. Meienhofer, J., Kuromizu, K.: Tetrahedron Lett., 3259 (1974); Kuromizu, K., Meienhofer, J.: J. Amer. Chem. Soc. *96*, 4978 (1974)

22. Bodanszky, M., Martinez, J., Priestley, G. P., Gardner, J. D., Mutt, V.: J. Med. Chem. *21*, 1030 (1978)
23. Schwyzer, R., Sieber, P.: Nature *199*, 172 (1963)
24. Schnabel, E., Klostermeyer, H., Berndt, H.: Liebigs Ann. Chem. *749*, 90 (1971)
25. Klausner, Y. S., Bodanszky, M.: Bioorg. Chem. *2*, 354 (1973)
26. Yajima, H., Ogawa, H., Fujii, N., Funakoshi, S.: Chem. Pharm. Bull. *25*, 740 (1977)
27. Kiso, Y., Ukawa, K., Akita, T.: J. Chem. Soc. Chem. Commun., 101 (1980)
28. Yajima, H., Watanabe, H., Okamoto, M.: Chem. Pharm. Bull. *19*, 2185 (1971)
29. Woodward, R. B., Heusler, K., Gosteli, J., Naegeli, P., Oppolzer, W., Ramage, R., Ranganathan, S., Vorbrüggen, H.: J. Amer. Chem. Soc. *88*, 852 (1966)
30. Windholz, T. B., Johnston, D. B. R.: Tetrahedron Lett., 2555 (1967)
31. Barany, G., Merrifield, R. B.: J. Amer. Chem. Soc. *99*, 7363 (1977)
32. Carpino, L. A., Han, G. Y.: J. Amer. Chem. Soc. *92*, 5748 (1970)
33. Jering, H., Schorp, G., Tschesche, H.: Hoppe-Seyler's Z. Physiol. Chem. *355*, 1129 (1974)
34. Wünsch, E.: in Houben-Weyl, Methoden der Organischen Chemie, Vol. 15, p. 493, Stuttgart: Thieme 1974
35. Rogers, S. J., Neilands, J. B.: Biochemistry *2*, 6 (1963)
36. Maurer, B., Keller-Schierlein, W.: Helv. Chim. Acta *52*, 388 (1969)
37. Bayer, E., Schmidt, K.: Tetrahedron Lett., 2051 (1973)
38. Bodanszky, M., Ondetti, M. A., Levine, S. D., Williams, N. J.: J. Amer. Chem. Soc. *89*, 6753 (1967)
39. Ondetti, M. A., Narayanan, V. L., von Saltza, M., Sheehan, J. T., Sabo, E. F., Bodanszky, M.: J. Amer. Chem. Soc. *90*, 4711 (1968)
40. Fruton, J. S., Bergmann, M.: J. Biol. Chem. *145*, 253 (1942)
41. Harington, C. R., Pitt-Rivers, R. V.: Biochem. J. *38*, 417 (1944)
42. Katsoyannis, P. G., Gish, D. T., du Vigneaud, V.: J. Amer. Chem. Soc. *79*, 4516 (1957)
43. Abderhalden, E., Bahn, A.: Hoppe-Seyler's Z. Physiol. Chem. *219*, 72 (1933)
44. Yamashiro, D., Li, C. H.: J. Org. Chem. *38*, 591 (1973)
45. Schnabel, E., Herzog, H., Hoffmann, P., Klauke, E., Ugi, I.: Liebigs Ann. Chem. *716*, 175 (1968)
46. Geiger, R., Jäger, G., Volk, A., Siedel, W.: Chem. Ber. *101*, 2189 (1968)
47. Jäger, G., Geiger, R., Siedel, W.: Chem. Ber. *101*, 2762 (1968)
48. Weygand, F., Steglich, W., Fraunberger, F., Pietta, P., Schmid, J.: Chem. Ber. *101*, 923 (1968)
49. Okawa, K., Tani, A.: J. Chem. Soc. Jpn. *75*, 185 (1950)
50. Okawa, K.: Bull. Chem. Soc. Jpn. *29*, 486 (1956)
51. Murase, Y., Okawa, K., Akabori, S.: Bull. Chem. Soc. Jpn. *33*, 123 (1960)
52. Hruby, V. J., Ehler, K. W.: J. Org. Chem. *35*, 1690 (1970)
53. Zahn, H., Diehl, J.: Z. Naturforsch. *12c*, 85 (1957)
54. Schröder, E.: Liebigs Ann. Chem. *670*, 127 (1964)
55. Callahan, F. M., Anderson, G. W., Paul, R., Zimmermann, J. E.: J. Amer. Chem. Soc. *85*, 201 (1963)
56. Wünsch, E., Fries, G., Zwick, A.: Chem. Ber. *91*, 542 (1958)
57. Erickson, B. W., Merrifield, R. B.: J. Amer. Chem. Soc. *95*, 3750 (1973)
58. Iselin, B.: Helv. Chim. Acta *45*, 1510 (1962)
59. Kiso, Y., Ukawa, K., Nakamura, S., Ito, K., Akita, T.: Chem. Pharm. Bull. *28*, 673 (1980)
60. Medzihradszky-Schweiger, H.: Ann. Univ. Sci. Budapestiensis de Rolando Eötvös Nom. Sectio Chim. *13*, 35 (1972)

61. Gosden, A., Macrae, R., Young, G. T.: J. Chem. Res. *1*, 22 (1977)
62. Tzougraki, C., Makofske, R. C., Gabriel, T. F., Michalewsky, J., Meienhofer, J., Li, C. H.: Int. J. Peptide Protein Res. *15*, 377 (1980)
63. Chen, F. M. F., Benoiton, N. L.: J. Org. Chem. *44*, 2299 (1979)
64. Hanson, R. W., Law, H. D.: J. Chem. Soc., 7297 (1965)
65. Stelakatos, G. C., Theodoropoulos, D. M., Zervas, L.: J. Amer. Chem. Soc. *81*, 2884 (1959)
66. Iselin, B., Feurer, M., Schwyzer, R.: Helv. Chim. Acta *38*, 1508 (1955)
67. Iselin, B., Schwyzer, R.: Helv. Chim. Acta *39*, 57 (1956)
68. Stewart, J. M.: in The Peptides, Vol. 3 (Gross, E., Meienhofer, J., Eds.), p. 170, New York: Academic Press 1981
69. Zahn, H., Schmidt, G.: Tetrahedron Lett., 5095 (1967); Liebigs Ann. Chem. *731*, 91, 101 (1970)
70. Swan, J. M.: Nature *180*, 643 (1967)
71. Weinert, M., Brandenburg, D., Zahn, H.: Hoppe-Seyler's Z. Physiol. Chem. *350*, 1556 (1969)
72. Inukai, N., Nakano, K., Murakami, M.: Bull. Chem. Soc. Jpn. *40*, 2913 (1967); *41*, 182 (1968)
73. Weber, U., Hartter, P.: Hoppe-Seyler's Z. Physiol. Chem. *351*, 1384 (1970)
74. Hartter, P., Weber, U.: Hoppe-Seyler's Z. Physiol. Chem. *354*, 365 (1973)
75. Zervas, L., Photaki, I.: J. Amer. Chem. Soc. *84*, 3887 (1962); *87*, 4922 (1965)
76. Zervas, L., Photaki, I., Ghelis, N.: J. Amer. Chem. Soc. *85*, 1337 (1963)
77. Berger, A., Noguchi, J., Katchalski, E.: J. Amer. Chem. Soc. *78*, 4483 (1956)
78. Zahn, H., Hammerström, K.: Chem. Ber. *102*, 1048 (1969)
79. Vandesande, F.: Bull. Soc. Chim. Belg. *78*, 395 (1969)
80. Photaki, I.: J. Chem. Soc. C, 2687 (1970)
81. Schnabel, E., Stolterfuss, J., Offe, H. A., Klauke, E.: Liebigs Ann. Chem. *743*, 57 (1971)
82. Muraki, M., Mizoguchi, T.: Chem. Pharm. Bull. *19*, 1708 (1971)
83. Guttmann, S.: Helv. Chim. Acta *49*, 83 (1966)
84. Storey, H. T., Beacham, J., Cernosek, S. F., Finn, F. M., Yanaihara, C., Hofmann, K.: J. Amer. Chem. Soc. *94*, 6170 (1972)
85. Jäger, G., Geiger, R.: in "Peptides 1972" (Hanson, H., Jakubke, H. D., Eds.), p. 90, Amsterdam: North Holland Publ. 1973
86. Sifferd, R. H., du Vigneaud, V.: J. Biol. Chem. *108*, 753 (1935)
87. Harington, C. R., Mead, T. H.: Biochem. J. *30*, 1598 (1936)
88. Berse, C., Boucher, R., Piché, L.: J. Org. Chem. *22*, 805 (1957)
89. Bodanszky, M., Ondetti, M. A.: Chem. Ind., 697 (1962)
90. Bachi, M. D., Ross-Petersen, K. J.: J. Org. Chem. *37*, 3550 (1972)
91. Akabori, S., Sakakibara, S., Shimonishi, Y., Nobuhara, Y.: Bull. Chem. Soc. Jpn. *37*, 433 (1964)
92. König, W., Geiger, R., Siedel, W.: Chem. Ber. *101*, 681 (1968)
93. Zervas, L., Photaki, I.: Chimia *14*, 375 (1960)
94. Hiskey, R. G., Adams, J. B., Jr.: J. Org. Chem. *30*, 1340 (1965)
95. Hanson, R. W., Law, H. D.: J. Chem. Soc. C, 7285 (1965)
96. Amiard, G., Heymés, R., Velluz, L.: Bull. Soc. Chim. France, 698 (1956)
97. Photaki, I., Taylor-Papadimitriou, J., Sakarellos, C., Mazarakis, P., Zervas, L.: J. Chem. Soc. C, 2683 (1970)
98. Kamber, B., Rittel, W.: Helv. Chim. Acta *51*, 2061 (1968)
99. Kamber, B., Brückner, H., Riniker, B., Sieber, P., Rittel, W.: Helv. Chim. Acta *53*, 556 (1970)
100. König, W., Kernebeck, K.: Liebigs Ann. Chem., 227 (1979)

101. Kamber, B., Rittel, W.: Liebigs Ann. Chem., 1928 (1979)
102. Caroll, F. I., Dickson, H. M., Wall, M. E.: J. Org. Chem. *30*, 33 (1965)
103. Hiskey, R. G., Tucker, W. P.: J. Amer. Chem. Soc. *84*, 4794 (1962)
104. Brois, S. J., Pilot, J. F., Barnum, H. W.: J. Amer. Chem. Soc. *92*, 7629 (1970)
105. Kamber, B.: Helv. Chim. Acta *56*, 1370 (1973)
106. Chimiak, A.: Roczniki Chem. *38*, 883 (1964)
107. Sakakibara, S., Shimonishi, Y., Kishida, Y., Okada, M., Sugihara, H.: Bull. Chem. Soc. Jpn. *40*, 2164 (1967)
108. Pastuszak, J. J., Chimiak, A.: Roczniki Chem. *51*, 1567 (1977)
109. Fontana, A., Scoffone, E., Benassi, C. A.: Biochemistry *7*, 980 (1968)
110. Nishimura, O., Kitada, C., Fujino, M.: Chem. Pharm. Bull. *26*, 1576 (1978)
111. Wünsch, E., Moroder, L., Gemeiner, M., Jaeger, E., Ribet, A., Pradayrol, L., Vaysse, N.: Z. Naturforsch. *35b*, 911 (1980)
112. Gosden, A., Stevenson, D., Young, G. T.: Chem. Commun., 1123 (1972)
113. Wieland, T., Sieber, A.: Liebigs Ann. Chem. *722*, 222 (1969)
114. Pimlott, P. J. E., Young, G. T.; Proc. Chem. Soc., 257 (1958)
115. Brownlee, P. J. E., Cox, M. E., Handford, B. O., Marsden, J., Young, G. T.: J. Chem. Soc. (C), 3832 (1964)
116. Holland, G. F., Cohen, L. A.: J. Amer. Chem. Soc. *80*, 3765 (1956)
117. Sakakibara, S., Nobuhara, Y., Shimonishi, Y., Kiyoi, R.: Bull. Chem. Soc. Jpn. *38*, 120 (1965)
118. King, F. E., Clarke-Lewis, J. W., Wade, R.: J. Chem. Soc. (C), 880 (1957)
119. Sheehan, J. C., Yang, D. D. H.: J. Amer. Chem. Soc. *80*, 1158 (1958)
120. Weygand, F., Steglich, W., Lengyel, I.: Acta Chim. Acad. Sci. Hung. *44*, 19 (1965)
121. Veber, D. F., Milkowski, J. D., Varga, S., Denkewalter, R. G., Hirschmann, R.: J. Amer. Chem. Soc. *94*, 5456 (1972)
122. Kamber, B.: Helv. Chim. Acta *54*, 927 (1971)
123. Wünsch, E.: in Houben-Weyl, Methoden der Org. Chem., Vol. 15/1, p. 735, Stuttgart: Thieme 1974
124. Hiskey, R. G.: in The Peptides, Vol. 3 (Gross, E., Meienhofer, J., Eds.), p. 137, New York: Academic Press 1981
125. Hiskey, R. G., Rao, V. R., Rhodes, W. G.: in "Protective Groups in Org. Chem." (McOmie, J. F. W., Ed.), p. 235, New York: Plenum 1973
126. Photaki, I.: in "The Chemistry of Polypeptides" (Katsoyannis, P. G., Ed.), p. 59, New York: Plenum 1973
127. Photaki, I.: Topics in Sulfur Chem. *1*, 113 (1976)
128. Wolman, Y.: in "The Chemistry of the Thiol Group" (Patai, S., Ed.), Part 2, p. 669, New York: Wiley 1974
129. Photaki, I., Yiotakis, A. E.: J. Chem. Soc. (Perkins I), 259 (1976)
130. Glass, J. D., Pelzig, M.: Int. J. Peptide Protein Res. *12*, 75 (1978)
131. Kossel, A., Kenneway, E. L.: Hoppe-Seyler's Z. Physiol. Chem. *72*, 486 (1911)
132. Bergmann, M., Zervas, L., Rinke, H.: Hoppe-Seyler's Z. Physiol. Chem. *224*, 40 (1934)
133. Hofmann, K., Rheiner, A., Peckham, W. D.: J. Amer. Chem. Soc. *75*, 6083 (1953)
134. Van Orden, H. O., Smith, E. L.: J. Biol. Chem. *208*, 751 (1954)
135. Paul, R., Anderson, G. W., Callahan, F. M.: J. Org. Chem. *26*, 3347 (1961)
136. Bodanszky, M., Sheehan, J. T.: Chem. Ind., 1268 (1960)
137. Bergmann, M., Koster, H.: Hoppe-Seyler's Z. Physiol. Chem. *159*, 179 (1926)

138. Pless, J., Guttmann, S.: in "Peptides 1966" (Beyerman, H. C., van de Linde, A., Maassen van den Brink, W., Eds.), p. 50, Amsterdam: North Holland Publ. 1967
139. Hayakawa, T., Fujiwara, Y., Noguchi, J.: Bull. Chem. Soc. Jpn. *40*, 1205 (1967)
140. Freidinger, R. M., Hirschmann, R., Veber, D. F.: J. Org. Chem. *43*, 4800 (1978)
141. Felix, A. M., Lambros, T. J., Tzougraki, C., Meienhofer, J.: J. Org. Chem. *43*, 4194 (1978)
142. Yajima, H., Kawasaki, K., Kinomura, Y., Oshima, T., Kimoto, S., Okamoto, M.: Chem. Pharm. Bull. *16*, 1342 (1968)
143. Tritsch, G. L., Woolley, D. W.: J. Amer. Chem. Soc. *82*, 2787 (1960)
144. Schafer, D. J., Young, G. T.: J. Chem. Soc. (C), 46 (1971)
145. Gros, C., de Garilhe, P., Costopanagiotis, A., Schwyzer, R.: Helv. Chim. Acta *44*, 2024 (1961)
146. Iselin, B. M.: in "Peptides 1963" (Zervas, L., Ed.), p. 27, Oxford: Pergamon Press 1965
147. Turán, A., Patthy, A., Bajusz, S.: Acta Chim. Acad. Sci., Hung. *85*, 327 (1975)
148. Felix, A. M.: J. Org. Chem. *39*, 1427 (1978)
149. Pless, J., Bauer, W.: Angew. Chem. *85*, 142 (1973)
150. Yajima, H., Kiso, Y., Ogawa, H., Fujii, N., Irie, H.: Chem. Pharm. Bull. *23*, 1164 (1975)
151. Yajima, H., Fujii, N., Ogawa, H., Kawatani, H.: J. Chem. Soc. Chem. Commun., 107 (1974)
152. Matsuura, S., Niu, C. H., Cohen, J. S.: J. Chem. Soc. Chem. Commun., 451 (1976)
153. Zervas, L., Winitz, M., Greenstein, J. P.: J. Org. Chem. *22*, 1515 (1957)
154. Zervas, L., Otani, T. T., Winitz, M., Greenstein, J. P.: J. Amer. Chem. Soc. *81*, 2878 (1959)
155. Paulay, Z., Bajusz, S.: Acta Chim. Acad. Sci. Hung. *43*, 147 (1965)
156. Bajusz, S.: Acta Chim. Acad. Sci. Hung. *44*, 31 (1965)
157. Guttmann, S., Pless, J.: Chimia *18*, 185 (1964)
158. Jäger, G., Geiger, R.: Chem. Ber. *103*, 1727 (1970)
159. Jäger, G., Geiger, R.: Liebigs Ann. Chem., 1928 (1973)
160. Milne, H. B., Peng, C. H.: J. Amer. Chem. Soc. *79*, 639 (1957)
161. Schwyzer, R., Li, C. H.: Nature *182*, 1669 (1958)
162. Nishimura, O., Fujino, M.: Chem. Pharm. Bull. *24*, 1568 (1976)
163. Yajima, H., Takeyama, M., Kanaki, J., Mitani, K.: J. Chem. Soc. Chem. Commun., 482 (1978)
164. Yajima, H., Takeyama, M., Kanaki, J., Nishimura, O., Fujino, M.: Chem. Pharm. Bull. *26*, 3752 (1978)
165. Yajima, H., Akaji, K., Mitani, K., Fujii, N., Funakoshi, S., Adachi, H., Oishi, M., Akazawa, Y.: Int. J. Peptide Protein Res. *14*, 169 (1979)
166. Kiso, Y., Satomi, M., Ukawa, K., Akita, T.: J. Chem. Soc. Chem. Commun., 1063 (1980)
167. Gazis, E., Bezas, B., Stelakatos, G. C., Zervas, L.: in "Peptides 1962" (Young, G. T., Ed.), p. 17, Oxford: Pergamon Press 1963
168. Takahashi, K.: J. Biol. Chem. *243*, 6171 (1968)
169. Pande, C. S., Pelzig, M., Glass, J. D.: Proc. Natl. Acad. Sci. USA *77*, 895 (1980)
170. Toi, K., Bynum, E., Norris, E., Itano, H.: J. Biol. Chem. *242*, 1036 (1967)
171. Patthy, L., Smith, E. L.: J. Biol. Chem. *250*, 557 (1975)

172. Christensen, H. N.: J. Biol. Chem. *160*, 75 (1945)
173. Habeeb, A. F. S. A.: Canad. J. Biochem. Physiol. *38*, 493 (1960)
174. Bodanszky, M., Ondetti, M. A., Birkhimer, C. A., Thomas, P. L.: J. Amer. Chem. Soc. *86*, 4452 (1964)
175. du Vigneaud, V., Behrens, O. K.: J. Biol. Chem. *117*, 27 (1937)
176. Tilak, M. A., Hollinden, C. S.: Tetrahedron Lett., 391 (1968)
177. Theodoropoulos, D. M.: J. Org. Chem. *21*, 1550 (1956)
178. Elliot, D. F., Morris, D.: Chimia *14*, 373 (1960)
179. Kalbag, S. M., Roeske, R. W.: J. Amer. Chem. Soc. *97*, 440 (1975)
180. Amiard, G., Heymés, R., Velluz, L.: Bull. Soc. Chim. France, 191 (1955)
181. Zervas, L., Theodoropoulos, D. M.: J. Amer. Chem. Soc. *78*, 1359 (1956)
182. Bosshard, H. R.: Helv. Chim. Acta *54*, 951 (1971)
183. Coyle, S., Young, G. T.: J. Chem. Soc. Chem. Commun., 980 (1976)
184. Losse, G., Krychowski, J.: Tetrahedron Lett., 4121 (1971)
185. Shaltiel, S.: Biochem. Biophys. Res. Commun. *29*, 178 (1967)
186. Shaltiel, S., Fridkin, M.: Biochemistry *9*, 5122 (1970)
187. Bell, J. R., Jones, J. H.: J. Chem. Soc. Perkin I, 2336 (1974)
188. Beyerman, H. C., Hirt, J., Kranenburg, P., Syrier, J. L. M., van Zon, A.: Rec. Trav. Chim. Pays-Bas *93*, 256 (1974)
189. Patchornik, A., Berger, A., Katchalski, E.: J. Amer. Chem. Soc. *79*, 6416 (1957)
190. Akabori, S., Okawa, K., Sakiyama, F.: Nature *181*, 772 (1958)
191. Sakiyama, F., Okawa, K., Yamakawa, T.: Bull. Chem. Soc. Jpn. *31*, 926 (1958)
192. Sakiyama, F.: Bull. Chem. Soc. Jpn. *35*, 1943 (1962)
193. Schaich, E., Fretzdorff, A. M., Schneider, F.: Hoppe-Seyler's Z. Physiol. Chem. *354*, 897 (1973)
194. Grønvald, F. C., Johansen, N. L., Lundt, B. F.: in "Peptides, Structure: Biol. Function" (Gross, E., Meienhofer, J., Eds.), p. 309, Pierce, Rockford, Ill. (1979)
195. Haas, W. L., Krumkalns, E. V., Gerzon, K.: J. Amer. Chem. Soc. *88*, 1988 (1966)
196. Jäger, G., Geiger, R., Siedel, W.: Liebigs Ann. Chem. *101*, 3537 (1968)
197. Sakakibara, S., Fujii, T.: Bull. Chem. Soc. Jpn. *42*, 1466 (1969)
198. Fujii, T., Sakakibara, S.: Bull. Chem. Soc. Jpn. *47*, 3146 (1974)
199. Fujii, T., Kimura, T., Sakakibara, S.: Bull. Chem. Soc., Jpn. *49*, 1595 (1976)
200. Kitagawa, K., Kitade, K., Kiso, Y., Akita, T., Funakoshi, S., Fujii, N., Yajima, H.: J. Chem. Soc. Chem. Commun., 955 (1979)
201. Jones, J. H., Ramage, W. I.: J. Chem. Soc. Chem. Commun., 472 (1978)
202. Weygand, F., Steglich, W., Pietta, P.: Chem. Ber. *100*, 3841 (1967)
203. Brown, T., Jones, J. H.: J. Chem. Soc. Chem. Commun., 648 (1981)
204. Geiger, R., König, W.: in "The Peptides", Vol. 3 (Gross, E., Meienhofer, J., Eds.), p. 70, New York: Academic Press 1981
205. Sivanandaiah, K. M., Gurusidappa, S.: J. Chem. Research (S), 108 (1979)
206. Guttmann, S., Boissonnas, R. A.: Helv. Chim. Acta *41*, 1852 (1958); ibid. *42*, 1257 (1959)
207. Weygand, F., Steglich, W.: Z. Naturforsch. *14b*, 472 (1959)
208. Lundt, B. F., Johansen, N. L., Vølund, A., Markussen, J.: Int. J. Pept. Protein Res. *12*, 258 (1978)
209. Irie, H., Fujii, N., Ogawa, H., Yajima, H., Fujino, M., Sinagawa, S.: J. Chem. Soc. Chem. Commun., 922 (1976); Chem. Pharm Bull. *25*, 2929 (1977)
210. Iselin, B.: Helv. Chim. Acta *44*, 61 (1961)
211. Bodanszky, M., Chandramouli, N.: Unpublished

212. Fujii, N., Sasaki, T., Funakoshi, S., Irie, H., Yajima, H.: Chem. Pharm. Bull. *26*, 650 (1978)
213. Houghten, R. A., Li, C. H.: Anal. Biochem. *98*, 36 (1979)
214. Landini, D., Modena, G., Montanari, F., Scorrano, G.: J. Amer. Chem. Soc. *92*, 7168 (1970)
215. Kunz, H.: Chem. Ber. *109*, 3693 (1976)
216. Bodanszky, M., Bednarek, M. A.: Int. J. Peptide Protein Res. *20*, 408 (1982)
217. Loffet, A., Dremier, C.: Experientia *27*, 1003 (1971)
218. Klausner, Y. S., Chorev, M.: J. Chem. Soc. Perkin I, 627 (1977)
219. Previero, A., Coletti-Previero, M. A., Cavadore, J. C.: Biochim. Biophys. Acta *147*, 453 (1967)
220. Ohno, M., Tsukamoto, S., Sato, S., Izumiya, N.: Bull. Chem. Soc. Jpn. *46*, 3280 (1973)
221. Yamashiro, D., Li, C. H.: J. Org. Chem. *38*, 2594 (1973)
222. Ohno, M., Tsukamoto, S., Makisumi, S., Izumiya, N.: J. Chem. Soc. Chem. Commun., 663 (1972); Bull. Chem. Soc. Jpn. *45*, 2852 (1972)
223. Löw, M., Kisfaludy, L.: Hoppe-Seyler's Z. Physiol. Chem. *360*, 13 (1979)
224. Akabori, S., Sakakibara, S., Shimonishi, Y.: Bull. Chem. Soc. Jpn. *34*, 739 (1961)
225. Shimonishi, Y., Sakakibara, S., Akabori, S.: Bull. Chem. Soc. Jpn. *35*, 1966 (1962)
226. Weygand, F., Steglich, W., Bjarnason, J., Akhtar, R., Chytil, N.: Chem. Ber. *101*, 3623 (1968)
227. König, W., Geiger, R.: Chem. Ber. *103*, 2041 (1970)
228. König, W., Geiger, R.: Chem. Ber. *105*, 2872 (1972)
229. Weygand, F., Steglich, W., Bjarnason, J.: Chem. Ber. *101*, 3642 (1968)
230. Pietta, P., Chillemi, F., Corbellini, A.: Chem. Ber. *101*, 3649 (1968)
231. Velluz, L., Amiard, G., Bartos, J., Goffinet, B., Heymes, R.: Bull. Soc. Chim. France, 1464 (1956)
232. Zaoral, M., Rudinger, J.: Collect. Czechoslov. Chem. Commun. *24*, 1933 (1959)
233. Liberek, B.: Chem. Ind., 987 (1961)
234. Rudinger, J.: Collect. Czech. Chem. Commun. *19*, 365 (1954)
235. Swan, J. M., du Vigneaud, V.: J. Amer. Chem. Soc. *76*, 3110 (1954)

V. Side Reactions in Peptide Synthesis

In two reviews the author, with J. Martinez, attempted a systematic discussion of side reactions encountered in peptide synthesis. The first of these articles [1] was organized according to the amino acids which cause or suffer changes, because the individuality of amino acids leads to side reactions which are characteristic for a particular residue. Of the twenty amino acid constituents of proteins only alanine and leucine are immune from characteristic side reactions, although not from racemization. The second, more comprehensive article [2] contains also a treatment of side reactions related to methods for protection and coupling. In this volume, in which an attempt is being made to discern the principles that govern peptide synthesis, side reactions proceeding through similar mechanisms are grouped together. Racemization will be treated among other undesired reactions.

The common cause of numerous side reactions is the abstraction of protons by tertiary bases present in the reaction mixture during activation and coupling. By discussing the various side reactions, including several pathways of racemization, which can be traced to such a common origin, we try to put emphasis on a principle, the omission of tertiary amines, whenever possible [3], and on the recommendation to use free amines rather than the combination of amine salts and tertiary bases as amino components. Side reactions caused by protonation require a separate section in this chapter, which might focus the attention on a second principle in this area, the need for weak rather than strong acids in acidolytic processes. A brief discussion of the unwanted reactions caused by overactivation will conclude this chapter.

A. Side Reactions Initiated by Proton Abstraction

Amino acids and their N-acyl derivatives are, in general, inert to bases. Abstraction of the acidic proton from the carboxyl group results in a carboxylate anion and this prevents, at least under the usual conditions of peptide synthesis, the formation of a second anionic center especially at the α-carbon atom:

$$\text{—HN—CHR—COOH} \xrightarrow[\text{(-BH)}]{\text{B}} \text{—HN—CHR—COO}^- \xrightarrow{\text{B}} \text{—HN—}\underset{\text{-}}{\text{C}}\text{R—COO}^-$$

In esters and in various activated derivatives of acylamino acids no such obstacle exists against proton abstraction. In fact the electron withdrawing forces present in the activating group "X" enhance the activity of the α-hydrogen and facilitate its abstraction:

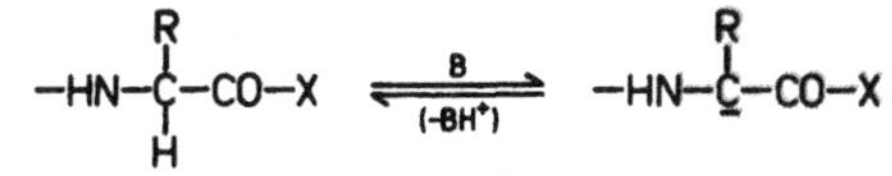

An obvious consequence of carbanion formation is the partial or total loss of chiral purity. Proton abstraction might be reversible and the equilibrium of the reaction might lie far to the left: gradually more and more molecules will pass through the stage of carbanion and suffer irreparable racemization. Therefore, the danger of racemization is inherent in peptide synthesis and in order to avoid it, it must be carefully considered. There are, however, side reactions in which proton abstraction occurs not at the α-carbon atom but at the amide nitrogen of an acylamino acid. The additional

unshared pair of electrons on the nitrogen atom renders the latter, in spite of the presence of the carbonyl substituent, a good nucleophile. Thus, it can participate in numerous side reactions, particularly in intramolecular attacks resulting in cyclizations. For instance, the formation of succinimide derivatives is usually preceded by proton abstraction from the amide nitrogen of an aspartyl amino acid residue:

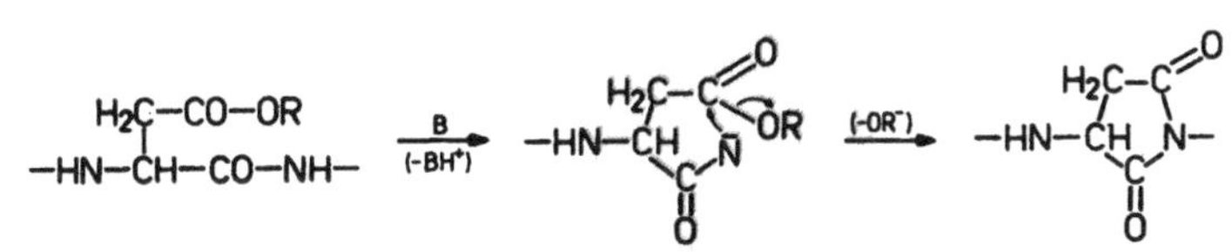

Analogous cyclization reactions and O-acylations initiated by proton abstraction will be discussed in separate sections.

1. Racemization

a. Mechanisms of Racemization

Understanding the mechanisms of racemization seems to be necessary for its prevention. Accordingly, a considerable amount of experimental work has been carried out in this area which is skillfully rendered in a review article by Kemp [4]. At this place we confine the discussion to the principal processes of base catalyzed racemization of activated acylamino acids. Three distinct pathways can be recognized:

a) direct abstraction of the α-proton,
b) racemization via reversible β-elimination and
c) racemization through azlactones [5(4H)-oxazolones].

The simple proton abstraction mechanism might be a contributor in several processes but it is the dominant pathway only in very special cases such as the rapid racemization of derivatives of phenylglycine, an amino acid which is not a constituent of proteins although it occurs in microbial peptides:

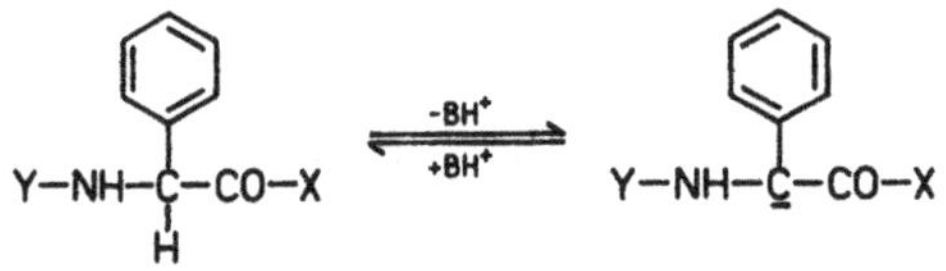

(where Y is a protecting group and X an activating group). The best known examples for racemization through reversible β-elimination are derivatives of S-benzyl-cysteine [5, 6] and β-cyanoalanine [7]:

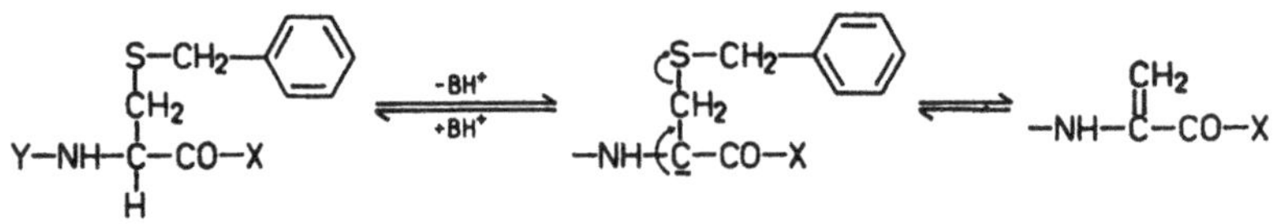

although an alternative pathway, a "conducted tour" mechanism can also be visualized and has certain experimental support [8].

The best studied and most important mechanism of racemization involves the formation of azlactones [9]:

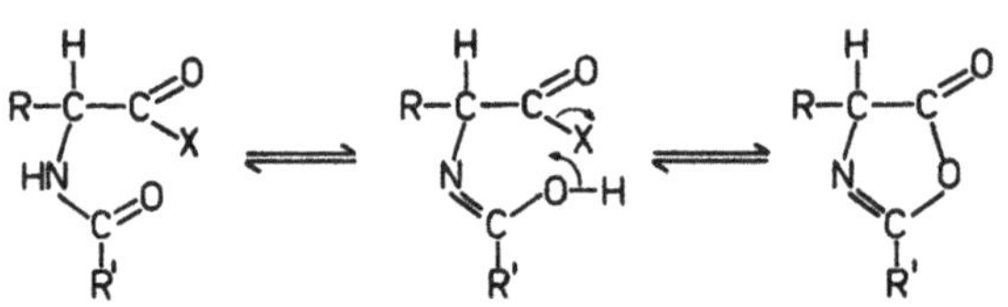

The explanation for the tendency for racemization of azlactones lies in the ease by which the acidic proton can be abstracted by bases from the chiral center due to resonance stabilization of the carbanion generated in the process:

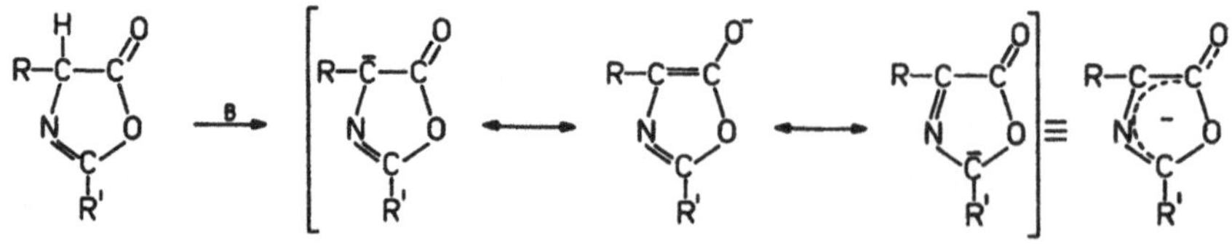

Azlactones are good acylating agents and could be useful for the activation of the carboxyl component. Yet, delocalization of the negative charge in the deprotonated intermediate provides them with sufficient lifetime to endanger the chiral purity of the product. The formation of an azlactone could be demonstrated [10] by its characteristic carbonyl frequency (1832 cm^{-1}) when benzoyl-L-leucine *p*-nitrophenyl ester was exposed to the action of tertiary amines

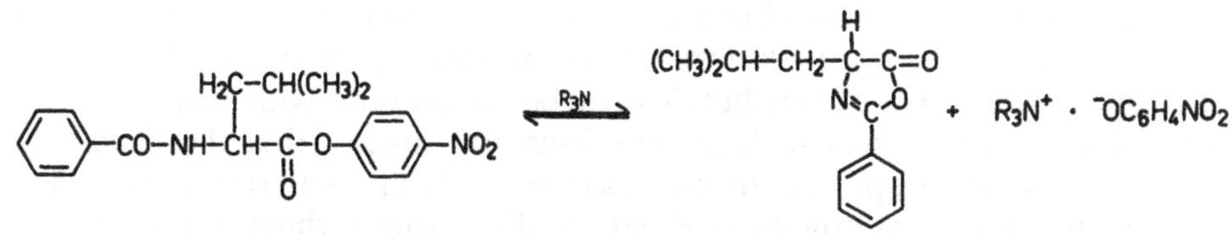

and equally convincing evidence incriminating the azlactone intermediate was found in the production of partially racemized benzoyl-leucyl-glycine ethyl ester when the reaction was completed with acylation of glycine ethyl ester. Characteristically, the unreacted portion of benzoyl-L-leucine *p*-nitrophenyl ester was recovered enantiomerically pure. Racemization through azlactone intermediates is influenced by several factors such as the nature of the amino acid involved, the solvent used in the reaction or the presence (or absence) of tertiary amines. The acyl group on the amine nitrogen, however, plays a decisive role in the conservation or loss of chiral purity. For instance, under identical conditions benzoylamino acids are more extensively racemized than acetylamino acids [11]. Such differences seem to be related to the electronic forces operating in the acyl group. Beyond the formation of azlactones the N-acyl substituents of the oxazolinone can also affect the acidity of the hydrogen atom on the chiral center. Expressed in another way: the stability of the anion produced in proton abstraction by bases is enhanced by electron withdrawing effects in the acyl group:

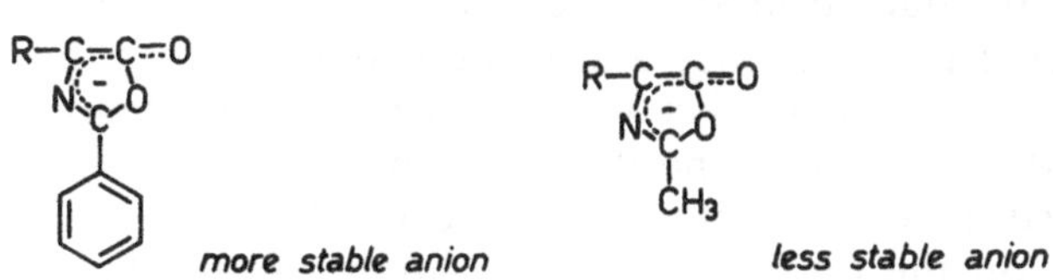

Probably not so much the formation of azlactones is of primary importance in determining the rate of racemization as are the electronic effects of the substituents of the oxazolinone, including those in the N-acyl group. Azlactones can be obtained in optically active form [12], and if immediately trapped by good nucleophiles [13], they can yield optically active products.

The influence of the N-acyl group on the stability of the anion generated through proton abstraction from the oxazolinone can range from extreme stabilization found in the formyl and trifluoroacetyl groups to pronounced destabilization shown by the benzyloxycarbonyl, tert. butyloxycarbonyl and other alkoxycarbonyl groups. In fact, trifluoroacetylamino acids yield an isomer [14] of the more common azlactones, an isomer in which the α-carbon atom is not a chiral center:

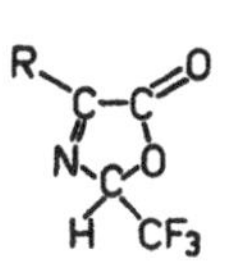

Until recently it was generally assumed that benzyloxycarbonylamino acids and, in general, amino acids protected by a urethane-type blocking group do not produce azlactones and hence they are resistant to racemization during activation and coupling. Isolation [15] of optically pure oxazol-(4H)-ones, e.g. from the reaction of tert.butyloxycarbonyl-L-valine with water soluble carbodiimides contradicts such assumptions and suggest that the beneficial effect of urethane type protecting groups rests on the electron release provided by them and on the ensuing destabilization of the anion which could form by proton abstraction:

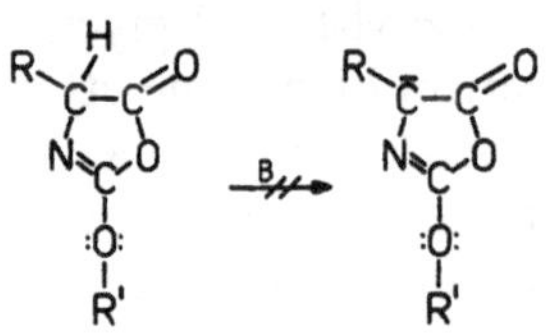

It seems now that the importance of azlactone formation in determining the rate of racemization was overemphasized in the past. The chiral stability of acylproline derivatives was, for instance, explained with the absence of an amide hydrogen in the N-acyl derivatives of this secondary amine. Without such an amide hydrogen no azlactone should form. In recent years, however, convincing evidence was found for the ready racemization of N-methylamino acids [16] during activation and coupling. Also, the absence of azlactone formation was used as the generally accepted rationale for the conservation of chiral purity in the couplings of peptides by the azide method. Unfortunately, reports on racemization that occured in the application of the azide procedure appear with increasing frequency and make it obvious that this, still valuable, procedure is less than perfect in conserving chiral integrity.

It is highly probable that in addition to the already mentioned mechanisms of base catalyzed racemization other pathways also exist. In the light of new evidence, particularly on the formation of oxazolinones from alkoxycarbonyl amino acids, the problem of racemization seems to require thorough reeximination.

The role of bases in at least some of the racemization processes is beyond doubt. For instance, time and again the advantage of free amines over a mixture of amine salts with tertiary bases was noted. Less attention was paid so far to the possibility of *intramolecular* base catalysis, although in several coupling methods the reactive intermediate contains a basic center and the latter could abstract the hydrogen from the chiral carbon atom. Since O-alkyl isoureas have pronounced basic character, it may not be farfetched to assume proton abstraction by a basic nitrogen atom in the O-acyl-isourea intermediates of carbodiimide mediated coupling reactions.

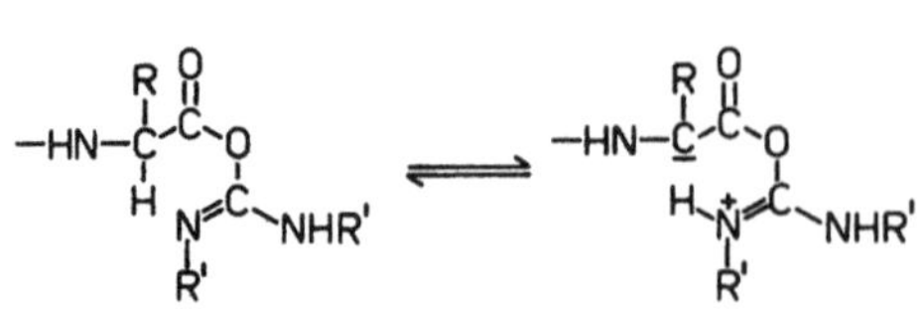

Alternatively hydrogen bond stabilized enols might play a role in such processes

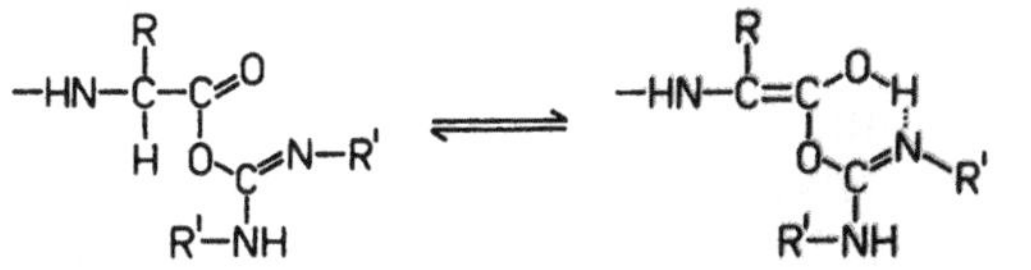

which would be then analogous to the effect of excess acetic anhydride on optically active amino acids. Here racemization probably proceeds through enolization of mixed anhydrides:

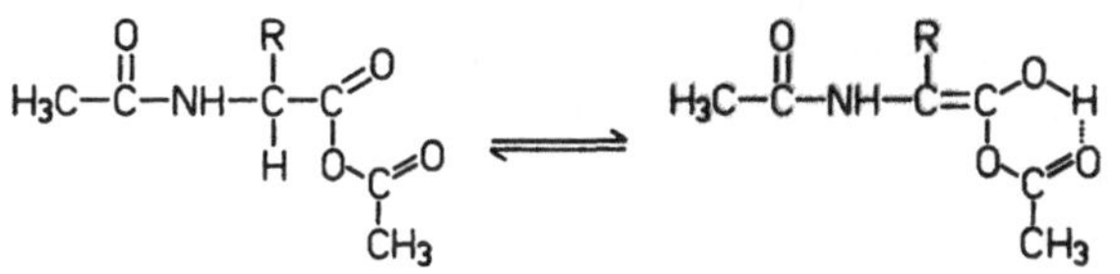

The possibility of enol stabilization by external bases, such as triethylamine or imidazole should also be considered:

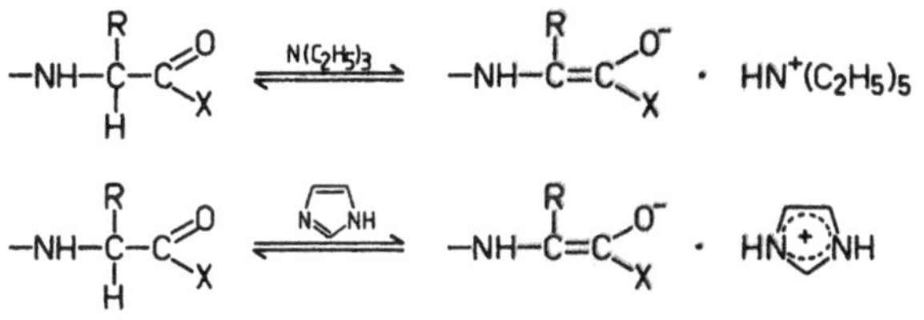

b. Models for the Study of Racemization

Numerous model systems have been proposed for the study of racemization. These systems are used to evaluate the effect of solvents, presence or absence of bases, temperature and other variables and last, but not least, the ability of different coupling methods to produce peptides without loss of chiral purity. The earliest suggestions came from Young's laboratory [17, 18] and involve the coupling of acetyl or benzoyl-L-leucine to glycine ethyl ester, followed by the examination of the optical rotation of the crude product. The results can be further refined by fractional crystallization and analysis of the fractions by weight, optical rotation and melting point. Because of the simplicity in its execution the Anderson-Callahan model [19] is more frequently used. The latter requires the coupling of benzyloxycarbonylglycyl-L-phenylalanine to glycine ethyl ester. If racemization occurs in the process the product contains benzyloxycarbonylglycyl-DL-phenylalanyl-glycine ethyl ester, which is rather insoluble in aqueous ethanol and can thus be separated and weighed. A word of caution is indicated here. This simple and useful method is reliable only if no by-products, other than the racemate, are formed in significant amount in the coupling reaction. Otherwise crystallization of the racemate might be impeded by the impurities and from the lack of crystallization the wrong conclusion, that there was no racemization, can be drawn. In principle, models should be so designed that the products of the test-experiment are not racemates but diastereoisomers and the conclusions are not based on negative evidence.

A more reliable, albeit also more time consuming, experiment is based on the coupling of benzyloxycarbonylglycyl-L-alanine to L-phenylalanylglycine ethyl ester (the "Kenner model") [20]. The diasteroisomers formed in the reaction are separated by countercurrent distribution. Somewhat less laborious are the methods introduced by Weygand and his associates [14, 21, 22], who condensed trifluoroacetyl-L-valine with L-valine methyl ester, or benzyloxycarbonyl-L-leucyl-L-phenylalanine with L-valine tert. butyl ester or trifluoroacetyl-L-prolyl-L-valine with L-proline methyl ester. The reaction products are examined with the help of vapor phase chromatography for the presence of diastereoisomers formed by racemization.

The test systems discussed so far are based on differences with respect to solubility or partition coefficient between diastereoisomers (or in the Anderson-Callahan test, between the racemate and the enantiomerically pure peptide derivative). An experimentally simple realization of the same principle is the examination of the products of model reactions by paper chromatography or thin layer chromatography [23]. Improvements in the realiability of the tests are also possible, e.g. the Young test can be perfected by the chromatographic separation of the products [24]. A more substantial simplification is, however, the use of the ubiquitous amino acid analyzer for the separation and quantative determination of the diasterosimers generated in the racemization tests. For instance coupling of acetyl-L-isoleucine [25] to glycine ethyl ester yields, in addition to the desired acetyl-L-isoleucylglycine ester, also acetyl-D-*allo*isoleucylglycine ethyl ester, if racemization occured in the reaction. Since alloisoleucine and isoleucine are routinely separated by the Spackman-Stein-Moore method [26] it is sufficient to

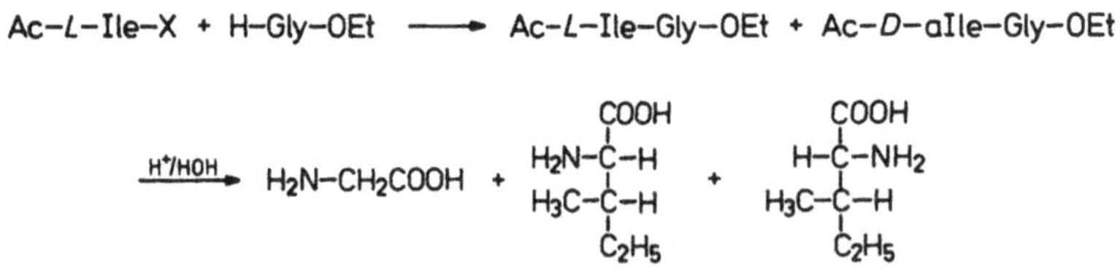

hydrolyze a small sample of the reaction mixture and to apply the hydrolysate to the analyzer. The main advantage of this model experiment is that no isolation of products is needed. This means a certain saving on time and effort, but more importantly the examination of the *crude* material assures that no distortion takes place in the isolation or separation of the products, thus no isomer is left in mother liquors, etc. The acetyl group has no major effect on the racemization of the amino acid to which it is attached, thus in this respect it can represent a peptide chain. This model can be applied for the study of the effect of coupling methods, solvents, tertiary amines added and also of the influence of the amino component, since glycine ethyl ester can be replaced by other nucleophiles. Yet, a certain limitation is caused, by the choice of isoleucine as the activated residue. It is a hindered amino acid and can suffer more loss in chiral purity than other less hindered residues which participate more readily in the desired reaction and allow therefore less time for unimolecular processes such as racemization.

The same principle, separation of diasteroisomers on the amino acid analyzer, appears also in the "Izumiya test" [27, 28] in which a benzyloxycarbonylglycyl amino acid is coupled to an optically active amino acid benzyl ester and the products examined after deprotection by hydrogenation. This model system allows variations with respect to the amino acid residue which is exposed to racemizing conditions. Thus, instead of Z-Gly-L-Ala one can couple Z-Gly-L-Phe, etc. to L-Leu-OBzl and the nucleophile can also be so selected that detection of the diastereoisomers cause no difficulty. The contributions of Benoiton and his associates [29, 30], who used N^{ε}-benzyloxycarbonyl-L-lysine benzyl ester for amino component, lie in this direction. The degree of racemization can be estimated, without deprotection and separation, through the examination of the nmr spectra of the coupling products. The model compounds acetyl-L-alanyl-L-phenylalanine methyl ester and acetyl-L-phenylalanyl-L-alanine methyl ester [31] allow the determination of the D-amino acid containing isomers by integration of the areas of the methyl protons of alanine while coupling of benzoylamino acids to N^{ε}-benzyloxycarbonyl-L-lysine methyl ester [32] permits a similar assessment of racemization through the examination of the methyl protons of the methyl ester group. In an interesting proposal [33] coupling of tert-butyloxycarbonyl-L-alanyl-L-methionyl-L-leucine to the tert. butyl ester of L-leucine is followed by acidolysis and then by a treatment with cyanogen bromide in aqueous acetic acid and by determination of the ratio of the two diastereoisomers, L-Leu-L-Leu and D-Leu-L-Leu with the help of the amino acid analyzer:

$$\text{Boc–Ala–Met–Leu} + \text{Leu–OBu}^t \longrightarrow \text{Boc–Ala–Met–Leu–Leu–OBu}^t \xrightarrow{CF_3COOH}$$

$$\text{H–Ala–Met–Leu–Leu–OH} \xrightarrow{BrCN} \text{Ala–Hse} + \textit{L}\text{–Leu–}\textit{L}\text{–Leu} + \textit{D}\text{–Leu–}\textit{L}\text{–Leu}$$

Hse = Homoserine

In a sophisticated and also very sensitive model experiment [34] benzyloxycarbonyl-L-alanyl-D-alanine is activated by the methods to be tested and coupled to L-alanyl-L-alanine p-nitrobenzyl ester. The crude product is deblocked by hydrogenation and the mixture of the two isomeric tetrapeptides L-Ala-D-Ala-L-Ala-L-Ala and L-Ala-L-Ala-L-Ala-L-Ala, is exposed to the action of leucineaminopeptidase. The enzyme will catalyze the complete hydrolysis of the all-L peptide, the product of racemization, but leaves the peptide in which the second position is occupied by a residue with D-configuration intact. With respect to sensitivty this method is surpassed by the isotope dilution techniques introduced into peptide chemistry by Kemp and his coworkers [35–37]. Radioactively labeled benzyloxycarbonylglycyl-L-leucine or benzoyl-L-leucine is coupled to glycine ethyl ester followed by dilution with "cold" racemate and fractional crystallization until products with constant count per mg are obtained. This yields reliable information on racemization and allows the detection of very slight racemization which would be left unnoticed in the original versions of the Anderson-Callahan or the Young tests (cf. above).

Some problems, e.g. the base catalyzed racemization of active esters of protected amino acids or peptides can be investigated simply by following the change of optical rotation with time [38]. The effect of solvents, protecting groups, temperature, activating groups, etc. can be studied in this simple manner. With well selected model compounds [39] it was possible to differentiate between various mechanisms of racemization and to determine the scope and limitations of hindered amines in preventing racemization.

c. Detection of Racemization. (Examination of Synthetic Peptides for the Presence of Diastereoisomers)

Racemization during the activation and coupling of protected amino acids cannot be excluded. It is even more likely to occur in the activation and coupling of protected peptides. Therefore, it is desirable and sometimes absolutely necessary to examine the synthetic products for the presence of unwanted diastereoisomers. Such contaminants, if they are only minor constituents in the crude synthetic material, might be lost in the isolation process or during purification but can also accompany the principal product through these steps. A simple and practical approach to the detection of diastereoisomers was devised by Manning and Moore [40]. A sample of the peptide is completely hydrolyzed with constant boiling hydrochloric acid and the mixture of liberated amino acids is acylated with an enantiomerically pure protected and activated amino acid, e.g. with L-leucine N-carboxy-anhydride. The resulting mixture of dipeptides is applied to the column of an automatic amino acid analyzer [26] which can separate dipeptides from their diastereoisomers. Accordingly, if racemization occurred at one or more residues then in addition to the peaks corresponding to the expected dipeptides (L-leucyl-L-amino acids) smaller satellite peaks will also appear on the recordings demonstrating the presence of L-leucyl-D-amino acids in the mixture. The areas under the peaks allow the quantitative determination of the amount of D-amino acids in the synthetic material. There is, of course, an inherent limitation in the examination of chiral integrity of a peptide through its hydrolysis with acids, since the process of hydrolysis itself is not unequivocal in this respect. In acid hydrolysates most amino acids appear more or less intact, but some, e.g. phenylalanine, suffer minor racemization during hydrolysis, while cystine is heavily contaminated with its D-isomer and also with mesocystine. Alkaline hydrolysis is even worse, it causes extensive racemization in several residues. Such details must be taken into consideration in the evaluation of the Manning-Moore analysis. The problem can be solved by using proteolytic enzymes for degradation.

The selectivity of proteolytic enzymes permits also their direct application for the study of optical homogeneity [41]. For instance complete digestability of a sample with leucine amino peptidase [42, 43] provides strong evidence for the absence of D-amino acid containing peptides. A comparison of the ratios of amino acids in hydrolysates obtained on digestion of a synthetic product with proteolytic enzymes with the ratios determined in a routine acid hydrolysate is probably one of the simplest and most reliable approaches for the study of chiral integrity.

The rates of hydrolysis in degradation with proteolytic enzymes are usually low at bonds following proline and glycine residues. Some aminopeptidases, e.g. aminopeptidase M, are less restrictive in this respect. Proline, a stumbling block in proteolysis, can be set free with the help of specific prolidases [44, 45]. In addition to aminopeptidases, carboxypeptidases A, B and Y, and dipeptidylaminopeptidases can also be adopted for the same purpose. Selective cleavage, e.g. with trypsin at the carboxyl side of arginine and lysine residues, provides useful information if these were the activated amino acids of carboxyl components. In general, the stereospecificity of enzyme catalyzed hydrolysis can serve in numerous ways the study of optical purity. Perhaps less reliable is an alternative approach in which one follows the disappearance of D-amino acids from hydrolysates on treatment with D-amino acid oxydases (e.g., from kidneys) or the elimination of L-amino acids by oxidation with enzymes from snake venoms. The evidence obtained in these oxidative processes should be trusted only if the catalytic effect of the enzyme preparation and the conditions used are shown to be operative in control experiments with mixtures containing both L and D amino acids.

Chromatographic procedures based on columns containing chiral supports [46, 47] can differentiate between D and L amino acids. This principle, perfected by the use of high pressure liquid chromatography, might become the standard control process for the detection of racemization that occurred in the synthesis of a peptide. Reversed phase high pressure chromatography is well suited [48] also for the implementation of the Manning-Moore procedure [40] because well selected columns can completely separate the diastereoisomers formed on acylation of the amino acids in a hydrolysate with an optically pure acylating agent.

d. Conservation of Chiral Purity

Chiral purity of activated residues is affected by several factors, such as the methods of activation and protection or the nature of the activated amino acid residue. It is influenced also by the solvent used in the reaction, the presence or absence of tertiary amines, and by the basic strength and bulk of the tertiary amine if one had to be added to the coupling mixture and, last but not least, by auxiliary nucleophiles (cf. Chapter 2). First and foremost of these factors seemed to be *the method of activation* and thus it received the most attention. The search for "racemization free" coupling methods is still actively pursued although gradually the impression can be gained that perhaps no such method may exist. Any increase in activation entails a similar increase in the acidity of the α-proton of the activated residue and hence its abstraction by bases is essentially unavoidable:

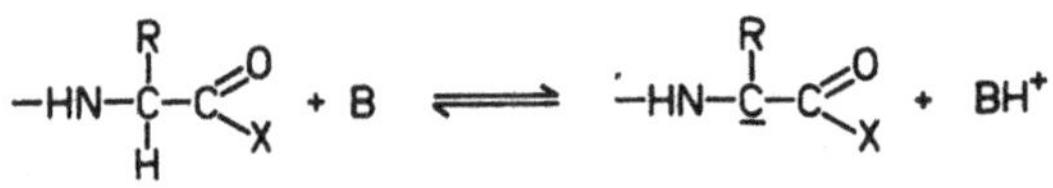

Therefore, it could be more profitable to focus the attention on each and every factor influencing racemization, rather than to try to develop perfect coupling methods which will yield chirally pure products under any conditions.

Through decades the strong belief prevailed that the azide method is free from racemization. Only in relatively recent years did we become aware of measurable racemization in azide coupling [21, 49, 50]. Those who observed no racemization in the preparation of peptides via azides (e.g., ref. [27]) knowingly or intuitively avoided the use of tertiary bases, or at least did not apply tertiary amines in excess [51]. By no means do we suggest that all methods are equal in this respect. The azide method still stands out as less conducive to racemization than many other procedures, but probably even the best methods can cause racemization under adverse conditions.

In the choice of coupling methods it is difficult to make positive recommendations, although some procedures, e.g. coupling via azides or with the help of EEDQ [52] have a fairly good record. It might be easier to point out coupling reagents which are notorious for their ability to cause racemization. Some of these, for instance the Woodward reagent [53], dicyclohexylcarbodiimide and other carbodiimides [54] caution the investigator by the structure of the reactive intermediates which contain a basic center, the potential cause of intramolecular proton abstraction from the chiral carbon atom:

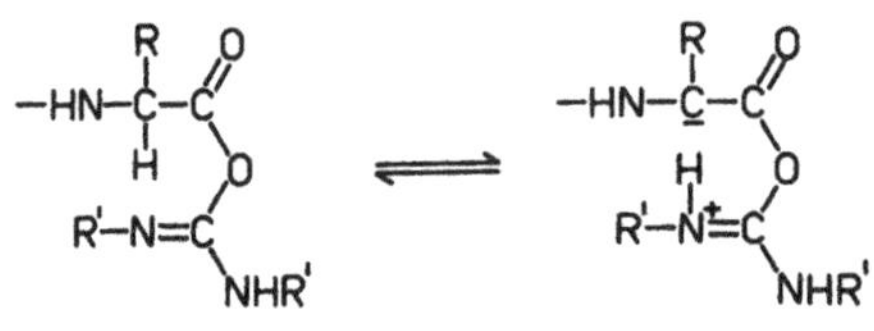

Similarly, among the various "push-pull acetylenes" [55–57] one with two basic centers [56]

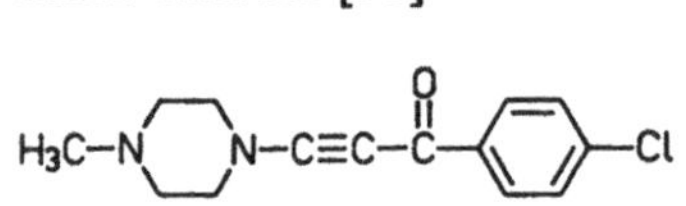

is more conducive to racemization than others with only a single proton abstracting site generated also in simpler ynamines [58]:

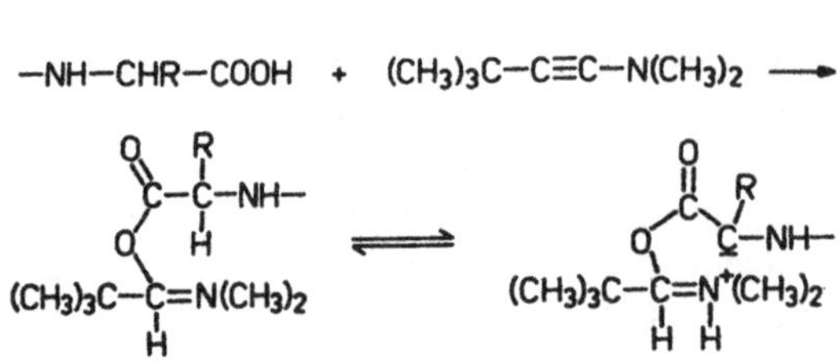

These considerations suggest that the lesser tendency of certain procedures to cause racemization is related to the absence of proton abstracting centers in the reactive intermediate and/or to the generation of materials which provide protons more readily than the chiral center of the activated residue. Thus, EEDQ [52] yields alcohol (and quinoline which has negligible basic strength):

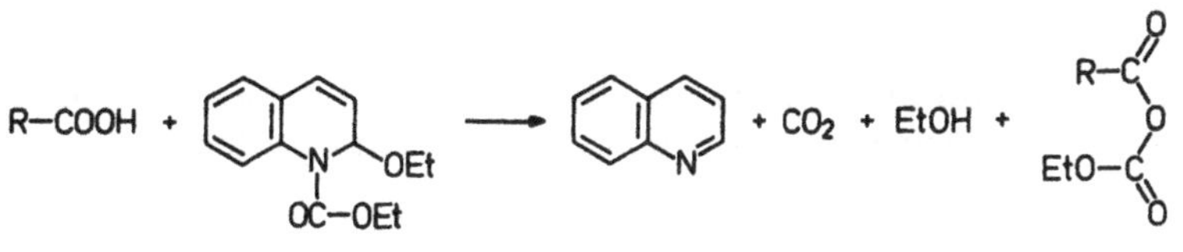

Also, in several other coupling methods substances are released which are not acidic enough to prevent acylation by protonation of the amino component, but which can, nevertheless, effectively compete with the chiral center if it comes to proton abstraction by bases. This is the situation with active esters which liberate substituted phenols or hydroxylamines during coupling.

In the base catalyzed racemization of reactive intermediates the amount and concentration of the base play an obvious role. The general principle to avoid basic conditions is supported by numerous reports and hardly requires further evidence. Thus, a free amine as nucleophile is preferable to a mixture of a salt of the amino component with a tertiary base. Weak acids, e.g. 1-hydroxybenzotriazole, do not interfere with acylation and coupling can be carried out without the addition of a tertiary amine [3]. Yet, over and above the *amount of the organic base* added to the reaction mixture its *chemical character* also has significant influence on the outcome of acylation. For instance in mixed anhydride reactions N-methylmorpholine causes less racemization [49] than the widely used triethylamine. In couplings via azides 1-diethylamino-2-propanol was found to be harmless [51] while triethylamine, N-methylmorpholine and diisopropylethylamine have, under certain conditions, an unfavorable effect on chiral purity. The last mentioned base prevents [39] the racemization of active esters of benzyloxycarbonyl-L-phenylglycine and of N-benzyloxycarbonyl-S-benzyl-L-cysteine, but had an almost as unfavorable effect on the optical purity of benzoyl-L-leucine *p*-nitrophenyl ester as other, less hindered, tertiary amines. Apparently steric hindrance in diisopropylethylamine is insufficient to interfere with proton abstraction from azlactone intermediates. Tribenzylamine seems to be more efficient in this respect. It is quite possible, however, that the influence of bases on racemization is determined not solely by their bulkiness but also by their basic strength [59, 60]. In this connection the racemization enhancing effect of the highly nucleophilic base *p*-dimethylaminopyridine [62, 63] should also be mentioned. On the other end of the scale, the weakly basic imidazole can also affect unfavorably the outcome of coupling reactions if its action is intramolecular. Thus, in acylation with activated derivatives of histidine significant racemization was observed [64], presumably caused by base catalyzed enolization or by cyclization and enolization.

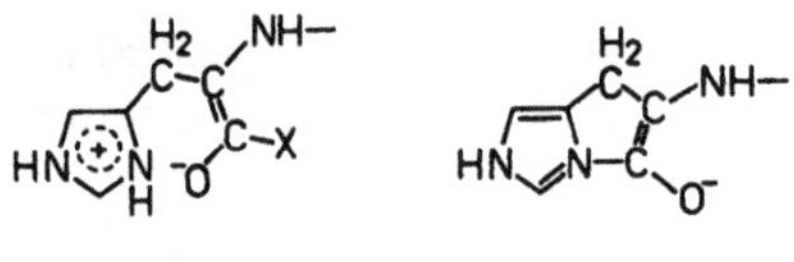

Substituents which reduce the basicity of the imidazole nucleus, e.g. the *p*-toluenesulfonyl group [65], reduce the extent of racemization as well [66]. Yet, a complete protection against loss of chiral purity of histidine residues can be expected only in derivatives in which the side chain protecting group (Y) is on the π-nitrogen atom of the imidazole:

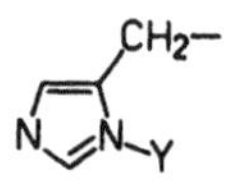

Among the factors which determine racemization the polarity of the *solvent* is quite important [4, 14]. In general, racemization is fast in highly polar solvents such as hexamethylphosphoramide, dimethylsulfoxide or dimethylformamide and less rapid in less polar solvents, e.g. pyridine, acetonitrile, chloroform, dichloromethane, tetrahydrofuran, dioxane or toluene. Unfortunately, most peptide intermediates are not sufficiently soluble in non-polar solvents and at this time the majority of acylation reactions are carried out in dimethylformamide. In solid phase peptide synthesis one applies solvents in which the peptidyl resin swells and a dissolution of the reactants is not needed. Thus, dichloromethane, which is not particularly conductive to racemization, can be used. An additional problem is created, however, by the solvent dependence of the rate of acylation of various activated intermediates. The most commonly used active esters react far better in polar solvents than in non-polar ones. These circumstances render the selection of solvents which would be favorable for acylation and yet cause little damage to chiral purity, rather difficult. A general remedy, which at least limits the extent of racemization, is to carry out the coupling reactions *at the highest possible concentration of the reactants* to ensure high coupling rates. This way the unimolecular and hence concentration independent racemization processes become less significant.

A better approach to the conservation of chiral purity is offered by the *protecting groups* which are available for the blocking of the α-amino function. Already at the time of the introduction of the benzyloxycarbonyl group its ability to protect against racemization during activation and coupling was noted and reported [67]. This unusual power to prevent the loss of chiral purity is absent from simple N-acyl groups such as the formyl, acetyl, trifluoroacetyl or benzoyl group and present only to some extent in the phthalyl group. On the other hand, several other amine protecting groups of the urethane type function equally well in this respect. Their ability to interfere with racemization was generally attributed to the lack of azlactone formation. The elimination of benzyl chloride and formation of N-carboxy-anhydrides from Z-amino acid clorides suggested [68] that

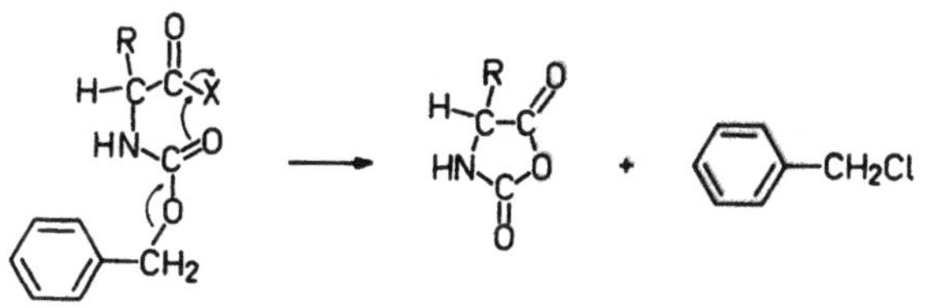

alkyloxycarbonylamino acids do not produce azlactones, the vulnerable intermediates. The recent discovery by Benoiton and Chen [15], the formation of both the symmetrical anhydride and the 5(4H) oxazolone from benzyloxy-carbonyl-L-valine and tert. butyloxycarbonyl-L-valine on reaction with water soluble carbodiimides demonstrates

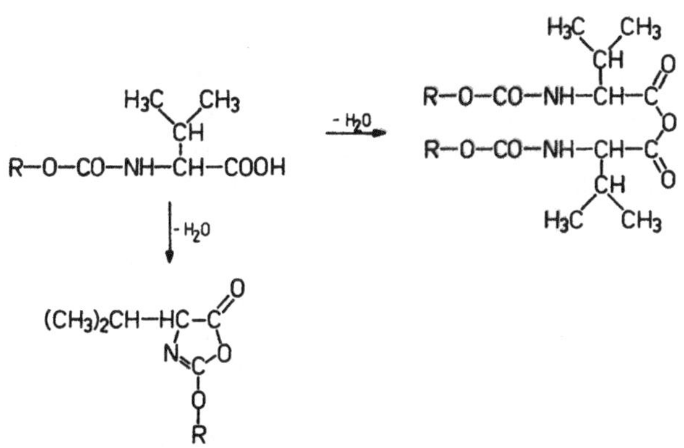

the imperfectness of this rationale. It seems now, that while amino acids provided with a urethane-type amine protecting group do form azlactones, the latter retain their chiral integrity even under basic conditions. Thus, the former explanation requires revision, but the empirical rule that the benzyl-oxycarbonyl group and other urethane-type amine blocking groups prevent the racemization of the residues to which they are attached, remains valid. Notable exceptions are the blocked derivatives of S-alkyl-cysteine, O-alkyl-serine and β-cyanoalanine. Some other amine masking groups, e.g. the *p*-toluenesulfonyl and the *o*-nitrophenylsulfenyl group, are similarly protective in this respect.

The *influence of the activated residue* on the extent of racemization can be considerable but it is not always fully understood. The benzylic character of the chiral carbon atom in phenylglycine or substituents which provide a good leaving group in the side chains of O-alkyl-serine and S-al' ,l cysteine offer simple explanations. It is less easy to interpret the somewhat reduced chiral stability of phenylalanine moieties, although this might be caused by the electron withdrawing effect of the aromatic nucleus even if it is separated by a carbon atom from the chiral center. On the other hand, tyrosine with a free phenolic hydroxyl was not racemized [69] in the coupling of Z-Val-Tyr via its azide in the presence of excess base, while the azide of Z-Val-His suffered considerable loss in chiral purity under similar conditions. An explanation might be found in the abstraction of a proton from the phenolic hydroxyl: the resulting anion interferes with the abstraction of a second hydrogen and therefore the chiral carbon does not become an anionic center. In general, formation of dianions requires stronger bases than those used in peptide synthesis:

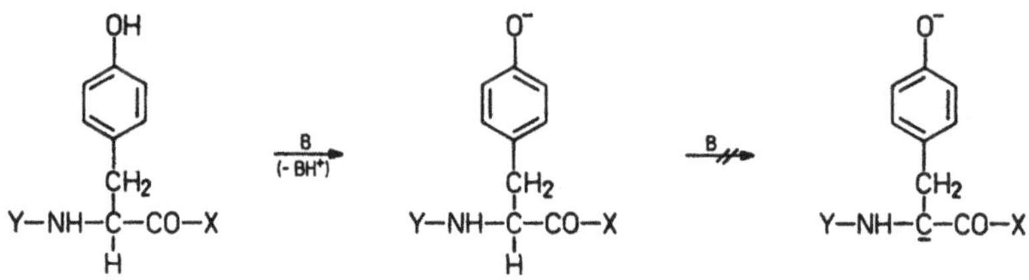

Racemization of activated valine and isoleucine residues occurs [70] in polar solvents. The electron release by the branched side chain should destabilize the anion which has to be assumed in base catalyzed racemization processes and thus an alternative rationale must be found. At this time we can suggest only the known assistance of bulky substituents in cyclizations and a cyclic intermediate is likely in the process of racemization. Chiral integrity is affected also by the residue(s) which precede the activated C-terminal amino acid in a peptide and also by the bulkiness of the N-terminal amino acid in the amino component [70]. The sequence dependence of racemization received, so far, only limited attention [71] and clearly requires further systematic studies.

Racemization of the C-terminal residue of amino components with a free C-terminal carboxyl is a recent and unexpected discovery [72]. This side reaction, which is enhanced by 1-hydroxybenzotriazole and suppressed by N-hydroxysuccinimide, is probably due to the transient activation of the unprotected carboxyl group through interaction with the acylating agent:

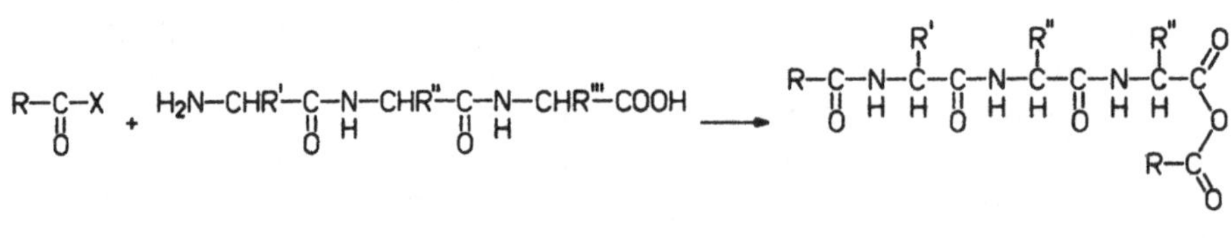

One of the most powerful methods for the preservation of chiral integrity is the use of *additives* or, perhaps more appropriately, of *auxiliary nucleophiles*. These can reduce the lifetime of overactivated, racemization-prone intermediates, such as O-acyl-isoureas. Also, the commonly applied additives have acidic hydrogens and thus can provide a proton which is more readily abstracted by bases than the proton from a chiral center. The best results, reported so far, were achieved with 1-hydroxybenzotriazole [73] (a), N-hydroxysuccinimide [74, 75] (b), 2-hydroximinocyanoacetic acid ethyl ester

[76] (c) and particularly with 3-hydroxy-3,4-dihydro-1,2,3-benzotriazole [77] (d).

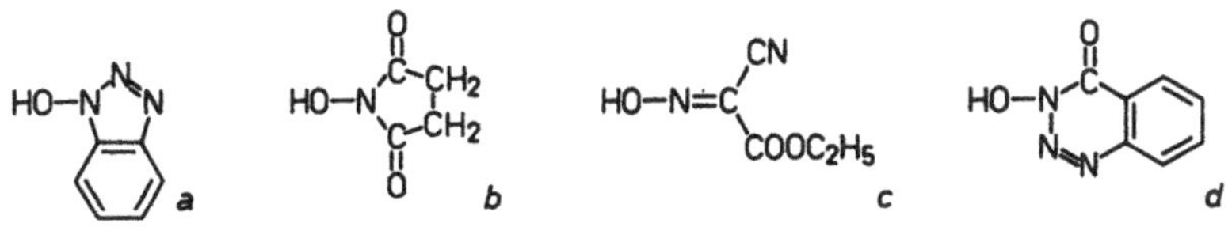

These racemization suppressing agents and several other potentially useful additives were compared by Izdebski [78].

From the foregoing discussion it is obvious that the extent of base catalyzed racemization is determined by a whole series of factors. An assessment of each of these in every coupling reaction is a demanding task and the results are probably not entirely satisfactory since not all the influences are known, or at least not well enough, to allow a quantitation of their contributions. Therefore, until the advent of truly racemization-free coupling methods, conservation of chiral integrity requires optimization in the choice of reagents, protecting groups, solvents, etc. Methods of activation which involve reactive intermediates containing a basic center should be used with caution. Overactivation, polar solvents should be avoided. The remaining choices are, however, not always conducive to an efficient formation of peptide bonds. Also, the selection of solvents is severely limited by the solubility of the intermediates. Hence, more weight has to be placed on the factors which provide some options and allow judicious decisions. For instance, the use of urethane-type amine protecting groups, attached to an amino acid rather than to a peptide, can greatly reduce the risk of racemization and the latter can be further diminished by avoiding the presence of tertiary bases in the reaction mixtures during activation and coupling. Last, but not least, the addition of well tested auxiliary nucleophiles creates conditions which no longer imperil chiral purity.

2. Undesired Cyclizations

Dipeptide esters readily cyclize to form *diketopiperazines*. Ring closure can take place spontaneously because the thermodynamic stability of the six-membered ring overcomes the energy barrier in the formation of a cis-peptide bond, but the reaction is accelerated by bases, e.g. ammonia:

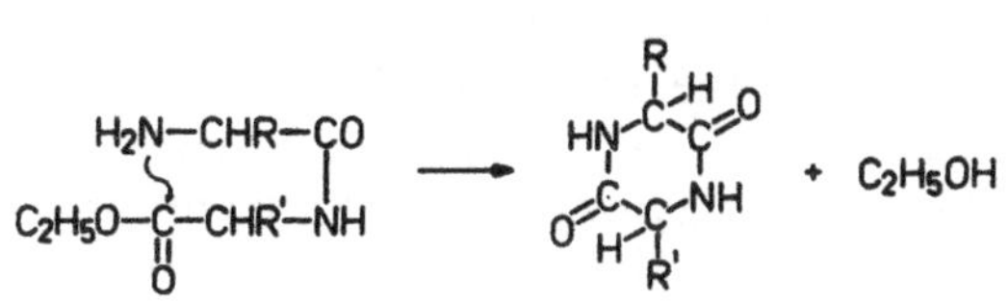

In solid phase peptide synthesis, where, at least in its original version [79], polymer bound benzyl esters are present, this side reaction can cause some premature cleavage of the chain from the insoluble support [80—83]:

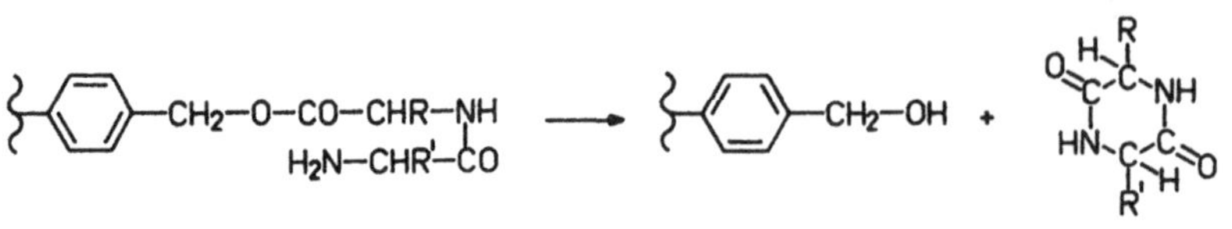

In most cases the losses suffered by diketopiperazine formation are minor, but certain residues, such as glycine, proline, N-methylamino acids, valine and isoleucine enhance the tendency for cyclization. Obviously, conformational factors provide further assistance in ring formation. Similarly, if one of the residues belongs to the L-family of amino acids while the other has the D-configuration, cyclization is accelerated because the amino acid side chains will lie on opposite sides of the general plane of the diketopiperazine ring. Of course, cyclization is less likely in dipeptide tert. butyl esters than in methyl, ethyl or benzyl esters which are more sensitive to nucleophilic attacks.

Protected and activated derivatives of glycyl-proline also cyclize to yield acyldiketopiperazines [84] under the influence of bases:

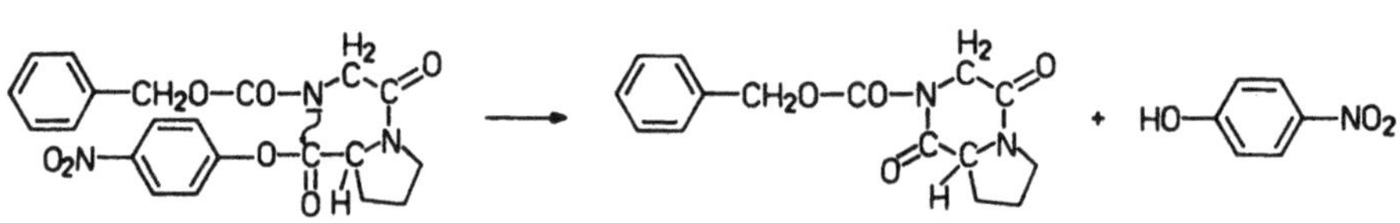

In the attempted ammonolysis of some benzyloxycarbonyl-dipeptide esters a different cyclization takes place: the formation of *hydantoins*:

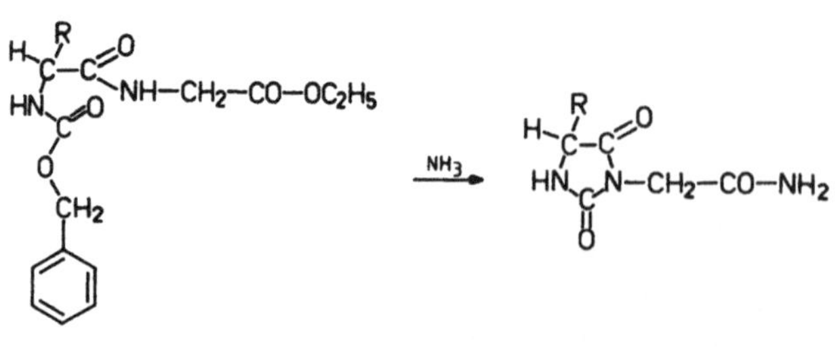

Such hydantoin formation was noted mainly in dipeptide derivatives which had glycine as the second residue in their sequence, but the preceding residue also had some influence on the course of the reaction. Thus, hydantoin formation was pronounced in the ammonolysis of benzyloxycarbonyl-L-phenylalanylglycine ethyl ester [85] and benzyloxycarbonyl-L-tryptophylglycine ethyl ester [86]. Cyclization to hydantoin derivatives followed by ring opening can accompany the saponification of esters by alkali in benzyloxycarbonyl peptides in which glycine is the second residue [87—90]:

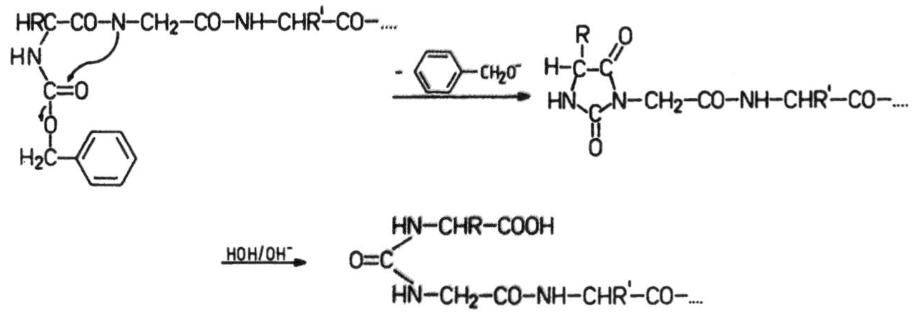

Ring closure in derivatives of aspartic acid leads to the formation of *aminosuccinimides*. This cyclization reaction takes place under various conditions and can also be base catalyzed. For instance, saponification of benzyloxycarbonyl-β-benzylaspartyl amino acids or peptides yields benzyloxycarbonylaminosuccinyl amino acids or peptides [91, 92]

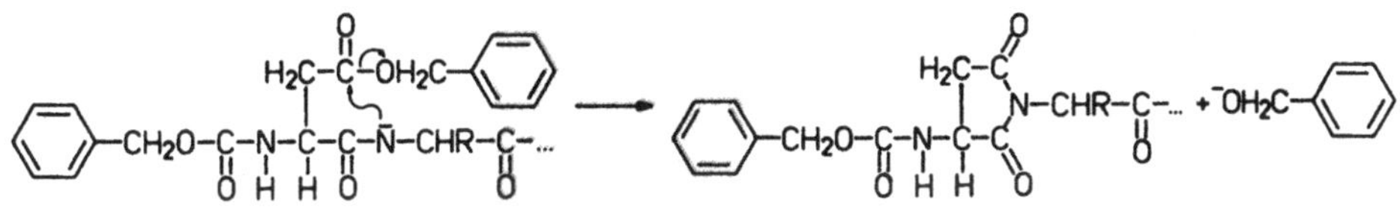

which, in turn, can open up to produce both α-aspartyl and β-aspartyl derivatives. Formation of the intermediate succinimides must also be assumed in the hydrolysis [93] and also in the hydrazinolysis [94] of tert.butyl esters attached to the β-carboxyl of aspartyl residues. Derivatives of β-cyclohexyl aspartate do not lend themselves to cyclization and hence provide more reliable protection for the side chain of aspartyl residues [95].

The production of aminosuccinyl derivatives, an often encountered side reaction in peptide synthesis, is not necessarily a major complicating factor: the base catalyzed cyclization of β-benzylaspartyl residues is quite pronounced if glycine is the nect residue in the sequence [96] but is extremely slow in peptides in which the amino group participating in ring closure belongs to an amino acid with a bulky side chain such as valine [97]:

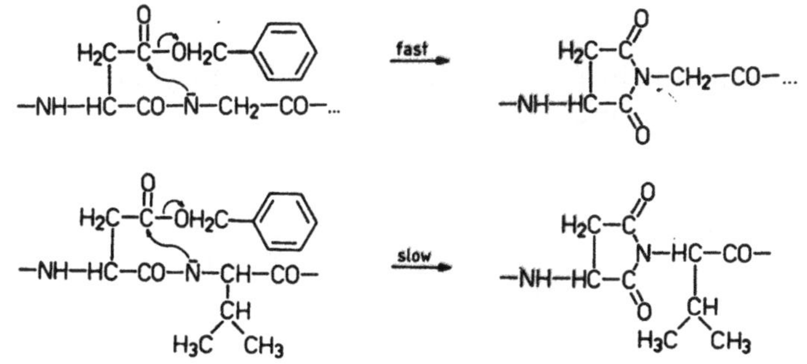

Also, the presence of an anion, like the one generated from the phenolic hydroxyl of tyrosine, will interfere with the formation of a new anionic center and hence with ring closure as well:

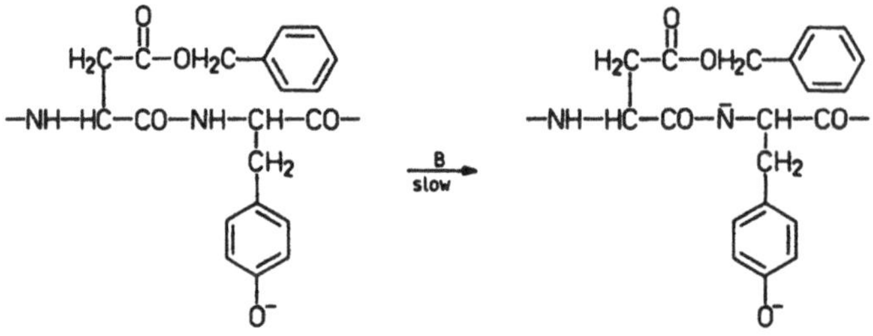

The base catalyzed ring closure observed in β-alkyl-aspartyl derivatives is less significant in γ-alkyl-glutamyl residues, but *glutarimides* do form in activated derivatives of N^{α}-acyl-glutamine [98, 99]. Intramolecular acylation of the poorly nucleophilic carboxamide nitrogen proceeds also under neutral conditions in case of powerful activation

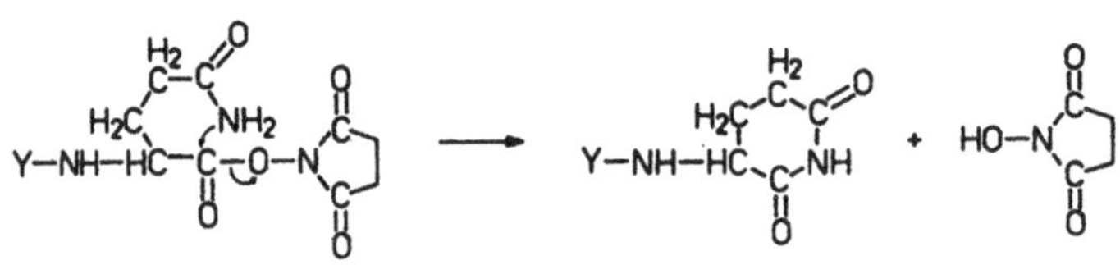

but requires the presence of a base where less potent activating groups are applied. Some glutarimide derivatives are produced in the coupling of peptides with C-terminal glutamine even by the azide method.

The alternative cyclization of glutamine moieties to pyroglutamyl residues is enhanced by weak acids more than by bases, but in activated derivatives of tosylglutamic acid ring closure to *pyrrolidones* can be preceded by the abstraction of the acidic hydrogen atom from the sulfonamide group [100]:

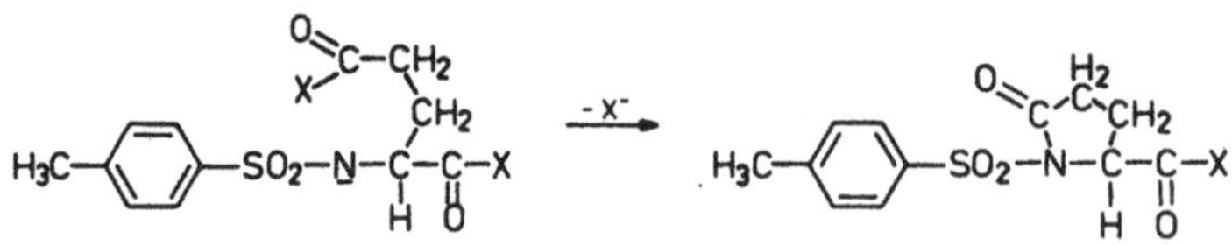

The analogous lactam formation in activated derivatives of N-tosyl-ornithine [38, 101] is similarly promoted by proton abstraction:

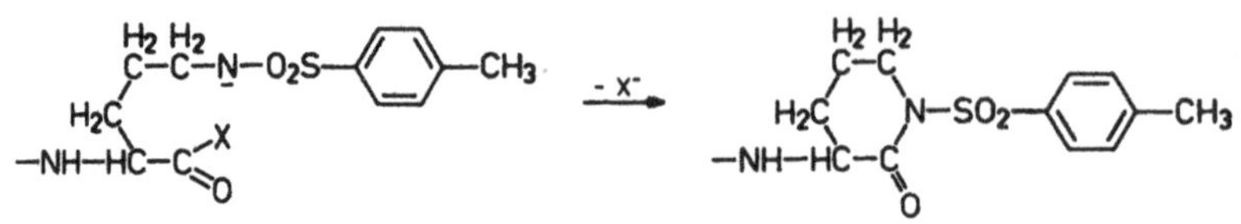

3. O-Acylation

The principal difference between an overactionated acylating agent and one which is more suitable for the selective acylation of amines is the reaction of the former with alcohols and phenols. For instance, anhydrides are attacked by hydroxyl groups while moderately active esters can be recrystallized from hot ethanol. Yet, this difference is greatly diminished in the presence of bases which convert the alcohols to alcoholates and phenols to phenolates. The formation of O-acyl derivatives when active esters were used in excess and with tertiary amines present in the solution has been known for a long time [102]:

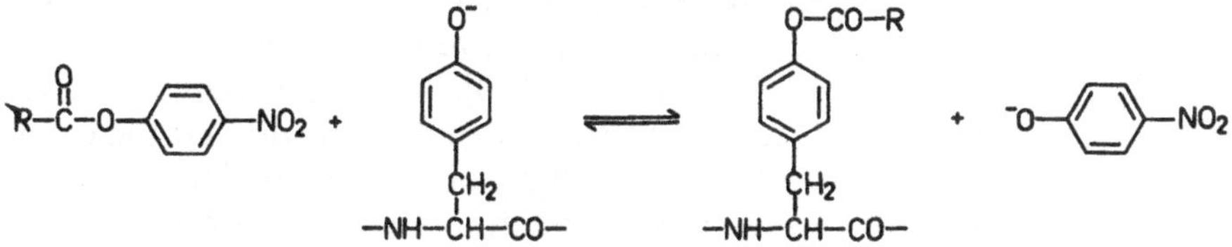

However, even in the absence of tertiary amines O-acylation occurs [103] in coupling reactions mediated by carbonyldiimidazole because the imidazole liberated in the process acts as proton abstractor:

The extensive O-acylation of serine residues noted [104] in histidine containing peptides[11] must be explained in the same way. Imidazole is an efficient catalyst in the transesterification of active esters and can be used in the preparation of substituted benzyl esters [106] or for the anchoring of protected amino acids to a hydroxymethyl polymer [107]:

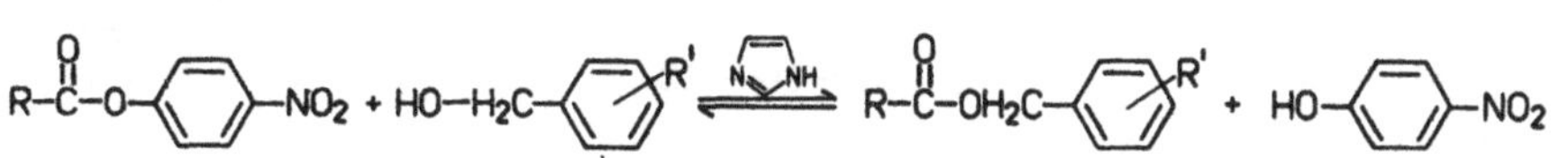

Undesired O-acylation of amino acid side chains can be suppressed by adding proton donors, particularly 2,4-dinitrophenol or pentachlorophenol to the reaction mixture [108]. Against the histidine catalyzed O-acylation the addition of 1-hydroxybenzotriazole (although in itself a catalyst of O-acylation) has similar beneficial effects [104].

---

[11] An analogous imidazole catalyzed ring opening of benzyloxycarbonyl-pyroglutamyl-histidyl peptides [105] yields, in methanol, benzyloxycarbonyl-γ-glutamyl-histidyl derivatives.

B. Side Reactions Initiated by Protonation

1. Racemization

Acid catalyzed racemization of amino acid derivatives probably involves protonation of the carbonyl oxygen followed by enolization:

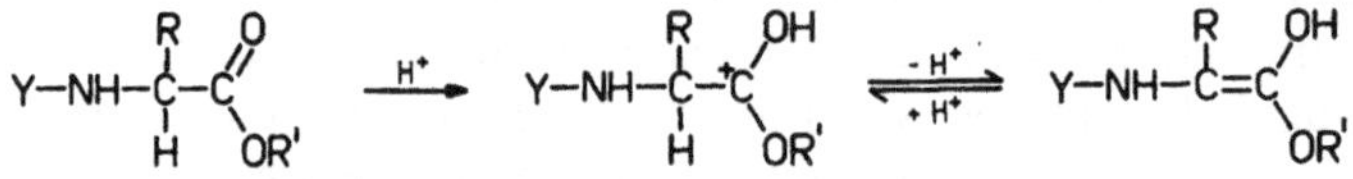

Such a process, however, requires strong acids because no protonation of an oxygen atom can be expected in weakly acidic media. Yet, acidolytic removal of protecting groups is often carried out with extremely strong acids, such as liquid hydrobromic [109] acid or liquid HF [110]. The good results achieved in the practical application of such powerful reagents indicate that the extent of racemization caused by them must be slight or negligible. This fortunate circumstance should be due to the low temperature at which strong acids are applied. Removal of protecting groups by acidolysis at elevated temperature [111] must cause concern because loss of chiral purity has been observed [112] under such conditions.

2. Undesired Cyclizations

Ring-closure of β-alkyl aspartyl residues leading to *amino-succinyl derivatives* has already been discussed among the side reactions which are initiated by proton abstraction. The same products are obtained, however, also under the influence of strong acids [96]. For instance protonation of β-benzyl-aspartyl peptides promotes their cyclization:

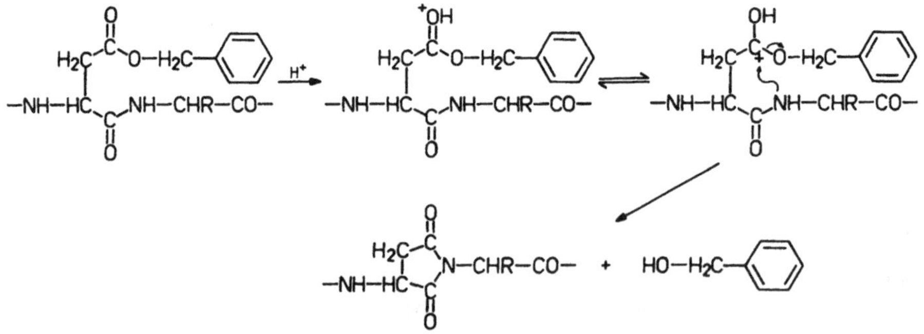

The rate of acid catalyzed cyclization is as dependent from the sequence as the base catalyzed version of this side reaction. Peptides which contain the Asp-Gly sequence are notoriously prone to cyclization [96]. The presence of aminosuccinyl moieties in the synthetic material can be reliably detected by digestion with aminopeptidases which do not catalyze the hydrolysis of the bond between the aminosuccinyl residue and the amino acid acylated by it.

From the numerous attempts for the suppression of this frequently encountered side reaction the use of the phenacyl group for the protection of the β-carboxyl [113] is certainly noteworthy. Somewhat disappointingly, prevention of acid catalyzed cyclization is counterbalanced by the propensity of the phenacyl group for intramolecular nucleophilic displacement under basic conditions [114]:

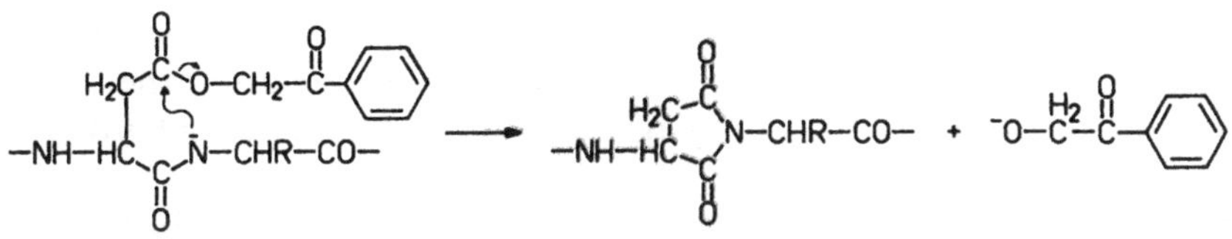

There are several other carboxyl protecting groups, e.g. benzyl esters with electron-withdrawing substituents which provide increased resistance to acids used in deblocking steps [115]. These are, however, conducive to ring closure during coupling, a step where basic conditions prevail. The more recently proposed β-cyclopentyl [116] and β-cyclohexyl [95] esters, though they require HF for their removal, are more auspicious.

In contrast to the acid catalyzed ring closure of β-alkyl-aspartyl residues, cyclization of glutamic acid derivatives takes place mainly under the influence of bases. On the other hand, formation of *pyroglutamyl peptides* in chains with N-terminal glutamine is accelerated by weak acids [117, 118].

3. Alkylation

Acidolytic removal of protecting groups is based on the formation of stable carbocations. Direct alkylation of nucleophilic centers by these reactive species, e.g. by the tert. butyl or benzyl cations formed in deblocking, is possible but not likely. The cations rather react with molecules of the solvent surrounding them. The compounds, however, generated in the interaction of the cations with the solvent are often good alkylating agents. For instance, benzyl trifluoroacetate is produced in the removal of benzyloxycarbonyl groups and O-benzyl groups with (hot) trifluoroacetic acid [111], while in the treatment of tert. butyloxycarbonyl derivatives, tert. butyl esters and tert. butyl ethers, tert. butyl trifluoroacetate forms (in the cold) [119]. Similarly, anisole, used as a scavanger of carbocations, can be the source of methylating agents like methyl fluoride which in turn can alkylate sensitive amino acid side chains [120]. Tyrosine is readily substituted in the ring on the carbon atom which is ortho to the phenolic hydroxyl. In addition to simple electrophilic aromatic substitution reaction, 3-benzyl-tyrosine derivatives can form also by intramolecular migration:

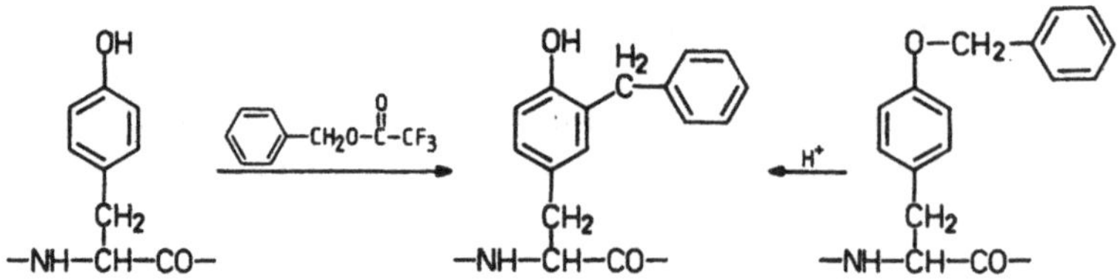

Substitution on the aromatic nucleus of tyrosine is facile in trifluoroacetic acid, but not in acetic acid, probably because benzyl acetate and other alkyl acetates are not as powerful alkylating agents as the corresponding trifluoroacetates.

Several other amino acids can suffer similar side reactions. Methionine is particularly prone to alkylation [120]. Methylation, benzylation and tert. butylation of the thioether sulfur atom leads to tert. sulfonium salts. The thioether function can be restored by the action of thiols or in the case of tert.butyl substitution simply by storage or gentle heating [121]:

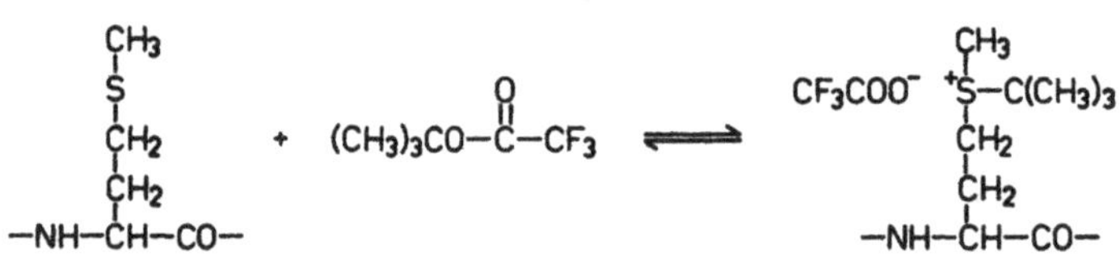

More serious damage is caused by benzylation of methionine residues, which could be remedied only by dealkylation of the S-benzylhomocysteine derivative with sodium in liquid ammonia followed by methylation with methyl iodide [122]:

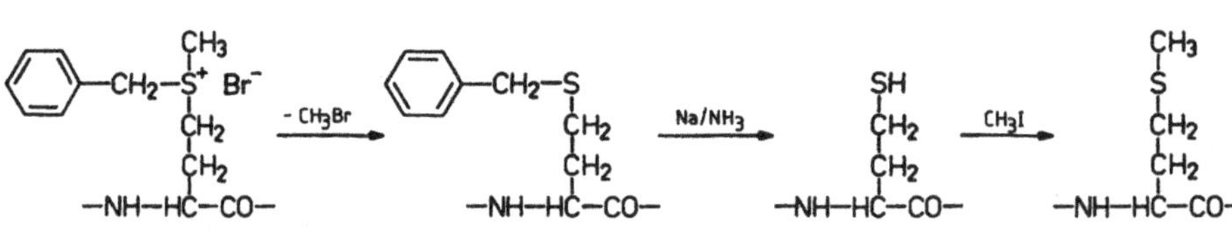

Therefore, alkylation of methionine side chains should be prevented. This can be achieved by the use of various dialkyl sulfides or thiols as scavangers or by the oxidation of methionine to the sulfoxide [123] and perhaps also by methylation with methyl *p*-toluenesulfonate [124]:

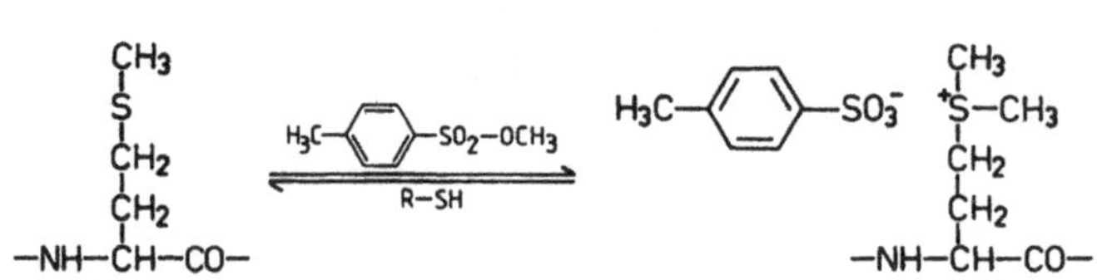

Alkylation of the side chain of tryptophan residues, overlooked for a long time, has been observed in several laboratories [125–127]. The principal byproducts, found after the removal of tert. butyloxycarbonyl or O-tert. butyl groups, are N^{in}-tert. butyl tryptophan derivatives

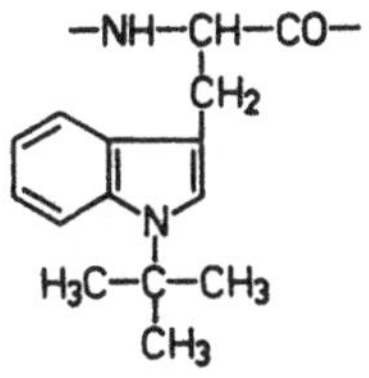

but the alkyl group can migrate from the nitrogen atom to ring positions 2,5 and 7 as well [128]. Since there is no available method for the removal of these alkyl groups, it is absolutely necessary to prevent alkylation of the indole. This can be done by the addition of scavangers such as thiols, dialkyl sulfides, indole or skatole and also by masking the indole nitrogen, e.g. with the formyl group, discussed in the preceding chapter. It might be even better to plan the synthesis of tryptophan containing peptides with consideration of their sensitivity to alkylation and of several other side reactions which occur under acidic conditions. Thus, schemes [129] based on protecting groups not requiring acidolysis are preferable to combinations in which acids are applied for deprotection.

Some alkylation, probably an intramolecular reaction, takes place during the removal of benzyloxycarbonyl groups from derivatives of benzyloxycarbonylglycine and N^{ε}-benzyloxycarbonyl-lysine [130]. This seldom observed side reaction

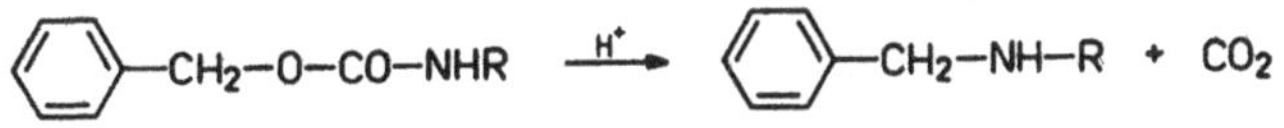

produced a few percent benzylamino acids when trifluoroacetic acid was used for acidolysis but only negligible amounts of byproducts with other acidic reagents, e.g. methanesulfonic acid [131].

4. Chain Fragmentation

The amide bonds linking amino acids to each other to create the backbone of a peptide chain are stable enough to withstand the usual rigors of peptide synthesis. In a few instances, however, this stability is reduced by special features of the amino acid residues participating in a peptide bond. Thus, under the influence of strong acids an acyl group attached to the nitrogen atom of a serine residue migrates to its hydroxyl oxygen. Such an N → O shift takes place also when the acyl group is a part of a peptide chain [132, 133].

This reaction which, in all likelihood proceeds via cyclic intermediates is easily reversed by treating the product with aqueous sodium bicarbonate

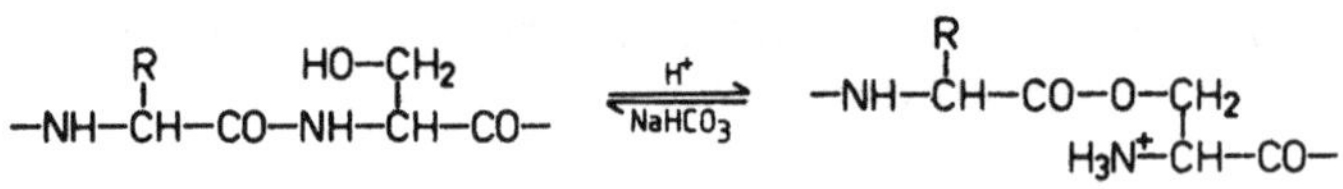

but partial hydrolysis of the sensitive ester bond will lead to fragmentation of the chain. Fortunately only few acids are strong enough to catalyze N → O acyl migration, but powerful acidic reagents such as hydrogen fluoride or trifluoromethanesulfonic acid should be used with caution, preferably for short times and below room temperature.

The amide bonds surrounding aspartyl residues are less resistant to acid catalyzed hydrolysis than other peptide bonds. Heating the solution of a peptide containing an aspartyl residue in 0.25 molar acetic acid is sufficient for the release of free aspartic acid [134]. One particular bond, the bond between aspartic acid and proline residues, is cleaved by aqueous acids

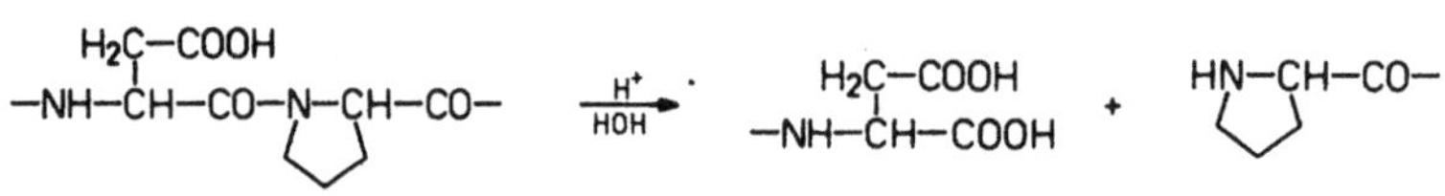

even at room temperature [135].

C. Side Reactions Due to Overactivation

The term "overactivation" [136] points to the ambiguities created in acylation reactions in which the activated derivative of the carboxyl component is too powerful to be selective and causes acylation not only of the amino group which is expected to form a peptide bond, but also of less good nucleophiles, e.g. hydroxyl groups. Anhydrides, both symmetrical and mixed, are obviously such overactivated derivatives, but the seemingly more subtle intermediates generated in the addition of carboxylic acids to carbodiimides are similarly overactivated. The O-acylisourea intermediates give rise to symmetrical anhydrides [137, 138] and azlactones [139] and react also with the weak but intramolecular nucleophilc center within the urea moiety:

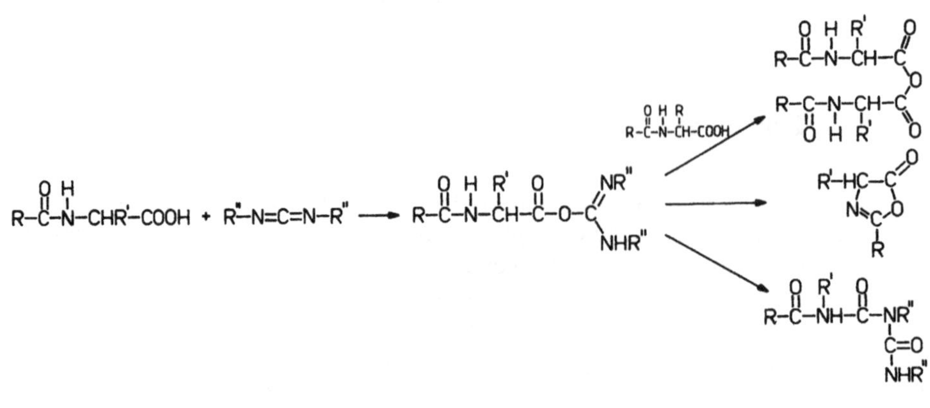

The last mentioned rearrangement to N-acylurea derivatives is still an extensively studied [140–142] side reaction. Not surprisingly, carbodiimides react with the unprotected sulfhydryl group of cysteine residues to form isothioureas which in turn yield, by β-elimination, dehydroalanine derivatives [143]:

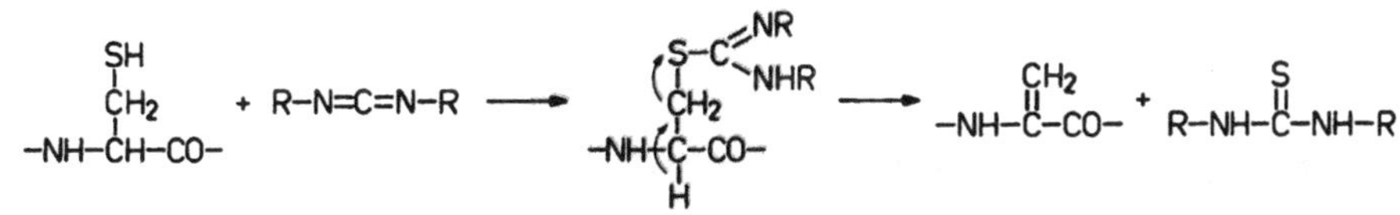

In an analogous manner the imidazole moiety in the histidine side chain can add to carbodiimides to produce substituted guanidines [144]:

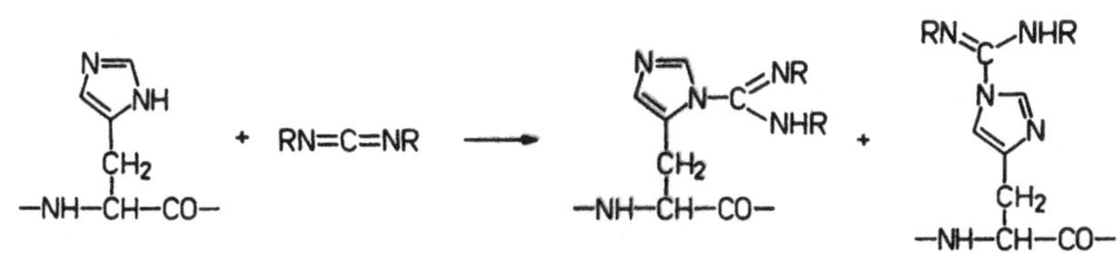

This side reaction can be reversed by acid catalyzed methanolysis, but

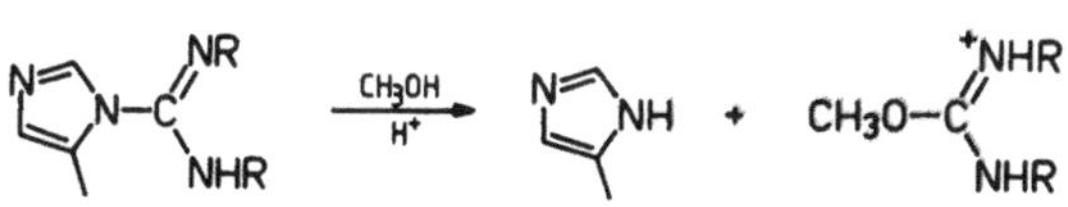

it serves as warning against the uncritical use of carbodiimides, particularly in excess.

It is not easy to avoid overactivation. Moderate activation can also be conducive to side reactions, e.g. to the formation of amides and urea derivatives in azide coupling [145, 146]. Active esters, which were thought for some time to be selective amine-acylating agents, were found to cause O-acylation if bases or catalysts such as imidazole are present [106]. Moreover, there is considerable difference in this respect between various active esters. For instance, the potent acyl derivatives of N-hydroxysuccinimide are more prone to O-acylation than the less reactive 2,4,5-trichlorophenyl esters. The rapid hydrolysis of 2,4-dinitrophenyl esters [147] and pentafluorophenyl esters [148] demonstrates that certain active esters should be considered overactivated. The ratio between O-acylation and N-acylation, a parameter probably more important than the absolute rate of either reaction, changes with the nature of the protected amino acid, the hydroxy-amino acid and with the experimental conditions [149].

It may seem unrealistic to expect reasonable coupling rates in dilute solutions with acylating agents which are completely devoid of overactivation.

Perhaps esters which are activated by the approaching nucleophile, e.g. esters of 1-hydroxypiperidine [150, 151]

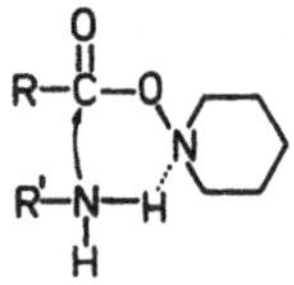

are the best approximations of such an ideal situation.

D. Side Reactions Related to Individual Amino Acid Residues

It would seem that amino acids which have no functional side chains should not be involved in specific side reactions, but this is true only for alanine and leucine among the amino acids which are constituents of proteins. In the case of *valine* and *isoleucine* branching of the side chain at the β-carbon atom leads to steric hindrance which lowers the rate of coupling reactions and can cause, therefore, an increase in the extent of unimolecular side reactions such as the formation of ureides from the O-acylisourea intermediates in condensations with carbodiimides:

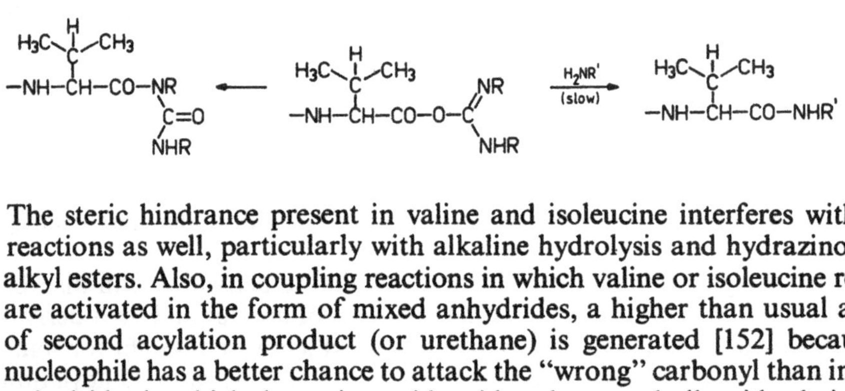

The steric hindrance present in valine and isoleucine interferes with other reactions as well, particularly with alkaline hydrolysis and hydrazinolysis of alkyl esters. Also, in coupling reactions in which valine or isoleucine residues are activated in the form of mixed anhydrides, a higher than usual amount of second acylation product (or urethane) is generated [152] because the nucleophile has a better chance to attack the "wrong" carbonyl than in mixed anhydrides in which the amino acid residues have no bulky side chains:

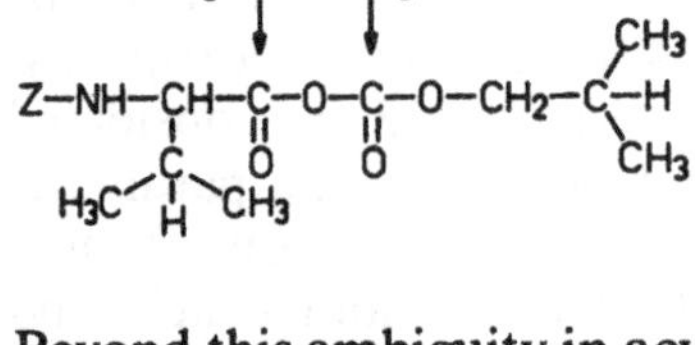

Beyond this ambiguity in acylation even the preparation of mixed anhydrides of protected valine and isoleucine can create problems, if the access of the activating reagent is limited by its own bulk. Thus the reaction of hindered amino acids with trimethylacetyl (pivaloyl) chloride is quite slow. In general the combination of two or more sterically hindered substances or groups can seriously impede a reaction. The matrix of insoluble polymers also interferes with the incorporation of hindered amino acids and it seems to be likely that acylation with trityl-isoleucine pentachlorophenyl ester cannot be brought to completion if the amino component is attached to an insoluble polymer.

Strangely enough the complete absence of steric hindrance in derivatives of *glycine* can also be the cause of side reactions. In other amino acids the inertness of the amide nitrogen is further enhanced by the bulky side chain (R) in its proximity:

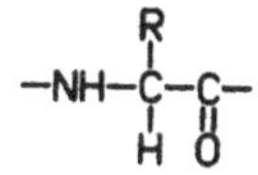

Since glycine has no side chain, its acylated amino group can accept a second acyl group. This indeed happens in reactions in which a powerful acylating agent is present, or with less potent derivatives of the carboxyl group, if the reaction is intramolecular. Thus, a diacylamide forms as a byproduct in the preparation of Z-Gly-Gly by the phosphoryl chloride method [153] and also in syntheses via mixed anhydrides [154, 155]:

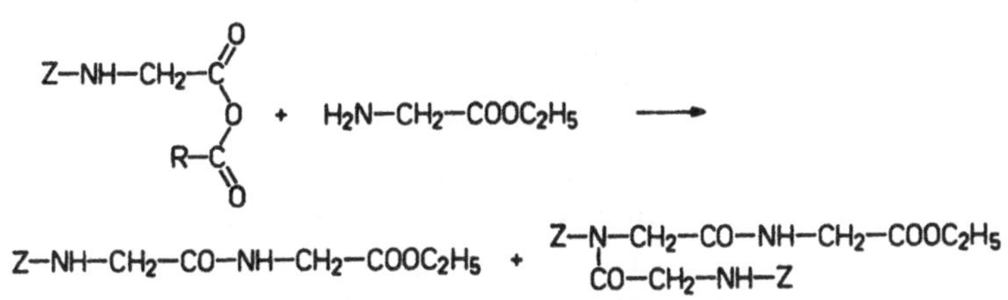

In a similar fashion p-toluenesulfonylglycine can form diacyl derivatives [156] in mixed anhydride reactions and the symmetrical anhydride of benzyloxycarbonylglycine undergoes a rearrangement yielding a diacylamide [157].

The already discussed base catalyzed ring closure reaction [84–90] of peptides with glycine as the second residue in their sequence leads to hydantoins, that is, to diacylamides

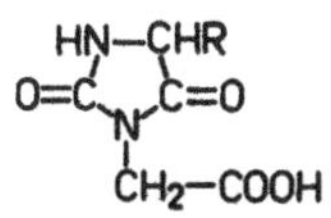

which are not readily formed if the amide nitrogen is sheltered by the bulk of an amino acid side chain. Similarly, base catalyzed cyclization of active esters of protected dipeptides [84] requires that the already acylated amino group of a glycine residue suffers intramolecular acylation; the product, an acyldiketopiperazine is, once again, a diacylamide:

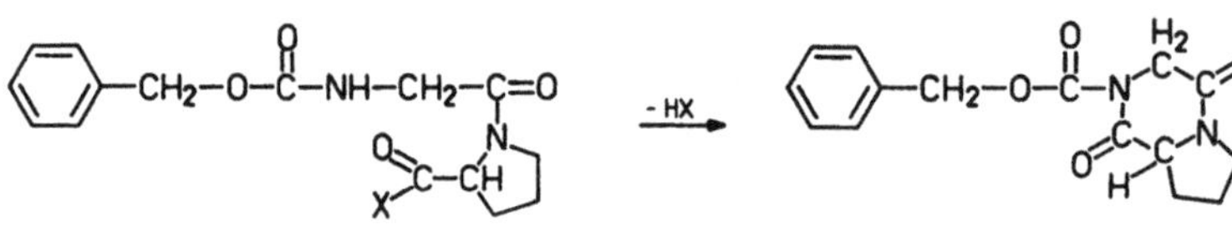

Because of this readiness of the amino group in glycine to accept two acyl substituents, serious consideration should be given to the blocking of the amide group of glycine residues. In this special case the otherwise superfluous masking of the amide, e.g. with benzyl [158–160], 4-methoxybenzyl [159] or 2,4-dimethoxybenzyl [159, 161] groups might be justified.

Proline with its cyclic and therefore relatively rigid "side chain" can be the source of several kinds of difficulties in coupling reactions. Thus, proline can attack one of the succinimide carbonyls instead of the active ester carbonyl in N-acyl-proline 1-hydroxysuccinimide esters [162], a side reaction already discussed in the section on coupling. Because of such spatial restrictions acylation with protected proline via carbodiimides can be less than satisfactory: rearrangement of the O-acylisourea intermediate to N-acylureas is often quite pronounced [163]. The geometry of proline residues is conducive also to folding of the chain. Hence, in the cyclization of benzyloxycarbonyl-glycyl-L-proline *p*-nitrophenyl ester the generation of an acyldiketopiperazine [84] is facilitated by the presence of a proline residue. In general: diketopiperazines readily form from dipeptides in which one of the residues is proline and ring closures yielding the Pro-Pro diketopiperazine are particularly facile.

Some additional problems arise from the circumstance that proline, unlike all other amino acid constituents of proteins, is not a primary but a secondary amine. This leads to the reductive cleavage of the peptide bond connecting proline with the preceding residue during reduction with sodium in liquid ammonia [164]:

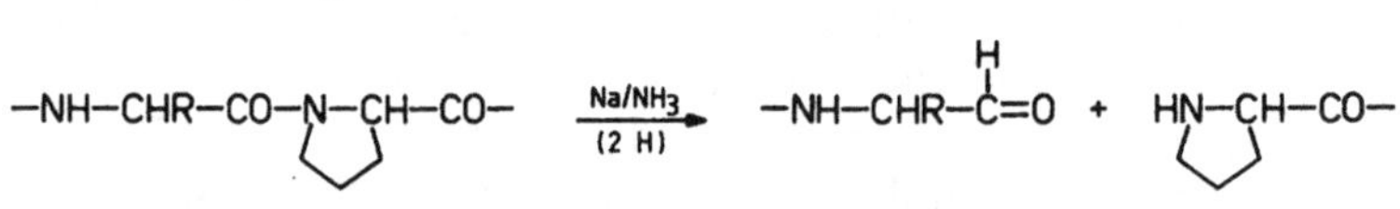

The reaction requires the presence of proton donors. Water or alcohols can be the source of the two hydrogen atoms, but these may stem also from amide groups, especially from *p*-toluenesulfonamides, or from the hydroxyls in amino acid side chains [165]. The acid catalyzed hydrolytic fission of the Asp-Pro bond has already been discussed in this chapter.

Base catalyzed acylation of the alcoholic hydroxyl groups on *serine* and *threonine* has been mentioned in the section on side reaction initiated by proton abstraction. The formation of esters is catalyzed also by acids. Hence, acidolytic removal of benzyl groups from serine containing peptides is usually carried out with hydrobromic acid in trifluoroacetic acid [166] because the classic reagent. HBr in acetic acid [167] causes partial acetylation of the hydroxyl group. In fact, even without a catalyst ester equilibrium is reached in a matter of weeks if solutions of serine containing peptides in acetic are stored at room temperature. The secondary hydroxyl group in threonine is generally less affected in these acylation reactions. This is true also for the acid catalyzed N → O acyl migration discussed among proton induced side reactions.

Alcoholic hydroxyls are good nucleophiles only in the presence of bases, when, in part, they are present as alcoholates. Yet, in intramolecular reactions they are reactive enough, even under neutral conditions, to be acylated by the activated carboxyl. Thus, carbodiimides, on reaction with N-protected serine derivatives produce lactones [73, 168, 169] which are, however, still reactive acylating agents:

The phenolic hydroxyl group in *tyrosine* side chains is at least as readily acylated, in the presence of base, as the alcoholic hydroxyl of serine. The acidity of the phenolic hydroxyl facilitates proton abstraction and the phenolate anion is an excellent nucleophile. The reactivity of the phenolic hydroxyl is, however, sufficient for the production of esters, even in the absence of proton abstracting reagents when powerful activating agents, e.g. carbodiimides are applied. Unwanted O-acyl groups can, fortunately, be readily removed from the tyrosine hydroxyl with nucleophiles such as ammonia, hydrazine or hydroxylamine. The damage caused by electrophilic aromatic substitution of the nucleus is more serious, since it cannot be repaired: 3-alkyltyrosines are stable compounds. In fact, 3-benzyl-tyrosine was first observed [170] in acid hydrolysates of tyrosine containing peptides. While alkyl migration (from O to C) can be, in part, intramolecular, the reaction proceeding through the collapse of an intimate ion pair [171, 172], it is important to realize that the substituent of carbon atom 3 can stem equally well from another residue of the chain or, in fact, from another molecule. The actual alkylating agents can be alkyl halogenides [170], alkyl trifluoroacetates [119] or alkyl *p*-toluenesulfonates [111]. Because of these possible intermolecular pathways it is not sufficient to choose for the protection of the phenolic hydroxyl such alkyl groups, as the O-isopropyl or the O-cyclohexyl group [172] which are unlikely to migrate to the ring carbon atom, but all acidolytically cleavable blocking groups on the chain must be carefully considered from this point of view. For instance, the mesitylene-2-sulfonyl group used for the masking of the guanidino group of arginine [173] is superior with respect to substitution on the tyrosine side chain to the p-methoxybenzenesulfonyl group [174].

The chemically inert side chain of *phenylalanine* is usually immune from side reactions but during catalytic hydrogenations the aromatic ring can be saturated and the amino acid residue converted to a hexahydrophenyl-

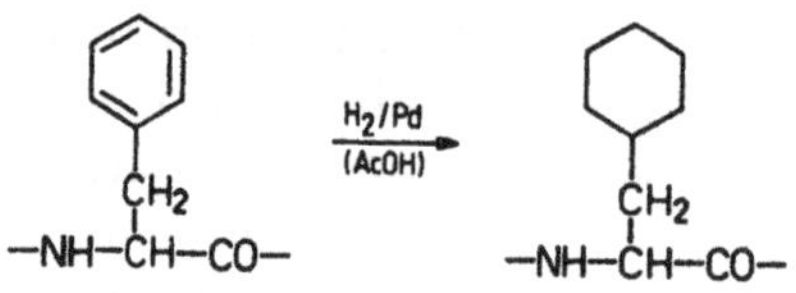

alanine (or cyclohexylalanine) moiety. This occurs, however, only as a minor side reaction except when reduction is carried out for a prolonged period of time [175, 176]. The formation of complexes [177–179] between alkali salts of benzyloxycarbonylphenylalanine and the protected amino acid has already been mentioned.

In contrast to phenylalanine, *tryptophan* is quite sensitive, particularly under acidic conditions. Its side chain can suffer oxidative degradation, dimerization, alkylation, substitution with sulfenyl chlorides, etc. A review on the chemistry of tryptophan [180] deals with a series of such reactions. Here we point only briefly to the already discussed [125–128] alkylation of the indole nucleus during acidolysis [181–184]. While this side reaction can be limited by formylation [185] or by the addition of scavengers such as thiols, indole, etc., it is more difficult to prevent oxidative decomposition, dimerization and additional ring formation in the tryptophan side chain. Some of these undesired reactions, e.g. the formation of carboline derivatives [186, 187]

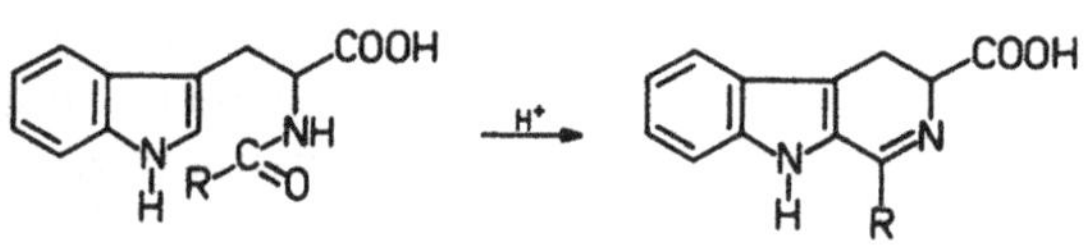

or dimerization [188, 189],

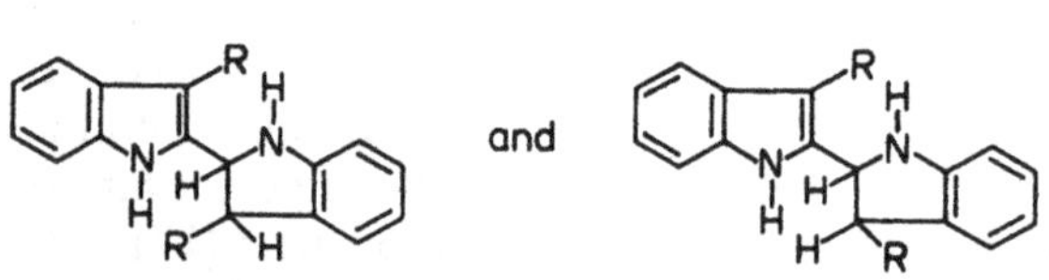

are favored by trifluoroacetic acid and by HCl in organic solvents or in water, while aromatic sulfonic acids and especially mercaptoethanesulfonic acid [190] seem to be less harmful.

A most pronounced undesired reaction in the synthesis of tryptophan containing peptides, substitution of the indole ring system by *o*-nitrophenylsulfenyl chloride [191],

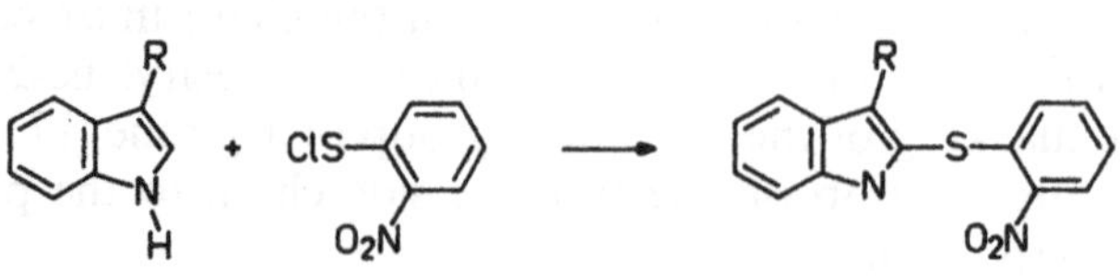

can be prevented by the application of nucleophiles rather than acids for the removal of the *o*-nitrophenylsulfenyl (Nps) group. In the choice of methods applied for deblocking a further limitation is caused by the imperfect resistance of the indole system to Pd-catalyzed hydrogenation. It can be partially [192] or completely [193] saturated.

The formation of a hydantoin derivative on ammonolysis of benzyloxycarbonyltryptophylglycine ethyl ester has been mentioned in connection with side reactions related to glycine [86].

Synthesis of *aspartic acid* containing peptides is seriously complicated by the tendency of β-alkyl-aspartyl residues to change to amminosuccinyl moieties. This ring closure is catalyzed both by acids and by bases and was accordingly treated in the appropriate sections of this chapter. The complexity is further increased by transpeptidation: the ready hydrolysis of the aminosuccinyl residue by alkali, leading to the formation of both α-aspartyl and β-aspartyl peptides. Yet, intramolecular nucleophilic attack on one of the aminosuccinyl carbonyls yielded a diketopiperazine derivative even under practically neutral conditions [194]:

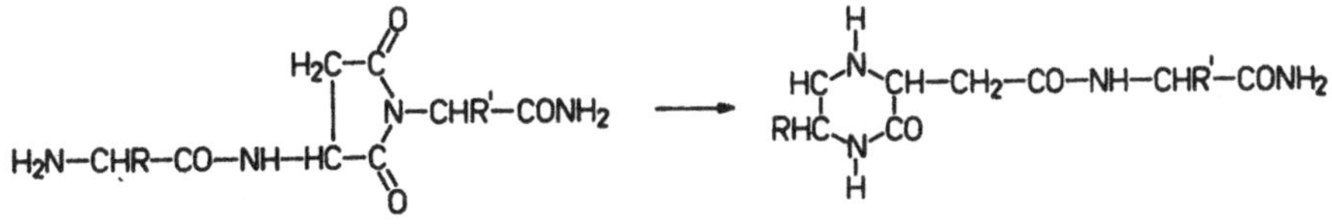

Ring closure can affect also peptides in which aspartyl residues with unmasked β-carboxyl groups are present [195]. Aminosuccinyl derivatives formed in this way during attempted purification in systems containing pyridine *and* acetic acid [194, 195]. The sensitivity of the peptide bond in the Asp-Pro sequence [135, 196] to aqueous acids should be considered both in synthesis and during purification.

Glutamic acid residues can yield both the six membered rings of glutarimides and the five membered cycles of pyroglutamyl (5-pyrrolidone-2-carboxylic acid) residues. The latter form mainly if an N-terminal glutamyl residue has an activated γ-carboxyl group, a situation which is usually not unintentional. Glutarimides, however, can be produced as by-products if the γ-carboxyl of a midchain glutamyl residue is left unprotected and thus becomes involved during the activation of the carboxyl component [20, 92, 197]. The glutarimide derivative can be the source of transpeptidation [198], a much studied side reaction:

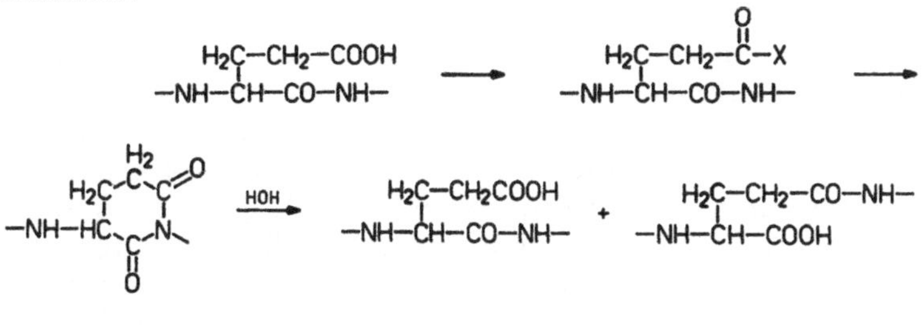

An analogous rearrangement occurs, obviously through a similar cyclic intermediate, during alkaline hydrolysis of γ-esters of glutamyl residues, but this can be prevented by the addition of copper(II) hydroxide to the reaction mixture [199]. Glutarimides are also the likely intermediates in the unexpected removal of γ-tert. butyl esters by hydrazine [200].

Irrevocable damage is caused to peptides through the Friedel Crafts acylation of anisole (added as a scavenger) by the γ-carboxyl of glutamyl residues, e.g. in liquid hydrogen fluoride [201, 202]. The side reaction, which probably proceeds via an acyl cation

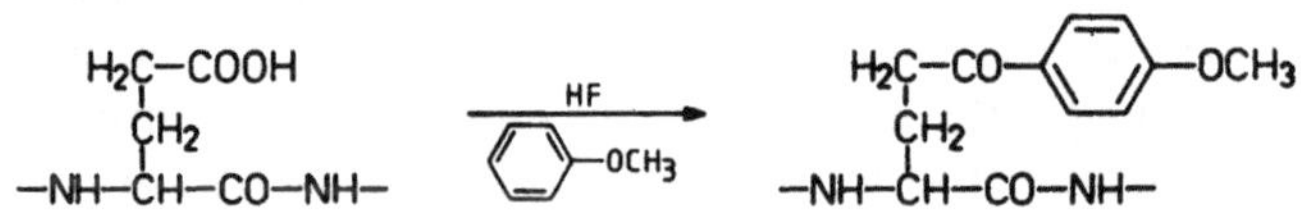

is eliminated if the γ-carboxyls remain blocked [203] during acidolysis in the form of phenacyl esters or *p*-nitrobenzyl esters which are resistant to liquid HF. The carboxyl masking groups are then removed in a separate operation.

The side reactions related to *asparagine*, e.g. the ready hydrolysis by alkali of the carboxamide group [204–206] or the saponification of N-acyl-asparagine methyl ester [91] generally involve cyclic intermediates as shown by the formation of both asparagine and isoasparagine derivatives:

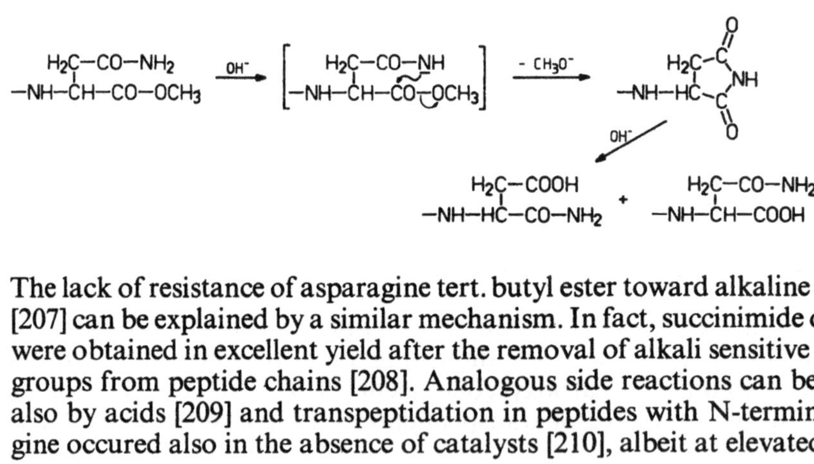

The lack of resistance of asparagine tert. butyl ester toward alkaline hydrolysis [207] can be explained by a similar mechanism. In fact, succinimide derivatives were obtained in excellent yield after the removal of alkali sensitive protecting groups from peptide chains [208]. Analogous side reactions can be catalyzed also by acids [209] and transpeptidation in peptides with N-terminal asparagine occured also in the absence of catalysts [210], albeit at elevated temperatures.

Cyclization to succinimides might also be one of the causes of poor yields experienced [211] in the synthesis of asparaginyl peptides, but a more firmly established and more common side reaction interfering with the incorporation of asparagine residues is the dehydration of the side chain carboxamide to a nitrile [212, 213]. Once an asparagine residue is part of a peptide chain its side chain suffers no more dehydration unless exceptionally potent reagents or drastic conditions are applied. Thus, the loss of water must occur in the reactive intermediate of the coupling reaction. A mechanism involving a cyclic cation

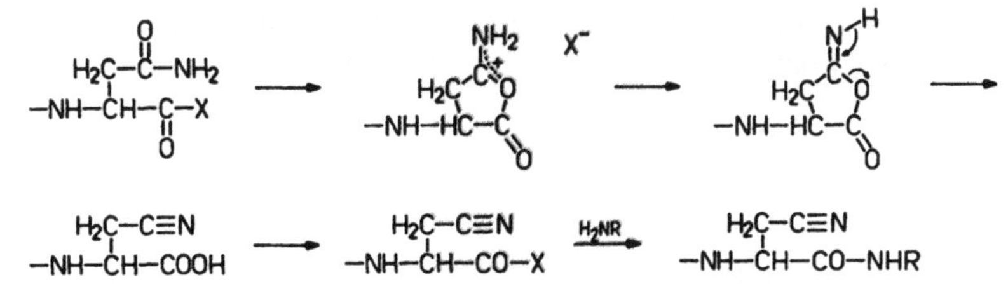

is plausible, although other related pathways have also been suggested [215, 216]. The method of activation has a certain influence on the extent of nitrile formation but it occurs, in addition to couplings with mixed anhydrides and carbodiimides, in which it was first noted, also on reaction of acylasparagine derivatives with phosphoryl chloride [217] or during the preparation of N-carboxyanhydrides with phosgene [218]. Nitrile formation can be prevented by the use of amide protecting groups mentioned in Chapter IV. It is suppressed by the addition of 1-hydroxybenzotriazole to the reaction mixture [219]. Alternatively, one can prepare active esters of N^{α}-protected asparagine derivatives and separate [220] the reactive derivative of asparagine from that of β-cyanoalanine:

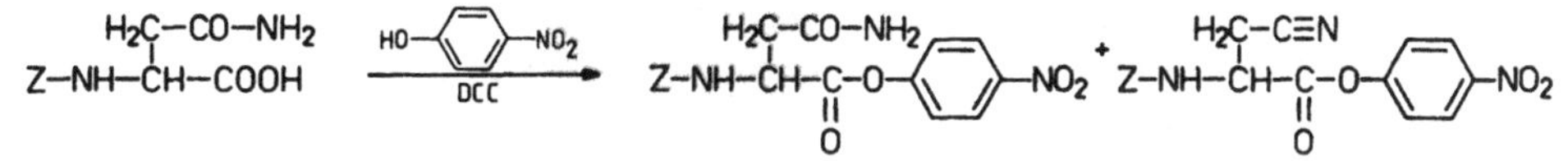

Acylation with nitrile-free active ester provides asparaginyl peptides in homogeneous form [221]. While the dehydration reaction might create serious problems, it is not "fatal". The nitrile in the side chain of β-cyanoalanine residues can be rehydrated with alkaline hydrogen peroxide [217] and is hydrated also during the final deprotection of peptides with hydrogen fluoride [219].

Dehydration of the carboxamide in the side chain of *glutamine* residues is less extensive [212, 220] than the analogous reaction in asparagine side chains and the remedies are similar. In glutamine, however, cyclizations to pyrrolidones and to glutarimides are quite pronounced. The ring closure which converts peptides with N-terminal glutamine to pyroglutamyl peptides was discussed as a side reaction initiated by proton abstraction. Here, we stress again formation of glutarimides in activated derivatives of N^{α}-acyl glutamine [98, 99]. Thus, in the preparation of the N-hydroxysuccinimide ester of tert. butyloxycarbonylglutamine with the help of carbodiimides cyclization interferes with the reaction and, even if the active ester is secured by carrying out the reaction at 0°, it is converted under the influence of tertiary amines (in dimethylformamide) to the N^{α}-acylglutarimide [99] and the same side reaction was observed also with o-nitrophenylsulfenylglutamine N-hydroxysuccinimide ester [222]. A general tendency toward glutarimide formation is indicated by the analogous cyclization of peptides in which a C-terminal glutamine residue is only moderately activated, as for instance in the azide [223].

The most important side reaction of *arginine* containing peptides is the formation of lactams (piperidones) in activated derivatives of arginine.

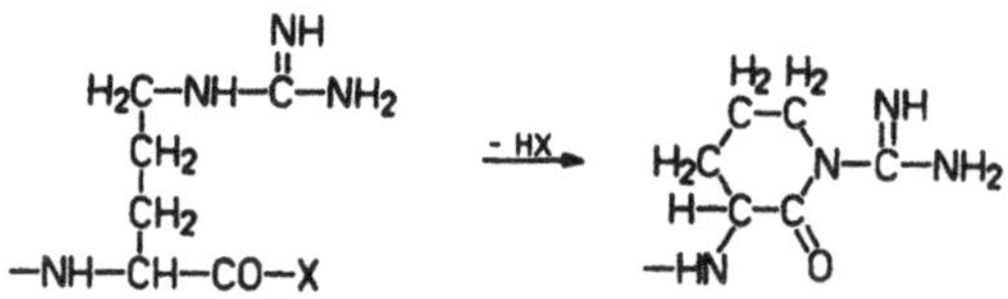

This often disturbing intramolecular acylation can be reduced by various protecting groups proposed for the blocking of the guanidino function, and discussed in Chapter IV, but for its complete prevention blocking of the guanidine, e.g. with two adamantyloxycarbonyl [224] or two isobornyloxycarbonyl [225] groups, is necessary. If this protection is lost during the unmasking of the α-amino function, this does not detract from its significance, because once the arginine residue is part of a peptide chain, its side chain is not acylated.

From the problems surrounding the numerous protecting groups applied for the blocking of the guanidine group the reduction in the stability of acylguanidines (when compared with resonance stabilized unacylated guanidines) must be reemphasized. For instance, while the arginine side chain is practically inert toward ammonia, nitroarginine containing peptides are ammonolyzed to yield both arginine and ornithine derivatives [226]:

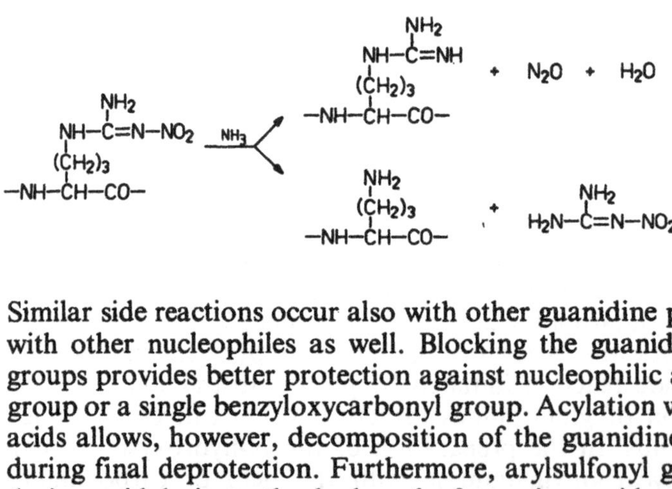

Similar side reactions occur also with other guanidine protecting groups and with other nucleophiles as well. Blocking the guanidine with arylsulfonyl groups provides better protection against nucleophilic attacks than the nitro group or a single benzyloxycarbonyl group. Acylation with aromatic sulfonic acids allows, however, decomposition of the guanidine in strong acids, e.g. during final deprotection. Furthermore, arylsulfonyl groups are transferred during acidolysis to the hydroxyl of tyrosine residues and, unfortunately, to some extent also to the aromatic nucleus of tyrosine.

In connection with *lysine* residues a general problem is posed by the partial loss of side chain protection during the unmasking of α-amino groups. This is the subject of a detailed discussion in the preceding chapter. Benzylation of the ε-amino group during the removal of the benzyloxycarbonyl group by acidolysis [130, 131] has been mentioned among acid catalyzed side reactions.

The *histidine* side chain can be the source of quite a few difficulties: imidazole is a weak base but strong enough to catalyze O-acylation of hydroxyamino acids or to cause racemization by intramolecular proton abstraction. Intramolecular nucleophilic attack by the imidazole nitrogen can result in lactam formation [168] and the histidine side chain can also catalyze the fission of peptide bonds [227]:

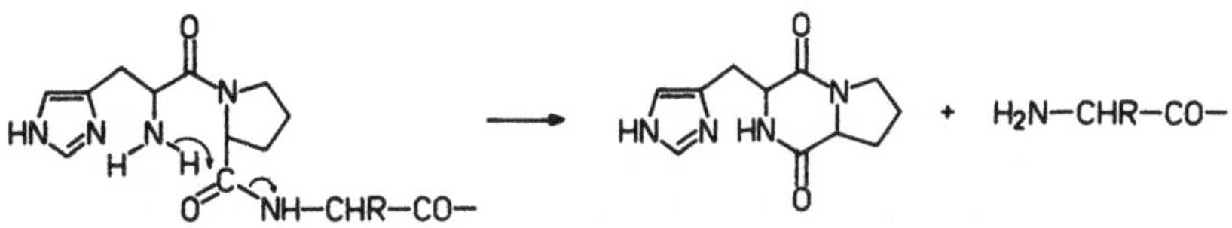

The reactivity of the imidazole nucleus leads to its alkylation by chloromethyl groups of the insoluble polymeric supports

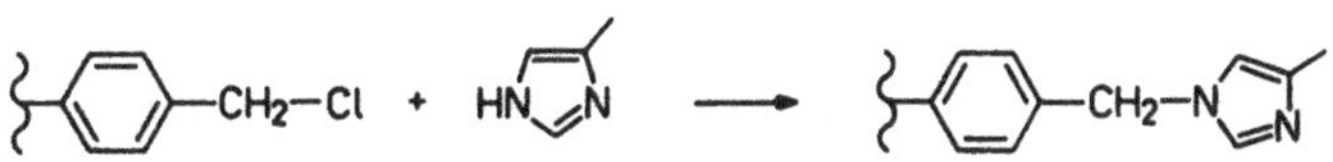

and also to an addition to carbodiimides [144]:

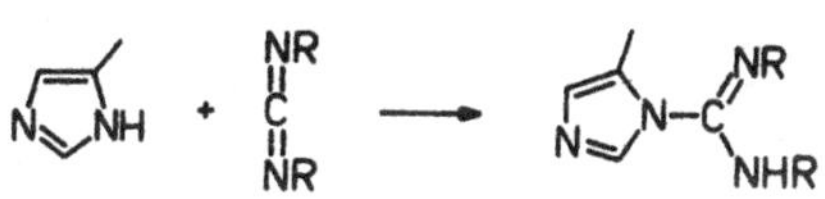

Masking the imidazole function remains a worthwhile objective, because some blocking groups (e.g. the classical benzyl protection) are not readily removed, while others are easily lost. The *p*-toluenesulfonyl group, for instance, is cleaved by acids and displaced [228] by nucleophiles such as 1-hydroxybenzotriazole. The problems surrounding histidine are further complicated by the lack of specificity in earlier literature about the position of substituents on the imidazole ring: generally no distinction was made between τ and π nitrogens:

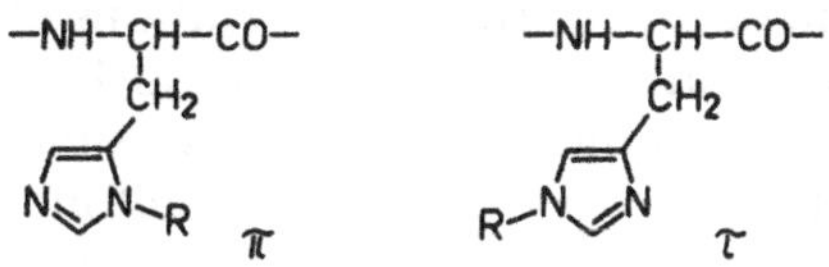

In order to avoid the mandatory protection of the sulfhydryl group, *cysteine* residues can be introduced in peptide chains in disulfide form [80]. Yet, cystine is also not exempt from side reactions. For instance, mixed disulfides suffer disproportionation to symmetrical disulfides [229]:

$$2\ R{-}S{-}S{-}R' \rightleftharpoons R{-}S{-}S{-}R + R'{-}S{-}S{-}R'$$

This disulfide interchange (or dismutation) can take place in strong acids [230] and also in neutral media [231] and is catalyzed by trace amounts of thiols.

Hence, it is more customary to apply S-alkyl or other side chain protected forms of cysteine. In this case, however, base catalyzed β-elimination appears as a complicating factor, especially in activated cysteine derivatives, and can be the cause of racemization as well:

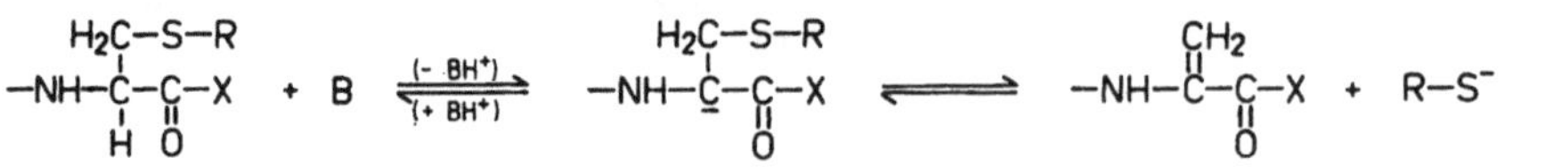

Cysteine both in free and in blocked form poisons platinum metal catalysts. Some methods proposed for the solution of this problem, e.g. catalytic reduction in liquid ammonia [232, 233] or the use of the 1,1-dimethyl-2-propynyloxycarbonyl group [234] for amine protection, have been mentioned in the preceding chapter in connection with the masking of the sulhydryl group. The enhanced tendency of S-benzylcysteine, activated as the azide, to form the amide as a by-product [235–237] can be suppressed [145] but still awaits explanation.

Similarly to cysteine the other sulfur containing amino acid, *methionine*, acts as a poison for palladium metal catalysts and this remains true for methionine containing peptides as well. It might be possible to find new hydrogenation catalysts which are not affected by thioethers, but the recently proposed [238] cobalt complex, $K_3[Co(CN)_5]$, still has to be examined with respect to its sensitivity to thioethers. The poisoning effect of methionine containing peptides in palladium catalyzed hydrogenolysis is reduced by the addition of boron trifluoride [239]. Also, removal of benzyloxycarbonyl groups can be carried out by catalytic hydrogenation in the presence of organic bases [240]; under the same conditions benzyl ethers are not cleaved [241]. Peptides which provide multiple ligands for palladium, e.g. compounds with more than one methionine residue, are not readily unmasked by hydrogenation even in the presence of base. Forced conditions, e.g. catalytic reduction for prolonged periods of time result in desulfurization and formation of α-aminobutyric acid residues [242]. Reduction with sodium in liquid ammonia remains a viable choice, but excess sodium demethylates the methionine side chain [243].

Oxidation of the thioether to a sulfoxide occurs during the operations of peptide synthesis or during purification, but can be prevented by working in an inert atmosphere. Fortunately, oxidation to the sulfoxide is reversible. A mild treatment with thiols will reduce a sulfoxide to the thioether. Sulfones cannot be reduced under mild conditions, but they also do not form from thioethers unless powerful oxidizing agents are used.

Alkylation of the sulfur atom in the methionine side chain readily occurs during the removal of blocking groups by acidolysis [111, 119]. Some alkylations are easily reversed; e.g. S-tert.butyl sulfonium salts decompose on standing or on warming with the regeneration of the thioether [121]. Alkylation by the benzyl group is a more serious side reaction because S-benzylmethionine (salts) give rise to a variety of products [244], among them S-benzylhomocysteine. Therefore, in reactions where alkylating agents are generated the thioether should be kept intact with the aid of scavengers. Alternatively the methionine side chain can be protected by oxidation to the sulfoxide [123] or by reversible alkylation with methyl *p*-toluenesulfonate [124]. Alkylation by chloromethyl groups of polymeric supports must be avoided.

References of Chapter V

1. Bodanszky, M., Martinez, J.: Synthesis, 333 (1981)
2. Bodanszky, M., Martinez, J.: in "The Peptides", Vol. V (Gross, E., Meienhofer, J., Eds.), p. 111, New York: Academic Press 1983
3. Bodanszky, M., Bednarek, M. A., Bodanszky, A.: Int. J. Peptide Protein Res. *20*, 387 (1982)
4. Kemp, D. S.: in "The Peptides", Vol I (Gross, E., Meienhofer, J., Eds.), p. 315, New York: Academic Press 1979
5. Iselin, B., Feurer, M., Schwyzer, R.: Helv. Chim. Acta *38*, 1508 (1955)
6. Iselin, B., Schwyzer, R.: Helv. Chim. Acta *43*, 1760 (1960)
7. Liberek, B.: Tetrahedron Lett., 1103 (1963); Liberek, B., Grzonka, Z.: ibid. 159 (1964)
8. Kovacs, J., Cortegiano, H., Cover, R. E., Mayers, G. L.: J. Amer. Chem. Soc. *93*, 1541 (1971)
9. Bergmann, M., Zervas, L.: Biochem. Z. *203*, 280 (1928)
10. Williams, M. W., Young, G. T.: J. Chem. Soc. *27*, 3409 (1962)
11. Neuberger, A.: Adv. Protein Chem. *4*, 344 (1948)
12. Cornforth, J. W.: in "The Chemistry of Penicillin" (Clarke, H., Johnson, J. R., Robinson, R., Eds.), p. 800, Princeton: Princeton Univ. Press 1949
13. Csonka, F. A., Nicolet, B. H.: J. Biol. Chem. *99*, 213 (1932)
14. Weygand, F., Prox, A., Schmidhammer, L., König, W.: Angew. Chem. *75*, 282 (1963)
15. Benoiton, N. L., Chen, F. M. F.: Canad. J. Chem. *59*, 384 (1981)
16. McDermott, J. R., Benoiton, N. L.: Canad. J. Chem. *51*, 2562 (1973)
17. Smart, N. A., Young, G. T., Williams, M. W.: J. Chem. Soc., 3902 (1963)
18. Williams, M. W., Young, G. T.: J. Chem. Soc., 881 (1963)
19. Anderson, G. W., Callahan, F. M.: J. Amer. Chem. Soc. *80*, 2902 (1958)
20. Clayton, D. W., Farrington, J. A., Kenner, G. W., Turner, J. M.: J. Chem. Soc., 1398 (1957)
21. Weygand, F., Prox, A., König, W.: Chem. Ber. *99*, 1451 (1966)
22. Weygand, F., Hoffmann, D., Prox, A.: Z. Naturforsch. *23b*, 279 (1968)
23. Taschner, E., Sokolowska, T., Biernat, J. F., Chimiak, A., Wasielewski, Cz., Rzeszotarska, B.: Liebigs Ann. Chem. *663*, 197 (1963)
24. Izdebski, J.: Roczniki Chemii *49*, 1097 (1975)
25. Bodanszky, M., Conklin, L. E.: Chem. Commun., 773 (1967)
26. Spackman, D. H., Stein, W. H., Moore, S.: Anal. Chem. *30*, 1190 (1958)
27. Izumiya, N., Muraoka, M.: J. Amer. Chem. Soc. *91*, 2391 (1969)
28. Izumiya, N., Muraoka, M., Aoyagi, H.: Bull. Chem. Soc. Jpn. *44*, 3391 (1971)
29. Benoiton, N. L., Kuroda, K., Cheung, S. T., Chen, F. M. F.: Canad. J. Biochem. *57*, 776 (1979)
30. Benoiton, N. L., Kuroda, K.: Int. J. Peptide Protein Res.: *17*, 197 (1981)
31. Halpern, B., Chew, L. F., Weinstein, B.: J. Amer. Chem. Soc. *89*, 5051 (1967)

32. Benoiton, N. L., Kuroda, K., Chen, M. F.: Int. J. Peptide Protein Res. *15*, 475 (1980)
33. Kitada, C., Fujino, M.: Chem. Pharm. Bull. *26*, 585 (1978)
34. Bosshard, H. R., Schechter, I., Berger, A.: Helv. Chim. Acta *56*, 717 (1973)
35. Kemp, D. S., Wang, S. W., Busby, G., Hugel, G.: J. Amer. Chem. Soc. *92*, 1043 (1970)
36. Kemp, D. S., Bernstein, Z., Rebek, J.: J. Amer. Chem. Soc. *92*, 4756 (1970)
37. Kemp, D. S., Rebek, J.: J. Amer. Chem. Soc. *92*, 5792 (1970)
38. Bodanszky, M., Birkhimer, C. A.: Chimia *14*, 368 (1960)
39. Bodanszky, M., Bodanszky, A.: Chem. Commun., 591 (1967)
40. Manning, J. M., Moore, S.: J. Biol. Chem. *243*, 5591 (1968)
41. Zuber, H.: Hoppe-Seyler's Z. Physiol. Chem. *349*, 1337 (1968)
42. Hill, R. L., Smith, E. L.: J. Biol. Chem. *228*, 577 (1957)
43. Hofmann, K., Woolner, M. E., Spühler, G., Schwartz, E. T.: J. Amer. Chem. Soc. *80*, 1486 (1958)
44. Sarid, S., Berger, A., Katchalski, E.: J. Biol. Chem. *234*, 1740 (1959); *237*, 2207 (1962)
45. Hill, R. L., Schmidt, W. R.: J. Biol. Chem. *237*, 389 (1962)
46. Gil-Av, E., Feibush, B., Charles-Sigler, R.: Tetrahedron Lett., 1009 (1966)
47. Bayer, E., Gil-Av, E., König, W. A., Nakaparskin, S., Oro, J., Parr, W.: J. Amer. Chem. Soc. *92*, 1738 (1970)
48. Takaya, T., Kishida, Y., Sakakibara, S.: J. Chromatography *215*, 279 (1981)
49. Anderson, G. W., Zimmermann, J. E., Callahan, F. M.: Ja. Amer. Chem. Soc. *88*, 1338 (1966)
50. Sieber, P., Riniker, B., Brugger, M., Kamber, B., Rittel, W.: Helv. Chim. Acta *53*, 2135 (1970)
51. Kisfaludy, L., Nyéki, O.: Acta Chim. Acad. Sci. Hung. *72*, 75 (1972)
52. Belleau, B., Malek, G.: J. Amer. Chem. Soc. *90*, 1651 (1968)
53. Woodward, R. B., Olofson, R. A.: J. Amer. Chem. Soc. *83*, 1007 (1961); Woodward, R. B., Olofson, R. A., Mayer, H.: *ibid. 83*, 1010 (1961)
54. Sheehan, J. C., Hess, G. P.: J. Amer. Chem. Soc. *77*, 1067 (1955)
55. Gais, H. F.: Angew. Chem. *17*, 597 (1978)
56. Neuenschwander, M., Lienhard, U., Fahrni, H. P., Hurni, B.: Helv. Chim. Acta *61*, 2428 (1978)
57. Neuenschwander, M., Fahrni, H. P., Lienhard, U.: Helv. Chim. Acta *61*, 2437 (1978)
58. Buyle, R., Viehe, G. H.: Angew. Chem. *76*, 572 (1964)
59. Sakakibara, S., Itoh, M.: Bull. Chem. Soc. Jpn. *40*, 656 (1967)
60. Williams, A. W., Young, G. T.: in "Peptides 1969" (Scoffone, E., Ed.), p. 52, North Holland Publ. Amsterdam (1971)
61. Atherton, E., Benoiton, N. L., Brown, E., Sheppard, R. C., Williams, B. J.: J. Chem. Soc. Chem. Commun., 336 (1981)
62. Steglich, W., Höfle, G.: Angew. Chem. Int. Ed. *8*, 981 (1969)
63. Wang, S. S., Kulesha, I. D.: J. Org. Chem. *40*, 1227 (1975); Wang, S. S.: *ibid. 40*, 1235 (1975)
64. Windridge, G. C., Jorgensen, E. C.: Intra Science Chem. Rep. *5*, 375 (1971)
65. Sakakibara, S., Fujii, T.: Bull. Chem. Soc. Jpn. *42*, 1466 (1969)
66. Terada, S., Kawabata, A., Mitsuyasu, N., Aoyagi, H., Izumiya, N.: Bull. Chem. Soc. Jpn. *51*, 3409 (1978)
67. Bergmann, M., Zervas, L.: Ber. dtsch. Chem. Ges. *65*, 1192 (1932)
68. Bodanszky, M., Ondetti, M. A.: in "Peptide Synthesis", p. 141, New York: Wiley-Interscience 1966

69. Sieber, P., Brugger, M., Rittel, W.: in "Peptides 1969" (Scoffone, E., Ed.), p. 60, Amsterdam: North Holland Publ. 1971
70. Benoiton, N. L., Kuroda, K., Chen, M. F.: in "Peptides 1978" (Siemion, I. Z., Kupryszewski, G., Eds.), p. 165, Wroclaw Univ. Press, Poland (1979)
71. Weygand, F., Steglich, W., Boracio de la Lama, X.: Tetrahedron, Suppl. 8 (1966)
72. Mihara, S., Takaya, T., Morikawa, T., Emura, J., Sakakibara, S.: in "Peptide Chemistry 1976" (Nakajima, T., Ed.), p. 36, Protein Res. Found., Osaka, Japan
73. König, W., Geiger, R.: Chem. Ber. *103*, 788 (1970)
74. Weygand, F., Hoffmann, D., Wünsch, E.: Z. Naturforsch. *21b*, 426 (1966)
75. Wünsch, E., Drees, F.: Chem. Ber. *99*, 110 (1966)
76. Itoh, M.: Bull. Chem. Soc. Jpn. *46*, 2219 (1973)
77. König, W., Geiger, R.: Chem. Ber. *103*, 2034 (1970)
78. Izdebski, J.: Polish J. Chem. *53*, 1049 (1979)
79. Merrifield, R. B.: J. Amer. Chem. Soc. *85*, 2149 (1963)
80. Lukenheimer, W., Zahn, H.: Liebigs Ann. Chem. *740*, 1 (1970)
81. Gisin, B. F., Merrifield, R. B.: J. Amer. Chem. Soc. *94*, 3102 (1972)
82. Khosla, M. C., Smeby, R. R., Bumpus, F. M.: J. Amer. Chem. Soc. *74*, 4721 (1972)
83. Rothe, M., Mazánek, J.: Liebigs Ann. Chem., 439 (1974)
84. Goodman, M., Steuben, K. C.: J. Amer. Chem. Soc. *84*, 1279 (1962)
85. Fruton, J. S., Bergmann, M.: J. Biol. Chem. *145*, 253 (1942)
86. Davis, N. C.: J. Biol. Chem. *223*, 935 (1956)
87. Goldschmidt, S., Wick, M.: Liebigs Ann. Chem. *575*, 217 (1952)
88. Wessely, F., Schlögl, K., Korger, G.: Nature *169*, 708 (1952)
89. Maclaren, J. A.: Austral. J. Chem. *11*, 360 (1958)
90. Bodanszky, M., Sheehan, J. T., Ondetti, M. A., Lande, S.: J. Amer. Chem. Soc. *85*, 991 (1963)
91. Sondheimer, E., Holley, R. W.: J. Amer. Chem. Soc. *76*, 2467 (1954)
92. Battersby, A. R., Robinson, J. C.: J. Chem. Soc., 259 (1955)
93. Bajusz, S., Lázár, T., Paulay, Z.: Acta Chim. Acad. Sci. Hung. *41*, 329 (1964)
94. Schwyzer, R., Iselin, B., Kappeler, H., Riniker, B., Rittel, W., Zuber, H.: Helv. Chim. Acta *46*, 1975 (1963)
95. Tam, J. P., Wong, T. W., Riemen, M. W., Tjoeng, F. S., Merrifield, R. B.: Tetrahedron Lett., 4033 (1979)
96. Ondetti, M. A., Deer, A., Sheehan, J. T., Pluscec, J., Kocy, O.: Biochemistry *7*, 4069 (1968)
97. Bodanszky, M., Kwei, J. Z.: Int. J. Peptide Protein Res. *12*, 69 (1978)
98. Zahn, H., Fölsche, E. T.: Chem. Ber. *102*, 2158 (1964)
99. Meyers, C., Havran, R. T., Schwartz, I. L., Walter, R.: Chem. Ind., 136 (1969)
100. Stedman, R. J.: J. Amer. Chem. Soc. *79*, 4691 (1957)
101. Zaoral, M., Rudinger, J.: Collect. Czechoslov. Chem. Commun. *24*, 1993 (1959)
102. Ramachandran, J., Li, C. H.: J. Org. Chem. *28*, 173 (1963)
103. Paul, R.: J. Org. Chem. *28*, 236 (1963)
104. Bodanszky, M., Fink, M. L., Klausner, Y. S., Natarajan, S., Tatemoto, K., Yiotakis, A. E., Bodanszky, A.: J. Org. Chem. *42*, 149 (1977)
105. Kurath, P., Thomas, A. M.: Helv. Chim. Acta *56*, 1656 (1973)
106. Stewart, F. H. C.: Austral. J. Chem. *21*, 477, 1639 (1968)
107. Bodanszky, M., Fagan, D. T.: Int. J. Pept. Protein Res. *10*, 375 (1977)

108. Martinez, J., Tolle, J. C., Bodanszky, M.: Int. J. Peptide Protein Res. *13*, 22 (1979)
109. Brenner, M., Curtius, H. C.: Helv. Chim. Acta *46*, 2126 (1963)
110. Sakakibara, S., Shimonishi, Y.: Bull. Chem. Soc. Jpn. *38*, 1412 (1965)
111. Weygand, R., Steglich, W.: Z. Naturforsch. *14b*, 472 (1959)
112. Taschner, E., Kupryszewski, G.: Bull. Acad. Pol. Sci. Ser. Chim. Geol. Geogr. *7*, 871 (1959)
113. Yang, C. C., Merrifield, R. B.: J. Org. Chem. *41*, 1032 (1976)
114. Bodanszky, M., Martinez, J.: J. Org. Chem. *43*, 3071 (1978)
115. Prestidge, R. L., Harding, D. R. K., Hancock, W. S.: J. Org. Chem. *41*, 2579 (1976)
116. Blake, J.: Int. J. Peptide Protein Res. *13*, 418 (1979)
117. Blombäck, B. E.: in "Methods in Enzymology", Vol. 11 (Hirs, C. H. W., Ed.), p. 398, New York: Academic Press 1967
118. Folkers, K., Chang, J. K., Curries, B. L.: Biochem. Biophys. Res. Commun. *39*, 110 (1970)
119. Lundt, B. F., Johansen, N. L., Vølund, A., Marcussen, J.: Int. J. Peptide Protein Res. *12*, 258 (1978)
120. Irie, H., Fujii, N., Ogawa, H., Yajima, H., Fujino, M., Shinagawa, S.: J. Chem. Soc. Chem. Commun., 922 (1976)
121. Noble, R. L., Yamashiro, D., Li, C. H.: J. Amer. Chem. Soc. *98*, 2324 (1976)
122. Brenner, M., Pfister, R. W.: Helv. Chim. Acta *34*, 2085 (1951)
123. Iselin, B.: Helv. Chim. Acta *44*, 61 (1961)
124. Bodanszky, M., Bednarek, M. A.: Int. J. Peptide Protein Res. *20*, 408 (1982)
125. Wünsch, E., Jaeger, E., Kisfaludy, L., Löw, M.: Angew. Chem. *89*, 330 (1977)
126. Löw, M., Kisfaludy, L., Sohár, P.: Hoppe-Seyler's Z. Physiol. Chem. *359*, 1643 (1978)
127. Masui, Y., Chino, N., Sakakibara, S.: Bull. Chem. Soc. Jpn. *53*, 464 (1980)
128. Löw, M., Kisfaludy, L., Jaeger, E., Thamm, P., Knof, S., Wünsch, E.: Hoppe-Seyler's Z. Physiol. Chem. *359*, 1637 (1978)
129. Bodanszky, M., Tolle, J. C., Bednarek, M. A., Schiller, P. W.: Int. J. Peptide Protein Res. *17*, 444 (1981)
130. Mitchell, A. R., Merrifield, R. B.: J. Org. Chem. *41*, 2015 (1976)
131. Fujii, N., Funakoshi, S., Sasaki, T., Yajima, H.: Chem. Pharm. Bull. *25*, 3096 (1977)
132. Shin, K. H., Sakakibara, S., Schneider, W., Hess, G. P.: Biochem. Biophys. Res. Commun. *8*, 288 (1962)
133. Sakakibara, S., Shin, K. H., Hess, G. P.: J. Amer. Chem. Soc. *84*, 4921 (1962)
134. Partridge, S. M., Davis, H. F.: Nature *165*, 62 (1950)
135. Piszkiewicz, D., Landon, M., Smith, E. L.: Biochem. Biophys. Res. Commun. *40*, 1173 (1970)
136. Brenner, M.: in "Peptides, Proc. 8th Europ. Peptide Symp." (Beyerman, H. C., van de Linde, A., Maassen van den Brink, W., eds.), p. 1, Amsterdam: North Holland Publ. 1967
137. Khorana, H. G.: Chem. Rev. *53*, 145 (1953); cf. also Smith, M., Moffatt, J. G., Khorana, H. G.: J. Amer. Chem. Soc. *80*, 6207 (1958)
138. Muramatsu, I., Hagitani, A.: J. Chem. Soc. Jpn. *80*, 1497 (1959)
139. Schnabel, E.: in Proc. Sixth Eur. Peptide Symp. Athens 1963 (Zervas, L., ed.), p. 71, Oxford: Pergamon Press 1965
140. Izdebski, J., Kubiak, T., Kunce, D., Drabarek, S.: Polish J. Chem. *52*, 539 (1978)
141. Izdebski, J., Kunce, D., Pelka, J., Drabarek, S.: Polish J. Chem. *54*, 117 (1980)

142. Izdebski, J., Kunce, D., Drabarek, S.: Polish J. Chem. *54*, 413 (1980)
143. Kisfaludy, L., Patthy, A., Löw, M.: Acta Chim. Acad. Sci. Hung. *59*, 159 (1969)
144. Rink, H., Riniker, B.: Helv. Chim. Acta *57*, 831 (1974)
145. Honzl, J., Rudinger, J.: Collect. Czech. Chem. Commun. *26*, 2333 (1961)
146. Schnabel, E.: Liebigs Ann. Chem. *659*, 168 (1962)
147. Bodanszky, M.: Nature *175*, 685 (1955)
148. Kisfaludy, L., Roberts, J. E., Johnson, R. H., Mayers, G. L., Kovács, J.: J. Org. Chem. *35*, 3563 (1970)
149. Girin, S. K., Shvachkin, Yu. P.: Z. Obschei Khimii *49*, 451 (1979)
150. Beaumont, S. M., Handford, B. O., Jones, J. H., Young, G. T.: Chem. Commun., 53 (1965)
151. Handford, B. O., Jones, J. H., Young, G. T., Johnson, T. F. N.: J. Chem. Soc., 6814 (1965)
152. Bodanszky, M., Tolle, J. C.: Int. J. Peptide Protein Res. *10*, 380 (1977)
153. Wieland, T., Heinke, B.: Liebigs Ann. Chem. *599*, 70 (1956)
154. Schellenberg, P., Ulrich, J.: Chem. Ber. *92*, 1276 (1959)
155. Kopple, K. D., Renick, R. J.: J. Org. Chem. *23*, 1565 (1958)
156. Zaoral, M., Rudinger, J.: Collect. Czech. Chem. Commun. *26*, 2316 (1961)
157. Kotake, H., Saito, T.: Bull. Chem. Soc. Jpn. *39*, 853 (1966)
158. Quitt, P., Hellerbach, J., Vogler, K.: Helv. Chim. Acta *46*, 327 (1963)
159. Weygand, F., Steglich, W., Bjarnason, J., Akhtar, R., Khan, N. M.: Tetrahedron Lett., 3483 (1966)
160. Stelakatos, G. C., Argyropoulos, N.: Chem. Commun., 271 (1966)
161. Weygand, F., Steglich, W., Bjarnason, J., Akhtar, R., Chytil, N.: Chem. Ber. *101*, 3623 (1968)
162. Merrifield, R. B.: Biochemistry *3*, 1385 (1964)
163. Savrda, J.: J. Org. Chem. *42*, 3199 (1977)
164. Hoffmann, K., Yajima, H.: J. Amer. Chem. Soc. *83*, 2289 (1961)
165. Marglin, A.: Int. J. Peptide Protein Res. *4*, 47 (1972)
166. Guttmann, S., Boissonnas, R. A.: Helv. Chim. Acta *42*, 1257 (1959)
167. Ben Ishai, D., Berger, A.: J. Org. Chem. *17*, 1564 (1952)
168. Sheehan, J. C., Hasspacher, K., Yeh, Y. L.: J. Amer. Chem. Soc. *81*, 6086 (1959)
169. Sheehan, J. C.: Ann. N.Y. Acad. Sci. *88*, 665 (1960)
170. Iselin, B.: Helv. Chim. Acta *45*, 1510 (1962)
171. Spanninger, P. A., von Rosenberg, J. L.: J. Amer. Chem. Soc. *94*, 1973 (1972)
172. Engelhard, M., Merrifield, R. B.: J. Amer. Chem. Soc. *100*, 3559 (1978)
173. Yajima, H., Takeyama, M., Kanaki, J., Mitani, K.: J. Chem. Soc. Chem. Commun., 482 (1978)
174. Nishimura, O., Fujino, M.: Chem. Pharm. Bull. *24*, 1568 (1976)
175. Schafer, D. J., Young, G. T., Elliott, D. F., Wade, R.: J. Chem. Soc. (C), 46 (1971)
176. Windrige, G. C., Jorgensen, E. C.: J. Amer. Chem. Soc. *93*, 6318 (1971)
177. Grassmann, W., Wünsch, E.: Chem. Ber. *91*, 462 (1958)
178. Goodman, M., Steuben, K. C.: J. Org. Chem. *24*, 112 (1959)
179. Grommers, E. P., Arens, J. F.: Rec. Trav. Chim. Pays-Bas *78*, 558 (1959)
180. Fontana, A., Toniolo, C.: in Progress in the Chemistry of Natural Products, Vol. 33 (Herz, W., Giesbach, H., Kirby, G. W., eds.), p. 309, New York: Springer 1976
181. Kessler, W., Iselin, B. M.: Helv. Chim. Acta *49*, 1330 (1966)

182. Sieber, P.: in "Peptides 1968" (Bricas, E., ed.), p. 236, Amsterdam: North Holland Publ. 1968
183. Alakhov, Yu. B., Kiryushkin, A. A., Lipkin, V. M., Milne, G. W. A.: J. Chem. Soc. Chem. Commun., 406 (1970)
184. Wünsch, E., Jaeger, E., Deffner, M., Scharf, R.: Hoppe-Seyler's Z. Physiol. Chem. *353*, 1716 (1972)
185. Previero, A., Colletti-Previero, M. A., Cavadore, J. C.: Biochim. Biophys. Acta *147*, 453 (1967)
186. Uphaus, R. A., Grossweiner, L. I., Katz, J. J., Kopple, K. D.: Science *129*, 641 (1959)
187. Previero, A., Prota, G., Coletti-Previero, M. A.: Biochim. Biophys. Acta *285*, 269 (1972)
188. Omori, Y., Matsuda, Y., Aimoto, S., Shimonishi, Y., Yamamoto, M.: Chem. Lett., 805 (1976)
189. Hashizume, K., Shimonishi, Y.: in "Peptide Chem. 1979" (Yonehara, H., ed.), p. 77, Protein Res. Foundation, Osaka, Japan (1980)
190. Loffet, A., Dremier, C.: Experientia *27*, 1003 (1971)
191. Anderson, J. C., Barton, M. A., Hardy, P. M., Kenner, G. W., McLeod, J. K., Preston, J., Sheppard, R. C.: Acta Chim. Acad. Sci. Hung. *44*, 187 (1965)
192. Löw, M., Kisfaludy, L.: Hoppe-Seyler's Z. Physiol. Chem. *359*, 1637 (1979)
193. Bajusz, S., Turán, A., Fauszt, I., Juhász, A.: in "Peptides 1972" (Hanson, H., Jakubke, H. D., eds.), p. 93, Amsterdam: North Holland Publ. 1973
194. Schön, I., Kisfaludy, L.: Int. J. Pept. Protein Res. *14*, 485 (1979)
195. Bodanszky, M., Sigler, G. F., Bodanszky, A.: J. Amer. Chem. Soc. *95*, 2352 (1973)
196. Fraser, K. J., Poulson, K., Haber, E.: Biochemistry *11*, 4974 (1972)
197. Battersby, A. R., Robinson, J. C.: J. Chem. Soc., 2076 (1956)
198. Clayton, D. W., Kenner, G. W.: Chem. Ind., 1205 (1953)
199. Bruckner, V., Kotai, A., Kovács, K.: Acta Chim. Acad. Sci. Hung. *21*, 427 (1959)
200. Shiba, T., Kaneko, T.: Bull. Chem. Soc. Jpn. *33*, 1721 (1960)
201. Sano, S., Kawanishi, S.: J. Amer. Chem. Soc. *97*, 3480 (1975)
202. Feinberg, R. S., Merrifield, R. B.: J. Amer. Chem. Soc. *97*, 3485 (1975)
203. Suzuki, K., Endo, N., Sasaki, Y.: Chem. Pharm. Bull. *25*, 2613 (1977)
204. Schwyzer, R., Iselin, B., Kappeler, H., Riniker, B., Rittel, W., Zuber, H.: Helv. Chim. Acta *41*, 1273 (1958)
205. Robinson, A. B.: Proc. Nat. Acad. Sci. U.S. *71*, 885 (1974)
206. Riniker, B., Schwyzer, R.: Helv. Chim. Acta *44*, 685 (1961)
207. Roeske, R.: J. Org. Chem. *28*, 1251 (1963)
208. König, W., Volk, A.: Chem. Ber. *110*, 1 (1977)
209. Sondheimer, E., Semeraro, R. J.: J. Org. Chem. *26*, 1847 (1961)
210. Riniker, B., Brunner, H., Schwyzer, R.: Angew. Chem. *74*, 469 (1962)
211. Boissonnas, R. A., Guttmann, S., Jaquenoud, P. A., Waller, J. P.: Helv. Chim. Acta *38*, 1491 (1955)
212. Gish, D. T., Katsoyannis, P. G., Hess, G. P., Stedman, R. J.: J. Amer. Chem. Soc. *78*, 5954 (1956)
213. Ressler, C.: J. Amer. Chem. Soc. *78*, 5956 (1956)
214. Stammer, J.: J. Org. Chem. *26*, 2556 (1961)
215. Paul, R., Kende, A. S.: J. Amer. Chem. Soc. *86*, 741 (1964)
216. Kashelikar, D. V., Ressler, C.: J. Amer. Chem. Soc. *86*, 2467 (1964)
217. Liberek, B.: Chem. Ind., 987 (1961)
218. Wilchek, R., Ariely, S., Patchornik, A.: J. Org. Chem. *33*, 1258 (1968)

219. Mojsov, S., Mitchell, A. R., Merrifield, R. B.: J. Org. Chem. *45*, 555 (1980)
220. Bodanszky, M., Denning, G. S., Jr., du Vigneaud, V.: Biochem. Prep. *10*, 122 (1963)
221. Bodanszky, M., du Vigneaud, V.: J. Amer. Chem. Soc. *81*, 5688 (1959)
222. Dewey, R. S., Barkemeyer, H., Hirschmann, R.: Chem. Ind., 1632 (1969)
223. Bodanszky, M., Yiotakis, A. E.: unpublished
224. Jäger, G., Geiger, R.: Chem. Ber. *103*, 1727 (1970)
225. Jäger, G., Geiger, R.: Justus Liebigs Ann. Chem., 1928 (1973)
226. Künzi, H., Manneberg, M., Studer, R. O.: Helv. Chim. Acta *57*, 566 (1974)
227. Mazur, R. H., Schlatter, J. M.: J. Org. Chem. *28*, 1025 (1963)
228. Fujii, T., Kimura, T., Sakakibara, S.: Bull. Chem. Soc. Jpn. *49*, 1595 (1976)
229. Sanger, F.: Nature *171*, 1025 (1953)
230. Benesch, R. E., Benesch, R.: J. Amer. Chem. Soc. *80*, 1066 (1958)
231. Ryle, P., Sanger, F.: Biochem. J. *60*, 535 (1955)
232. Meienhofer, J., Kuromizu, K.: Tetrahedron Lett., 3259 (1974)
233. Kuromizu, K., Meienhofer, J.: J. Amer. Chem. Soc. *95*, 4978 (1974)
234. Southard, G. L., Zaborowski, B. R., Pettee, J. M.: J. Amer. Chem. Soc. *93*, 3302 (1971)
235. Hegedüs, B.: Helv. Chim. Acta *31*, 737 (1948)
236. Holland, G. F., Cohen, L. A.: J. Amer. Chem. Soc. *80*, 3765 (1958)
237. Roeske, R., Stewart, F. H. C., Stedman, R. J., du Vigneaud, V.: J. Amer. Chem. Soc. *78*, 5883 (1956)
238. Losse, G., Stiehl, H. U.: Z. Chem. *21*, 188 (1981)
239. Yajima, H., Kawasaki, K., Kinomura, Y., Oshima, T., Kimoto, S., Okamoto, M.: Chem. Pharm. Bull. *16*, 1342 (1968)
240. Medzihradszky-Schweiger, H., Medzihradszky, K.: Acta Chim. Acad. Sci. Hung. *50*, 339 (1966)
241. Medzihradszky-Schweiger, H.,: Acta Chim. Acad. Sci. Hung. *76*, 437 (1973)
242. Dekker, C. A., Taylor, S. P., Jr., Fruton, J. S.: J. Biol. Chem. *180*, 155 (1949)
243. Stekol, J. A.: J. Biol. Chem. *140*, 827 (1941)
244. Dekker, C. A., Fruton, J. S.: J. Biol. Chem. *173*, 471 (1948)

VI. Tactics and Strategy in Peptide Synthesis

Over and above the problems of activation, coupling, protection and removal of protecting groups there are some more general aspects of peptide synthesis which, in order to be treated in a systematic manner, need to be identified and defined. Thus, schemes for the *combination* of various protecting groups have to be developed for syntheses in which certain blocking groups, e.g. those applied for the masking of the α-amino function, must be removed after coupling, while others are expected to stay intact throughout the chain building process. Considerations which govern the selection of protecting groups and coupling methods can be designated as *tactics.* A separate, although not independent, part of the plan of a synthesis of a large peptide is the general design of the synthetic scheme. Decisions such as the construction of a long chain from larger segments or from single residues, form the *strategy* of synthesis. A further category, the *techniques* of peptide synthesis, encompasses methods of *facilitation,* the choice between synthesis carried out in solution and chain building in which the peptide is anchored to an insoluble support. Experimental devices which simplify the isolation of intermediates or dispense with their isolation will also be discussed among the varieties of techniques proposed for peptide synthesis. It seems to us, that for the sake of exact communication between peptide chemists, these concepts [1–3] have to be clearly distinguished and consistently described with appropriate terms.

A. Tactics

1. Combinations of Protecting Groups

In the synthesis of dipeptides the problem of protecting group combinations does not arise. For instance a peptide bond can be formed between a benzyloxycarbonylamino acid and an amino acid benzyl ester and both protecting groups can be removed from the resulting dipeptide derivative in a single operation:

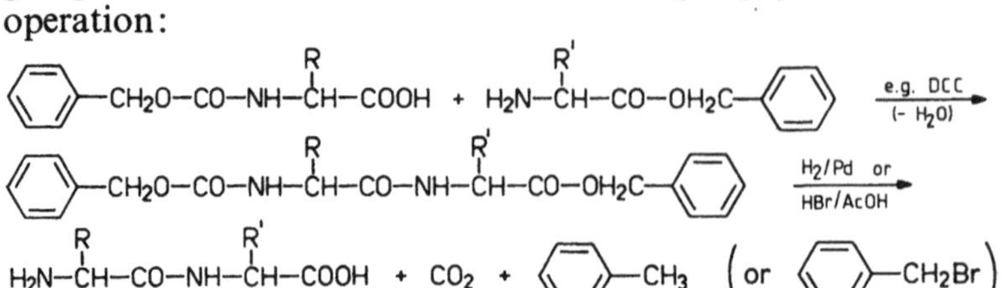

Should the amino component or the carboxyl component have a functional group in the side chain, this can be similarly masked, e.g. in the form of benzyl ester or benzyl ether. The task of chosing protecting groups is, however, radically different in the preparation of tripeptides or longer peptide chains. Such syntheses require at least two kinds of protecting groups. One of these is used for the blocking of the α-amino function and should be readily removable to allow its acylation in the following couplings step. The protecting groups on the C-terminal carboxyl and the side chain functions should remain intact during these operations. Thus, we have to differentiate between the *transient protection* of α-amino groups and the *semipermanent protection* of all other functions. The semipermanent protecting groups are removed after completion of the chain-lengthening process but their removal should require only reagents and conditions which do not endanger the chemical or chiral integrity of the product. For illustration we point to a synthesis of oxytocin [4] in which the α-amino groups were blocked by the benzyloxycarbonyl group and the sulfhydryl function in the side chain of the two cysteine residues by the benzyl group. Removal of the benzyloxycarbonyl protection after each coupling step did not affect the S-benzyl groups since these are inert toward hydrobromic acid in acetic acid, the reagent used for unmasking the α-amino groups. Therefore, the S-benzyl groups can perform their function throughout the synthesis. They are cleaved, at the conclusion of chain building, by reduction with sodium in liquid ammonia. Since the benzyloxycarbonyl group is removed also by reductive methods, the protected nonapeptide derivative

```
  Bzl                 Bzl
   |                   |
Z—Cys—Tyr—Ile—Gln—Asn—Cys—Pro—Leu—Gly—NH2
```

was unmasked in a single operation[12]. This is an early and simple example of a principle in the tactical selection of protecting groups. In such schemes, later designated as *orthogonal protection* [5], the transient protecting groups used for the blocking of α-amino groups and the semipermanent blocking groups selected for the remaining functions are cleaved in completely different reactions, by different reagents. An alternative approach can be exemplified by syntheses [6] in which all protecting groups are cleaved by acidolysis and selectivity is achieved by the use of acids of widely different strength. Such a combination of acid sensitive protecting groups was used in a synthesis of insulin [7]: triphenylmethyl (trityl) groups were removed under very slightly acidic conditions, biphenylisopropyloxycarbonyl) (Bpoc) groups with hydrochloric acid at pH2 (as measured on a glass electrode) in 90% trifluoroethanol and tert. butyloxycarbonyl (Boc) groups with trifluoroacetic

[12] Protection of N-terminal amino groups deserves special consideration. Since they need not be unmasked selectively, their protection can be similar to that of side chain functions. For instance, in oxytocin blocking of the terminal amine by the *p*-toluenesulfonyl group is equally satisfactory.

acid. The selectivity afforded by these conditions was sufficient for the preparation of homogeneous intermediates. It is noteworthy in this method, that the combination of protecting groups is based on an acid sensitive group (Boc) and others (Bpoc and Trt) which are even more sensitive to acids. The often applied alternative, in which acid resistant groups are used in combination with the Boc group and the side chain functions are unmasked at the end of the synthesis with very strong acids, such as hydrogen fluoride or trifluoromethanesulfonic acid, has inherent problems: under extremely acidic conditions many side reactions occur and their prevention requires numerous countermeasures. Nevertheless, the obviously more attractive combination of acid sensitive protecting groups also has certain limitations. During the cleavage of an acid sensitive group such as the Bpoc group some loss in the more resistant Boc blocking must also occur. Such losses might be acceptable in the synthesis of shorter chains but not in a long series of similar operations. The cumulative effects of minor imperfections should cause serious difficulties in the isolation and purification of the final product. In this respect the orthogonal principle is more auspicious.

It is tempting to present useful orthogonal combinations. e.g. in the form of a table, but on reexamination of a table composed some years ago (p. 170 in ref. 1) we find that new methods introduced for the removal of "old" protecting groups render such tables soon obsolete. Therefore, only a few, frequently used, combinations will be mentioned here. An often applied and usually successful approach is the blocking of α-amines with groups removable by catalytic reduction and the application of acid sensitive groups for all other functions. Such a combination was realized in the first synthesis [8] of porcine corticotropin in which the α-amines were blocked with the benzyloxycarbonyl group, while the side chain amino functions were protected with the tert. butyloxycarbonyl group and the carboxyls masked in the form of tert. butyl esters. The same methods of deprotection, acidolysis and catalytic reduction, were applied but in an opposite manner, in the synthesis of porcine secretin [9, 10]: acid sensitive groups were used for the protection of α-amines, and benzyl groups for the side chain functions. In the final deprotection, by hydrogenolysis, nitro groups, attached to the guanidino groups of arginine residues were also removed.

In recent years the 9-fluorenylmethyloxycarbonyl (Fmoc) group [11] considerably enriched the possibilities of orthogonal combinations, because it is quite resistant to acids but is readily cleaved by secondary amines, such as piperidine, under conditions which do not affect most other protecting groups. Thus, it became practical to use the Fmoc group for α-amine protection and to rely on acid sensitive groups for the masking of side chains. Such combinations are particularly useful in solid phase peptide synthesis, because they permit an acid sensitive anchoring of the peptide to the insoluble polymeric support and allow facile cleavage of the completed peptide from the resin concomitantly with the complete unmasking of the peptide. By reserving acidolysis for the final deprotection, formation of alkylating agents at each deprotection step is also circumvented. In syntheses carried out in solution, blocking the terminal amino group of segments [12] by the Fmoc group offers the distinct advantage that on deblocking the peptide appears as the free amine rather than a salt and thus is available in its full amount as amino component in the following condensation. In stepwise chain building, however, a certain limitation is caused by the Fmoc group. If, namely, some of the protected intermediates are poorly soluble in neutral solvents, such as dimethylformamide, then both the coupling reactions and removal of the transient blocking (Fmoc) group have to be performed in suspensions or in gels rather than in solution, and it can be difficult to bring these reactions to completion [13]. An important advantage of deprotection with trifluoroacetic acid is that this reagent, in addition to being a suitable acid, is, at the same time, a uniquely general solvent of both free and protected peptides. (The similarly powerful solvent, hexafluoroisopropanol, is too dangerous for everyday use.) The reverse combination, protection of α-amino functions by the tert. butyloxycarbonyl group and those in the side chains by groups derived from the 9-fluorenylmethyl group, remains an interesting possibility [14].

Further combinations can be designed when new protecting groups are discovered, particularly if these require specific reagents and only mild conditions for their removal. A good example for such developments is the introduction of the fluoride sensitive trimethylsilylethyloxycarbonyl group [15] in peptide synthesis and the subsequent adaptation of the same chemistry for the blocking of carboxyl functions in the form of trimethylsilylethyl esters [16]. Novel methods of removal of well-established protective groups, like cleavage of the benzyloxycarbonyl group, benzyl esters and benzyl ethers with trifluoroacetic acid in the presence of thioanisole [17], similarly broaden the range of tactical choices.

2. Final Deprotection

Removal of the remaining blocking groups from a completed peptide chain is a delicate and sometimes disappointing operation. Not infrequently the seemingly homogeneous protected material gives rise to a mixture instead of a single free peptide. This can be due to the properties of the intermediates in which the functional groups are masked. Because of the limited solubility of such materials their examination by chromatography is less incisive than the scrutiny of the deprotected product. In addition to chromatography the latter can be studied also by other analytical methods such as electrophoresis, sequencing by Edman degradation or via mass spectrometry and last but not least through hydrolysis catalyzed by specific proteolytic enzymes. Therefore, impurities hidden in the blocked material appear loud and clear after deprotection. Furthermore, the reagents and conditions used in the final deprotection are often drastic enough to generate new impurities. Hence, the protecting groups for the functional groups in the amino acid side chains and for the C-terminal carboxyl must be selected with care and forethought, bearing in mind the risks to be encountered in their removal. Protection of carboxyl groups in the form of alkyl esters might serve as an example. Methyl or ethyl esters are ideal in several respects. They are stable toward acids and hydrogenation and can be kept, therefore, intact through numerous steps in which α-amino protecting groups are cleaved by reduction or by acidolysis. Yet, the practice of cleaving alkyl esters by *saponification* with aqueous alkali should be regarded with suspicion. The effect of strong bases on peptide derivatives is far from harmless. Thus, in addition to racemization of the C-terminal residue [18] partial hydrolysis of side chain carboxamides, β-elimination in substituted cysteine and serine side chains, formation of hydantoins in certain sequences, transpeptidation (via imide formation) in aspartyl and glutamyl peptides, etc. might occur. A series of alternative methods for the cleavage of methyl esters (cf. Chapter III) remains to be tested in praxis, but until some process other than alkaline hydrolysis becomes well established, carboxyl protection in the form of methyl or ethyl esters remains of questionable value. A notable exception is the application of methyl esters for the blocking of C-terminal carboxyls of peptides which serve subsequently as carboxyl components in the condensation of segments. These esters are not hydrolyzed, but hydrazinolyzed and converted to azides: a sophisticated and reliable approach, which has often been applied with satisfactory results.

Acidolysis with *strong acids* can be regarded as counterpart to base catalyzed hydrolysis and it is similarly risky. Quite a few of the side reactions discussed in Chapter V are caused by excessive protonation or by the alkylating agents generated in acidolysis. Therefore such powerful reagents as hydrogen fluoride or trifluoromethanesulfonic acid must be used with considerable care. They are attractive tools which can cleave most of the commonly used protecting groups and also the bond between a completed chain and its polymeric support, but their efficiency is not a pure blessing and the convenience in final deprotection is often paid for by the necessity of extensive purification of the deblocked material. At least some of the side reactions (e.g., acid catalyzed N → O shift) can be avoided if *moderately strong acids*, such as trifluoroacetic acid, are applied for final deprotection. This was the case in the synthesis of the biologically active N-terminal 24-peptide of porcine corticotropin [8] where the last protected intermediate carried only acid sensitive masking groups:

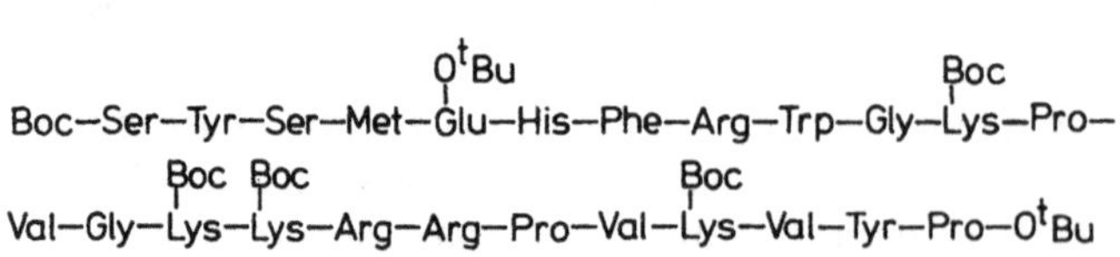

which could be cleaved in a single operation.

Reduction with sodium in liquid ammonia was used for the simultaneous removal of all masking groups in the synthesis of arginine vasopressin [19]. The protected nonapeptide intermediate

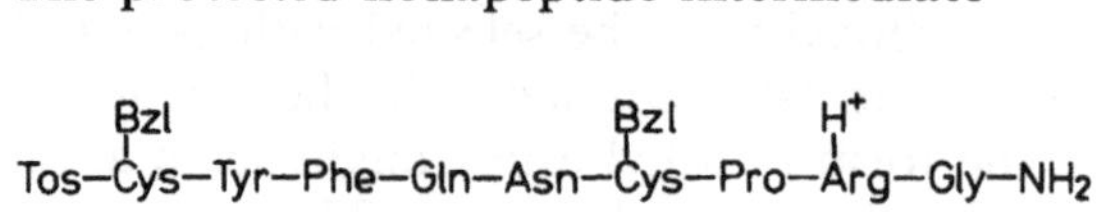

yielded the reduced form of the hormone which was oxidized by air to the active material. Of course, even this elegant method is not free from possible complications. Excess sodium, in the presence of proton donors, splits the peptide bond between proline and the preceding residue, cysteine. With proper care, however, reasonably pure peptides can be obtained by this method.

Catalytic hydrogenation was applied for the complete unmasking of the blocked 27-peptide intermediate

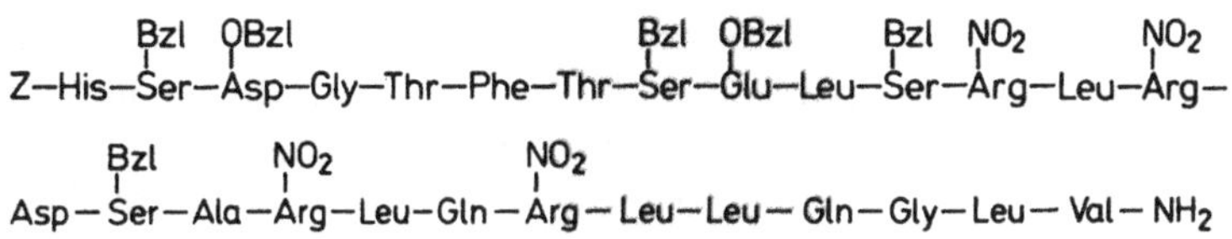

in the first synthesis [10] of porcine secretin. In spite of the relatively innocuous process, the prolonged periods of hydrogenation needed for the complete reduction of the nitro groups caused saturation of the aromatic ring in the phenylalanine side chain, albeit only to a slight extent.

Although we could point to possible side reactions in connection with most methods applied for final deprotection, these illustrate a trend toward processes which do not give rise to byproducts. Further enrichment of the armament of the peptide chemist can be expected from blocking groups which allow final deprotection by specific reagents, under mild conditions. For instance, simultaneous cleavage by secondary amines of the 9-fluorenylmethyloxycarbonyl (Fmoc) [11] group, the O- and S-9-fluorenylmethyl (Fm) groups [20] and 9-fluorenylmethyl esters [21] might prove to be superior to final deblocking by acidolysis. Deprotection with fluoride ions could turn out more than just imaginative and may lend importance to the trimethylsilylethyloxycarbonyl group [15] and trimethylsilylethyl esters [16]. Deblocking with the aid of specific enzymes (cf., e.g. ref. [22]) might be the ultimate goal in the search for methods for final deprotection.

Simultaneous removal of all protecting groups is an obviously attractive idea, but unmasking the final product in two or more steps might have certain advantages. Thus, an amine protecting group which can be selectively cleaved allows the eventual continuation of the synthesis, if this would turn out to be desirable and similar flexibility is provided by selectively removable blocking of the C-terminal carboxyl. Also, a partially deprotected peptide could be purified by methods which are not applicable for the completely blocked intermediate nor for the free peptide. Finally, in the tactical planning, of a synthesis, schemes which include certain flexibility and thus allow an alternative approach in case of unforeseen difficulties, are particularly valuable.

B. Strategies

1. Segment Condensation

For a considerable period of time it seemed obvious to build peptide chains through the condensation of segments[13] of the target compound. A memorable example of this strategy is the first synthesis of oxytocin by du Vigneaud and his associates [24]. The C-terminal tetrapeptide portion of the molecule was combined with the central tripeptide (both used in partially protected form) and the resulting heptapeptide derivative further lengthened after appropriate partial deblocking, by condensation with the properly masked N-terminal dipeptide, to yield a nonapeptide derivative with the complete sequence of the hormone:

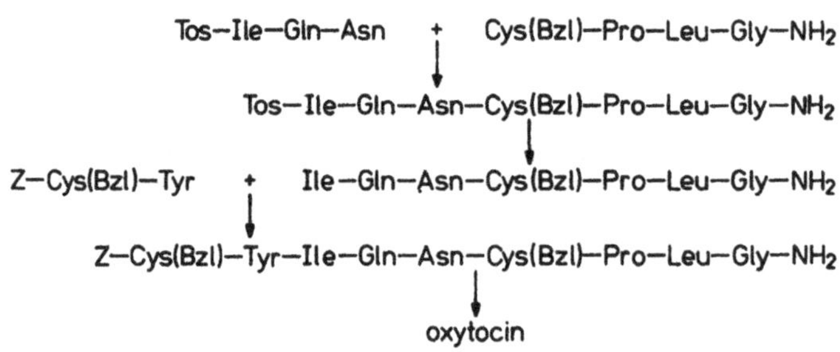

This scheme, which can be expressed, in a shorthand version, as a $3 + 4 = 7$; $2 + 7 = 9$ segment condensation, had several favorable features. It required relatively few operations of protection and deprotection, permitted the isolation and purification of the segments, which were peptides of moderate size, prior to their condensation to more complex molecules and last, but not least, allowed the distribution of the task between smaller teams of a larger research group. Because of such obvious advantages other early syntheses of biologically active peptides, e.g. angiotensin [25, 26] α-melanotropin [27] and corticotropin [28] were based on the strategy of segment condensation. In order to illustrate the strategy of segment condensation we sketch in scheme 2 a synthesis of an analog of angiotensin [25], in a self-explanatory manner:

[13] Instead of the earlier used term "fragment condensation" the more exact expression "segment condensation" [23] is adopted in this volume.

Scheme 2. A Synthesis of an Analog of Angiotensin [25] by Segment Condensation

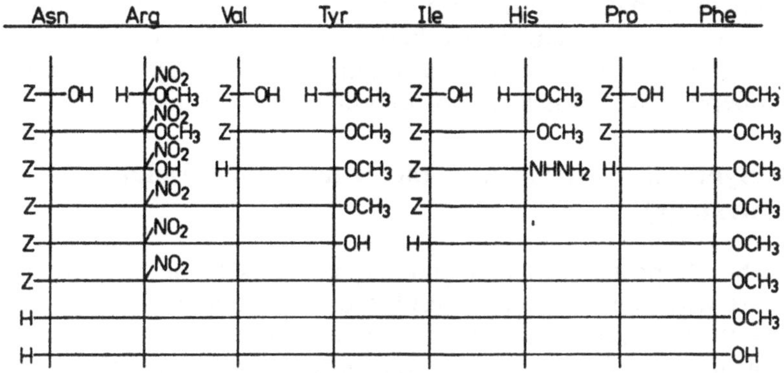

Condensation of large segments can create major difficulties, if for no other reason, because of the low molar concentration of the components to be coupled. The use of one component in excess can enhance the rate of coupling and lead to improved yields. In an effort toward the synthesis of ribonuclease A the last coupling reaction [29] involved a 44-peptide carboxyl-component and an amino-component of 60 residues. The azide of the protected 44-peptide was applied in about four-fold excess but even in this way the 104-peptide (the "S-protein") formed only in low yield and did not lend itself to isolation. The fact that the components carried only the mandatory protecting groups and were, therefore, reasonably soluble in the solvent, was in itself not sufficient for satisfactory results. Even more troublesome situations can arise when large segments are completely blocked: the ensuing poor solubility of the reaction components [30] might completely prevent the desired coupling. In a later, more successful synthesis of ribonuclease A [31] small segments were activated (in the form of azides) and incorporated into the growing peptide chain. The activated, low molecular weight intermediates were used in considerable excess and this excess increased with the increase in the molecular weight of the amino components. Thus, the *principle of excess*, initially proposed for stepwise chain-lengthening with single amino acid residues [32] could be successfully applied for the construction of the molecule of a protein from relatively small peptide segments.

A general disadvantage of the segment condensation strategy is the possible racemization of the activated C-terminal residue of the carboxyl component. A major advantage is offered, however, by the significant difference in the properties of the starting materials and the expected product. Such differences should alleviate problems in the isolation and purification of the desired material.

An interesting subclass of the segment condensation approach can be envisaged in the successive addition of dipeptide units in the form of their acetone derivatives [33]:

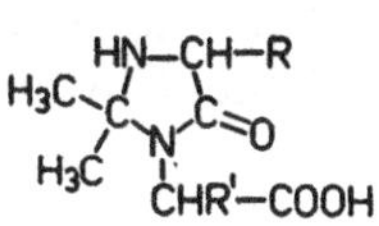

The highly original four center condensation (4CC) method of Ugi [34] is a special case of segment condensation. It has been discussed already in Chapter II, but we should reemphasize here that with selected chiral components this method can produce optically pure peptides [35]. Its practical application in peptide synthesis has also been demonstrated [36].

2. Stepwise Synthesis Starting with the N-terminal Residue (N → C Strategy)

Since Nature builds proteins in this manner, it seems to be an attractive thought to start with the N-terminal residue of a peptide and to continue its synthesis by the incorporation of single amino acid residues. This strategy requires the protection of only one α-amino group, that of the N-terminal residue and if the amino acids to be added can be applied without a blocking group at their carboxyl, then only the side chain functions have to be masked:

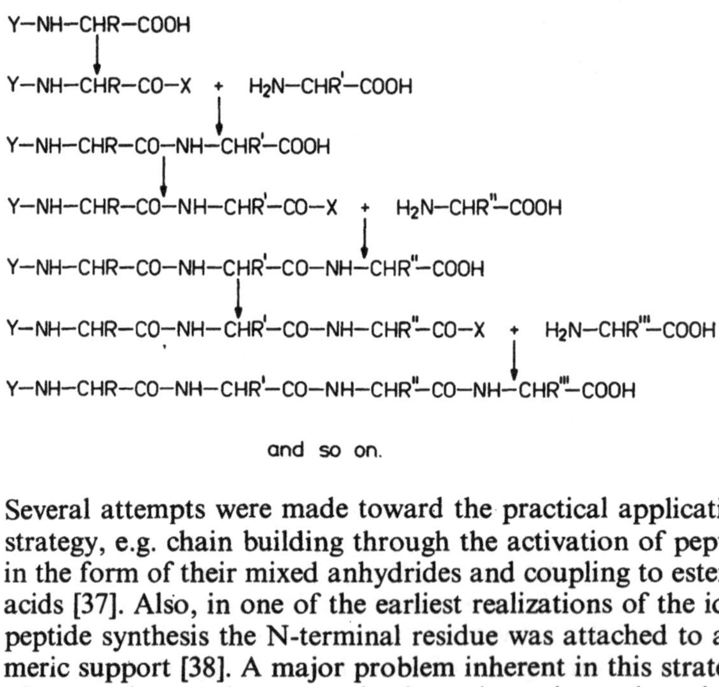

Several attempts were made toward the practical application of the N → C strategy, e.g. chain building through the activation of peptide intermediates in the form of their mixed anhydrides and coupling to esters of single amino acids [37]. Also, in one of the earliest realizations of the idea of solid phase peptide synthesis the N-terminal residue was attached to an insoluble polymeric support [38]. A major problem inherent in this strategy is the absence of protection against racemization, since the activated residues are not provided with a urethane-type amine protecting group. The use of the azide method, which is least conducive to racemization, for coupling [39], could not solve this dilemma, because other side reaction, such as Curtius rearrangement, interfere with the process. The extent of this rearrangement can be underestimated if the products are exposed to acidolytic conditions prior to their examination: strong acids cause decomposition [40] of the urea derivatives generated through the isocyanate byproducts of the Curtius method.

An interesting suggestion [22] for chain building is to start with the carboxymethyl derivative of polyethyleneglycol (CM-PEG) and to attach to it, with the help of water soluble carbodiimides, first glycine, then methionine, both added as ethyl esters and deprotected after coupling with the immobilized form of the proteolytic enzyme, carboxypeptidase Y. The peptide chain is lengthened toward its C-terminus in the same manner:

$$HO{-}CH_2{-}CH_2{-}O{-}(CH_2{-}CH_2{-}O{-})_n CH_2{-}CH_2{-}O{-}CH_2COOH \quad = \quad CM{-}PEG$$

$$CM{-}PEG{-}Gly{-}OEt \xrightarrow{\text{carboxypeptidase Y}} CM{-}PEG{-}Gly{-}OH$$

$$CM{-}PEG{-}Gly{-}OH \;+\; Met{-}OEt \xrightarrow{\text{carbodiimide}} CM{-}PEG{-}Gly{-}Met{-}OEt$$

$$CM{-}PEG{-}Gly{-}OMet{-}OEt \xrightarrow{\text{carboxypeptidase Y}} CM{-}PEG{-}Gly{-}Met{-}OH$$

$$CM{-}PEG{-}Gly{-}Met{-}OH \;+\; H_2N{-}CHR{-}CO{-}OEt \xrightarrow{\text{carbodiimide}} CM{-}PEG{-}Gly{-}Met{-}NH{-}CHR{-}CO{-}OEt$$

and so on.

Once chain lengthening is completed, the peptide bond between the methionine residue and the next amino acid is selectively cleaved with cyanogen bromide. A critical evaluation of the new approach will be possible only after it has been applied in the synthesis of complex peptides.

The N → C strategy can be used if some racemization, which occurs during most of the coupling reactions, is accepted [41] in the expectation of subsequent separation of the diastereoisomers. So far no general application of such a process can be found in the literature. There is, however, a somewhat neglected version of the N → C approach: one of the two insertion methods proposed by Brenner and his associates [42, 43]. In the aminodiacylhydrazine insertion procedure [43] the chain is built by the addition of single residues in the N → C direction, e.g.

$$Z{-}Gly{-}Gly{-}NH{-}NH_2 \longrightarrow Z{-}Gly{-}Gly{-}NH{-}NH{-}Phe \longrightarrow Z{-}Gly{-}Gly{-}Phe{-}NH{-}NH_2 \quad \text{etc.}$$

This intriguing process might lend itself to further development and could reach practicality.

3. Stepwise Synthesis Starting with the C-terminal Residue (C → N Strategy)

In contrast to the economy associated with the N → C strategy, the opposite direction of chain building, the approach which starts with the C-terminal residue is quite demanding. Thus, if an octapeptide consisting of residues A, B, C, D, E, F, G, H is constructed in that manner

```
              H
            G H
          F G H
        E F G H
      D E F G H
    C D E F G H
  B C D E F G H
A B C D E F G H
```

then after having provided the amino acid H with a semipermanent blocking group (Y), residue G has to be incorporated in protected and activated form (Y′—G—X) and after the coupling reaction the N-protecting group must be removed from the dipeptide intermediate (Y′—G—H—X) in order to allow the incorporation of residue F. The latter, however, has to be protected and activated prior to its introduction into the peptide. Blocking the α-amino function of each residue before, and partial unmasking of the coupling product after each chain lengthening step requires a considerable number of synthetic operations. Yet, one of the principal advantages of the C → N strategy can be found in the same circumstance. Incorporation of single amino acids provided with urethane-type amine protection does not involve significant racemization[14]. From this point of view the C → N strategy is unsurpassed. Also, the C → N approach readily lends itself to the application of the "principle of excess" [32], since the excess acylating agent, an amino acid derivative, is usually quite different with respect to solubility from the product, a protected peptide and, therefore, their separation creates no problems. In the case of longer chains simple washing with judiciously selected solvents, such as ethyl acetate or ethanol, removes the excess acylating agent together with the byproduct formed from the leaving group and the salts of the tertiary amine generated in the coupling reaction: the protected peptide is readily obtained in pure form.

In the first application of the C → N strategy [4], the stepwise synthesis of oxytocin (Scheme 3) *active esters* were used, because they were considered the acylating agents of choice. Since mixed anhydrides, at that time the most popular reactive intermediates, give two acylation products, they were thought to be unsuited for this approach. In later years several laboratories [45—47] chose *mixed anhydrides* for stepwise chain lengthening. The identification of permanently blocked intermediates among the products of the process [47] indicates that the earlier concerns were not unwarranted. Better results should be expected from the application of *symmetrical anhydrides* in the C → N strategy [48], because, at least in principle, only a single acylation product can form from them.

---

[14] Notable exceptions are the carboxyl-activated derivatives of O-alkyl serine and S-benzyl-cysteine even if they are protected with urethane-type amine blocking groups. In these cases, however, base catalyzed racemization can be suppressed by the use of hindered tertiary amines in the coupling reaction [44].

Scheme 3. Stepwise Synthesis of Oxytocin [4]

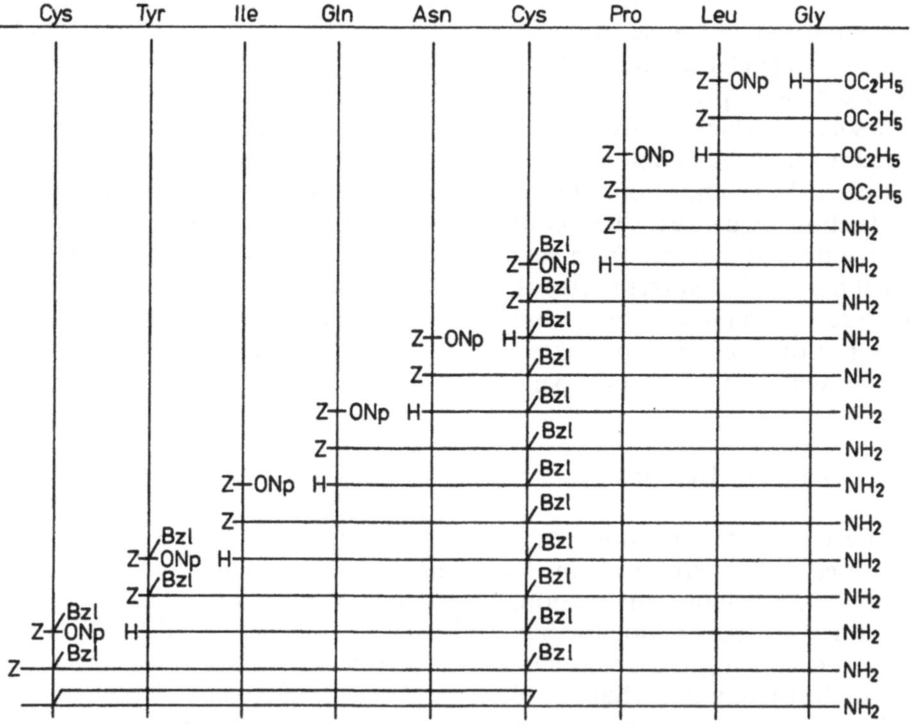

In addition to mixed and symmetrical anhydrides and various active esters [49, 50], N-carboxyanhydrides [51] and their sulfur analogs (2,5-thiazolidinediones) [52] could also be applied for stepwise chain building in C → N direction. Also, in most instances, the synthesis of peptides anchored to a polymeric support followed the same strategy. Chain lengthening was usually accomplished with the aid of carbodiimides as condensing agents and could be carried through many steps, leading to the incorporation of well over 100 residues, for instance in the synthesis of ribonuclease A [53].

A less obvious implementation of the C → N approach is the amino acid insertion method via O-acyl-salicylamides [42]:

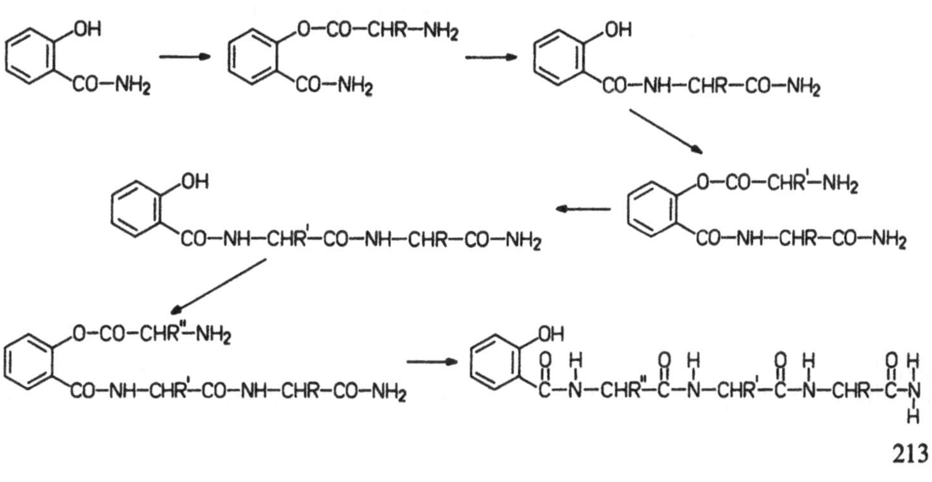

Thus, the C → N strategy can be implemented with various methods of coupling and, of course, also with a considerable choice of protecting groups. It does not lend itself so readily to team-work as the condensations of segments and can cause problems if the protected intermediates are poorly soluble in the solvents which are useful in peptide synthesis. Also, the C → N approach requires that the coupling reactions and the unmasking of α-amino groups be carried to completion, otherwise peptides from which a residue is missing ("deletion sequences") will contaminate the final product. The separation of the desired material from such rather similar byproducts can be an overwhelming task. Nevertheless, the repetitiveness of the operations renders the process conductive to mechanization and automation [54] and the virtual absence of racemization makes the C → N strategy an attractive approach in the synthesis of complex peptides.

C. Disulfide Bridges

The architecture of peptides and proteins is determined in part by the structural features of the peptide chain. Next neighbor interactions, hydrogen bonds, ionic and non-polar interactions define the geometry of the molecule. Yet, disulfide bridges provide further rigidity to a chain and can also link two (or more) chains together. Intrachain disulfides will be discussed in the section on cyclization in this Chapter. Here we deal with interchain disulfide bridges which can produce symmetrical and asymmetrical cystinyl peptides.

A peptide chain with a single cysteine residue can be readily converted to the corresponding *symmetrical disulfide*. Oxidation by air is often sufficient for this purpose. In an early example [55] an aqueous solution of the barium salt of L-tyrosyl-L-cysteine was oxidized by a stream of air to the cystinyl

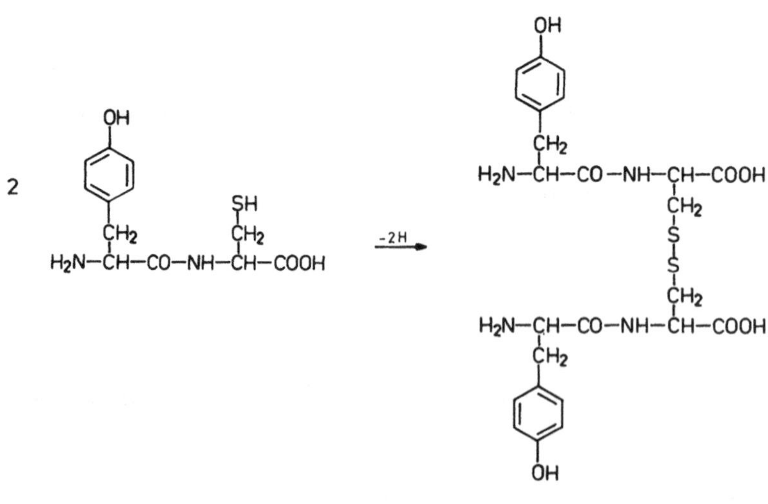

peptide, but other oxidizing agents, such as hydrogen peroxide [56], iodine [56] or diiodoethane [57] have also been used for the same purpose. The rate of disulfide formation depends, however, not solely upon the redox properties of the oxidizing agent, but also upon the concentration of the sulfhydryl derivative. High concentrations obviously favor the bimolecular reaction. Symmetrical disulfides can be obtained from S-protected peptides also by the oxidative removal of the sulfhydryl blocking group. The S-trityl group, for instance, is cleaved by iodine in methanol [58] with the concomitant formation of the disulfide [59]:

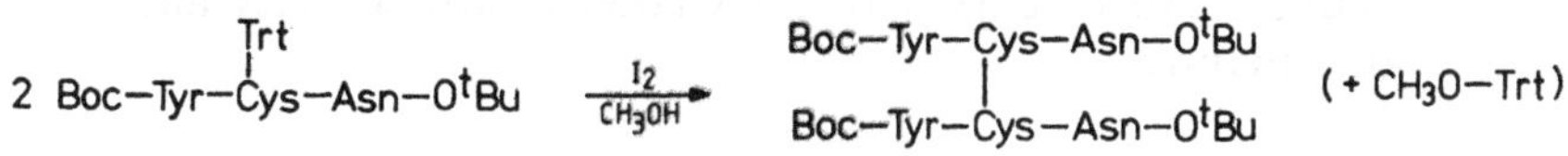

Several other thiol-protecting groups (cf. Chapter IV) e.g., the S-acetamidomethyl group or the S-diphenylmethyl group, are similarly treated in the preparation of symmetrical cystinyl peptides.

Preparation of symmetrical disulfides from chains which contain two or more cysteine residues can pose serious problems. Unless the sulfhydryl groups in the cysteinyl residues are blocked with different protecting groups and can be, therefore, selectively unmasked, several cystinyl peptides will form, amog them a single chain compound with an intramolecular disulfide bridge.

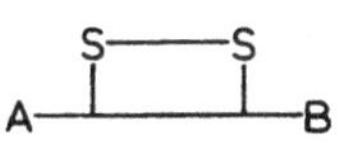

a parallel and an antiparallel dimer

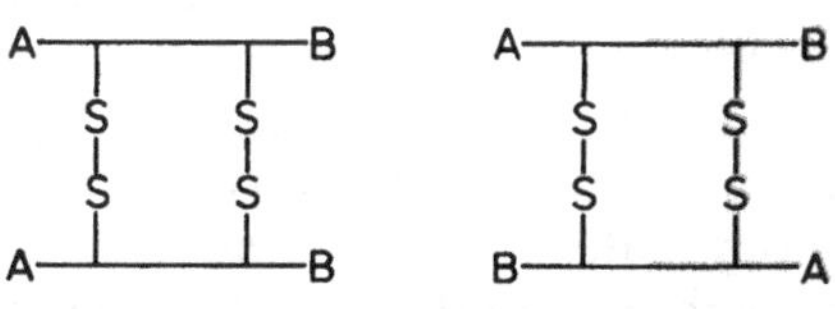

and, of course, also various polymers.

With two different sulhydryl protecting groups, each of which can be removed independently from the other, the problem is readily solved:

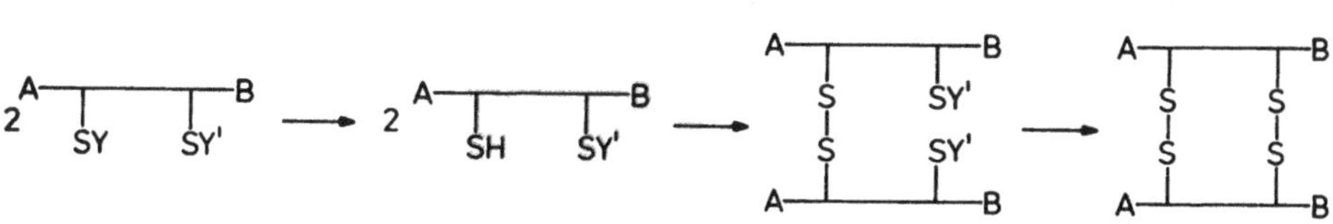

In this way only parallel dimers are formed. Polymerization can be suppressed by closing the second disulfide bridge in dilute solution. (Cf. the section on cyclization in this Chapter.)

An alternative approach to symmetrical cystinyl peptides starts with protected and activated derivatives of cystine, in a classical example [60] with bisbenzyloxycarbonyl-L-cystine dichloride:

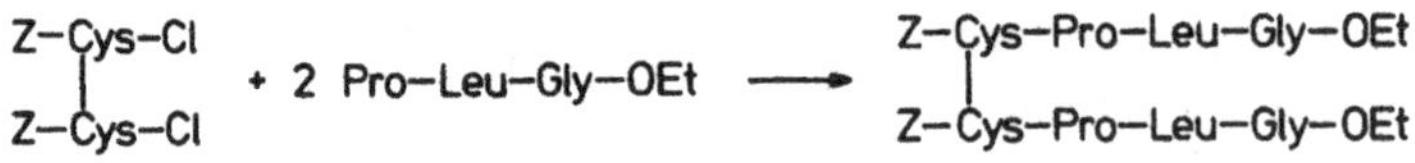

Obviously, difficulties can be expected if a second cystine moiety has to be added in the same manner:

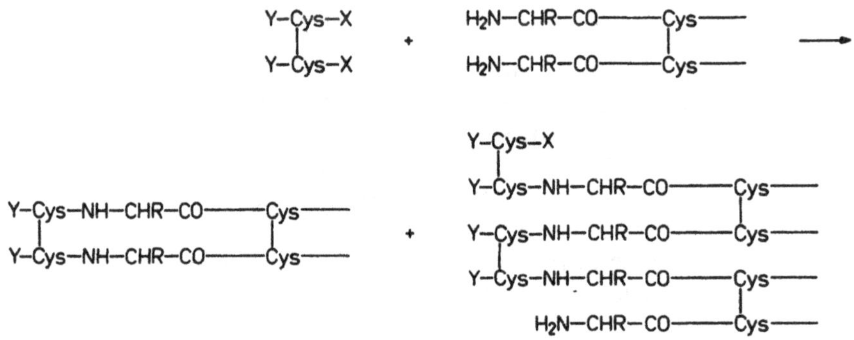

etc.

Similar problems are inherent in the synthesis of *asymmetrical cystinyl peptides*. For a systematic discussion of these problems, which transcend the limits of a small volume, we refer to the work of Wünsch [61]. Comprehensive articles dealing with the complex questions surrounding the formation of specific disulfides were written by Photaki [62] and more recently by König and Geiger [63]. The preparation of asymmetrical cystinyl peptides was extensively investigated by Hiskey and his associates [64–67] who recognized that *thiocyanogen* reacts both with thiols and with certain thioethers, e.g. derivatives of S-diphenylmethyl-, S-trityl-, S-tetrahydropyranyl- or S-methoxymethylene cysteine. The sulfenylthiocyanates formed in the removal of these S-protecting groups react, in turn, with a second thioether to yield an unsymmetrical cystinyl pepetide. For instance:

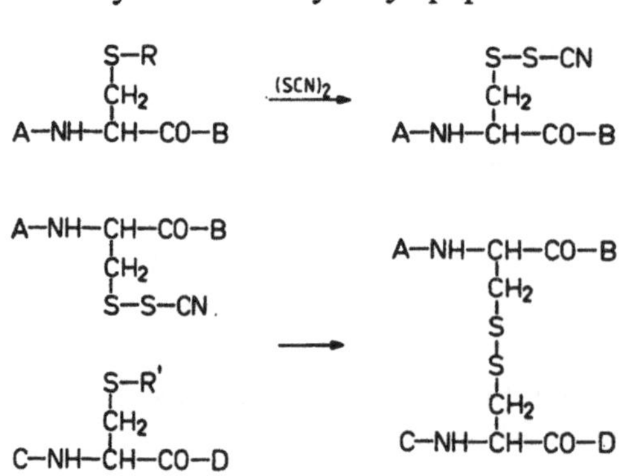

Protection of the sulfhydryl function by the benzyl groups resists thiocyanogen but not the more potent methoxycarbonylsulfenylchloride [68–70]:

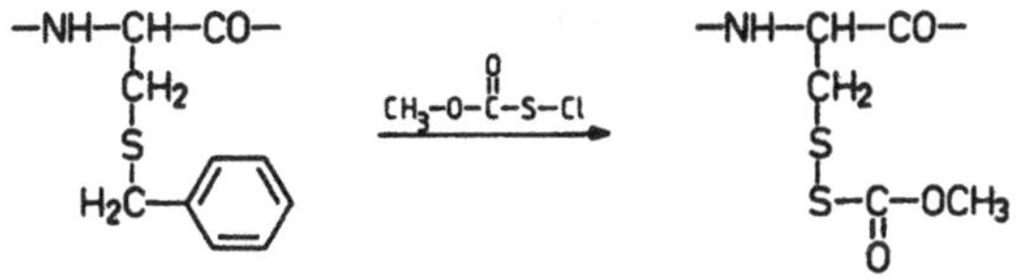

The methoxysulfenylthiocarbonates resulting from the removal of S-alkyl groups are stable toward acids and can be kept intact through various operations of peptide synthesis. Yet, they do react with mercaptanes to produce mixed disulfides:

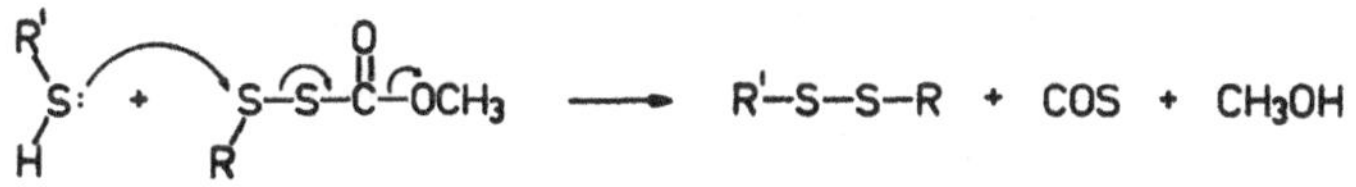

An impressive realization of these principles, developed for the synthesis of asymmetrical cystinyl peptides, is the construction of segments of insulin [59] and finally of (human) insulin itself, an endeavor [7] which required the formation of three disulfide bridges and the condensation of large building blocks via amide bonds.

D. Synthesis of Cyclic Peptides

Cyclic peptides play a major role in nature. The majority of microbial peptides, including peptide antibiotics, have ring structure [71]. One can distinguish between two main classes of compounds: *homodetic* and *heterodetic cyclopeptides* [72]. In members of the former group the constituent amino acids are connected through peptide bonds, while in heterodetic peptides other functions, such as disulfides, ester (lactone) groups, ethers or thioethers also contribute to ring formation. In *depsipeptides* or *peptolides* some of the constituents are hydroxyacids rather than amino acids and accordingly the building components are linked to each other both by ester and by amide bonds. From the reviews dealing with various aspects of cyclic peptides we point to an article of Kopple [73] which describes the methods of ring closure leading to cyclopeptides.

1. Homodetic Cyclopeptides

In the synthesis of homodetic cyclic peptides most methods of peptide bond formation are applicable for ring closure. Nevertheless, cyclization requires special considerations. The readiness of open chain compounds (or linear peptides) to cyclize is a function of several properties, first and foremost among these is the size of the ring to be closed. Usually no difficulties arise in the cyclization of peptides with six or more residues, but pentapeptides are often not the best starting materials in cyclization and ring closure is even more hampered in most tetra- and tripeptides. Such small rings can be closed only if at least one of the peptide bonds has cis rather than the more stable trans geometry:

In cyclic dipeptides (diketopiperazines) both amides are in cis arrangement. In this case the thermodynamic stability of the six-membered ring compensates for the energy required for the trans → cis rearrangement in which the barrier created by the partial double bond character of the amide bond has to be overcome. Therefore, diketopiperazines are produced without difficulty and they form spontaneously without particular activation of the carboxyl of dipeptides, e.g. from their alkyl esters:

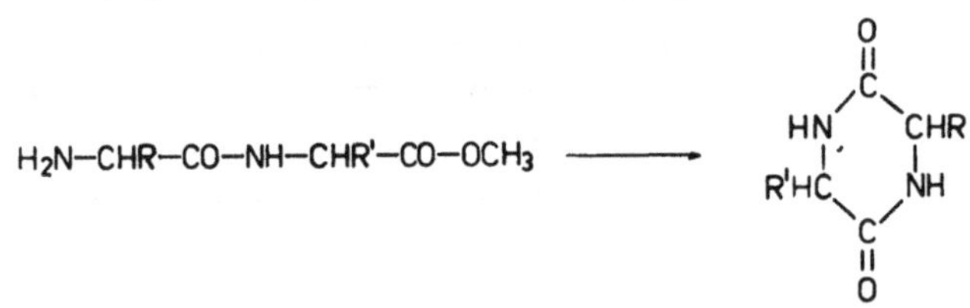

Cyclization of tripeptides is facilitated by the presence of proline and/or glycine residues. The former provides a geometry which is favorable for ring closure, while glycine, because of the absence of a side chain, creates at least no obstacle. Tri-, tetra-, and pentapeptides are prone to cyclodimerization, that is to the formation of rings of twice the size than obtained in simple cyclization.

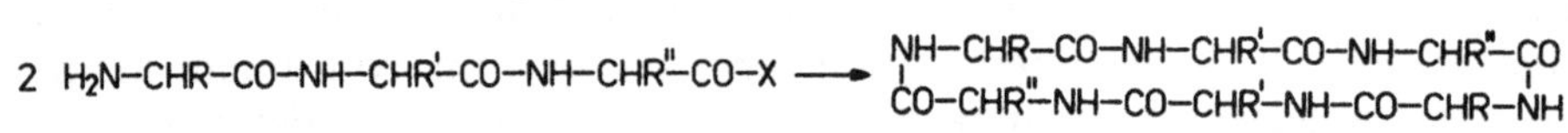

This phenomenon was noted in the synthesis of gramicidin S [74], the first naturally occurring cyclic peptide synthesized in the laboratory [75]. The best explanation for this conspicuous tendency for cyclodimerization is the assumption of an antiparallel arrangement of two pentapeptide derivatives prior to cyclization. The two chains are held together by hydrogen bonds in a β-sheet like conformation.

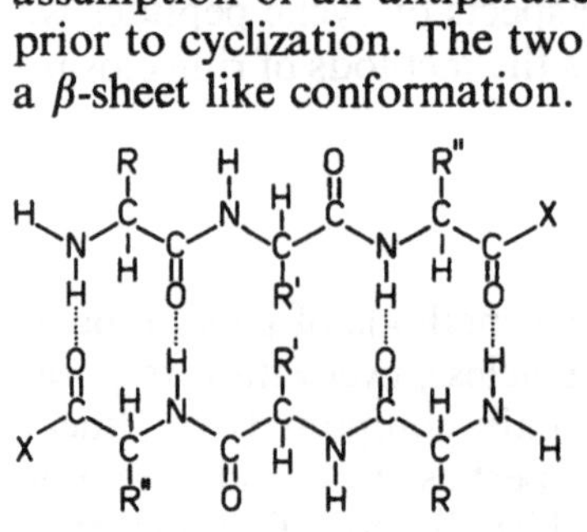

While no cyclic monomer was obtained in the early experiments [74] in subsequent studies conditions were found [76] which were more conducive to cyclization than to cyclodimerization. The coupling method applied for ring closure also plays some role, e.g. cyclodimerization is less pronounced when the open chain precursor is activated in the form of its azide [77] than in cyclizations by other procedures. A more decisive influence is exerted, however, by the concentration of the peptide subjected to cyclization. The *principle of dilution* [78] must be adopted if cyclization, a unimolecular reaction, is our aim. At high concentration of the activated peptide bimolecular reactions, dimerization and polymerization emerge as serious competitors. This principle has been applied in the synthesis of gramicidin S [75]. The solution of a salt of the activated decapeptide was added, in a thin stream, to pyridine:

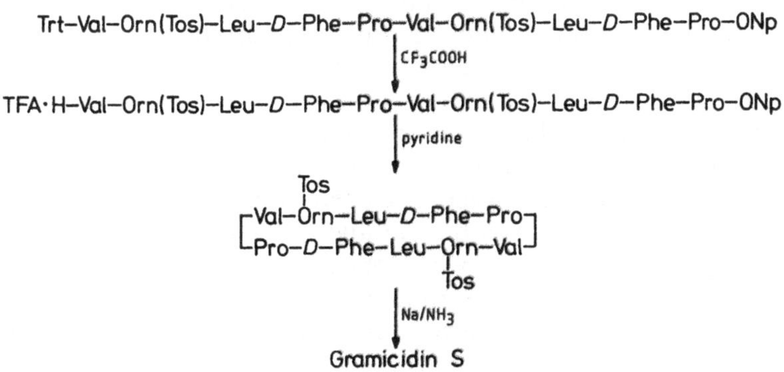

The same solvent, pyridine, which functions also as the base which abstracts a proton from the salt of the activated intermediate, and the same dilution technique have been adopted in the preparation of numerous cyclopeptides.

In the alternative approach, synthesis of the cyclodecapeptide through cyclodimerization of a pentapeptide derivative [79], H-Val-Orn(Z)-Leu-D-Phe-Pro-ONp, if the ring closure was carried out at 3×10^{-3} molar concentration, about one-third of the cyclic products consisted of the cyclopentapeptide (cyclosemigramicidin S) while two-thirds of the cyclic material was the cyclodecapeptide gramicidin S. About equal amounts of the monomeric and dimeric products formed if the reaction mixture was further diluted, to about 3×10^{-4} molar. The sequence has also a major influence on the outcome of the cyclization. When valine in the pentapeptide derivative was replaced [79] by glycine no cyclic decapeptide could be detected in the product, only the cyclopentapeptide formed. Also, practically no cyclodimerization took place in the ring closure of a pentapeptide with alternating L and D residues [80], because β-turns in the open chain precursor stabilize a ring like conformation [81].

In the selection of coupling methods which are suitable for ring closure through an amide bond two alternative tactics can be followed. In the more unequivocal pathway activation of the carboxyl and coupling are clearly separated. The N-terminal amino group of the open chain precursor remains blocked during activation, the amine protection is then removed and the ring is closed, preferably in dilute solution:

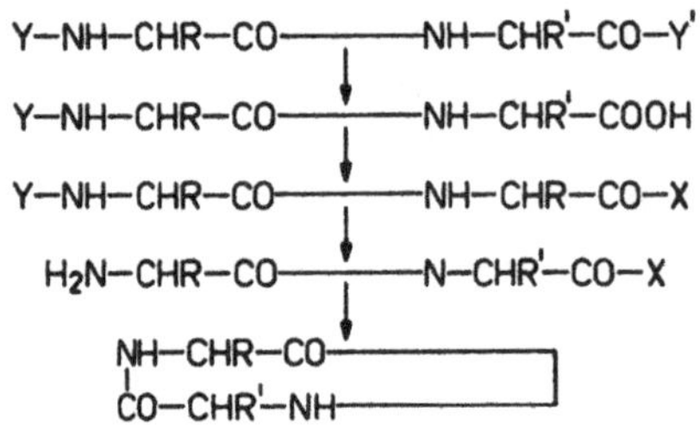

Activation, a bimolecular reaction, requires high concentration of the reactants, but, if the removal of the amine protecting group is carried out by acidolysis or by hydrogenolysis in acid solution, a salt of the amine is obtained and its acylation is still hampered. Therefore, it is possible to dilute the solution of the salt intermediate and to render it basic afterwards to initiate the ring closure reaction. This way the formation of dimers and polymers is effectively diminished. Such considerations led to the use of pyridine, which can function both as a diluent and as a base [72]. Of course, only certain methods of activation can be applied in this approach, since the reactive intermediates must be stable enough to withstand the unmasking of the N-terminal amino group. Active esters are probably best suited [72, 74, 75].

Interestingly, the azide method also permits the separation of the steps of activation and coupling [82]. The reaction of acid·hydrazides with nitrous acid or alkyl nitrites is much faster than deamination of the N-terminal amino acid residue by the same reagents. Thus the amino group remains essentially intact during the conversion of the hydrazide to the azide (with the calculated amount of nitrite!). Also, because of protonation in the acidic medium of this reaction the amino group is inert toward the azide, it is possible and practical [80] to dilute the solution of the azide and to expose only afterwards the amino group to acylation:

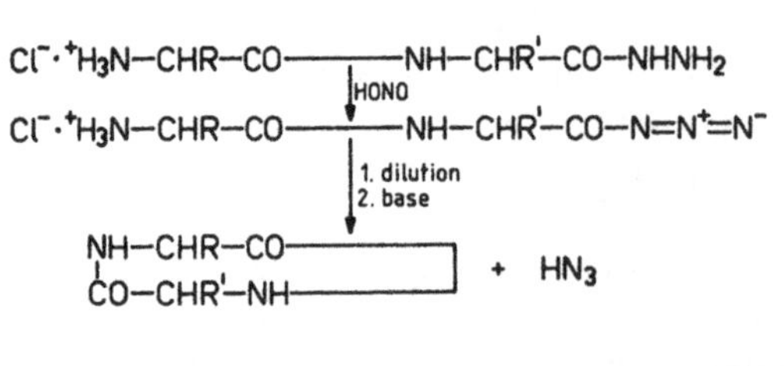

It is more problematic to follow the second alternative, to activate the carboxyl group of a peptide with an unprotected amino group. In this case the methods of activation must be selected with special care. For instance, the reaction of free peptides with alkyl chlorocarbonates should, in all probability, produce not only the desired mixed anhydrides but also urethanes, in which cyclization is blocked:

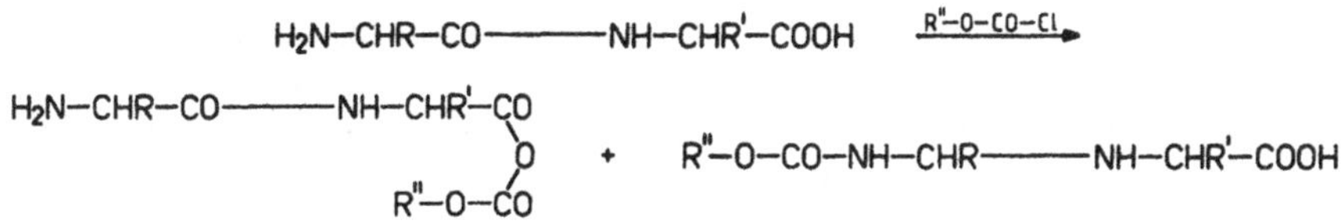

Toward such potent acylating agents as mixed anhydrides protonation of the amine does not provide complete protection. In spite of this complication, cyclization via mixed anhydrides can provide valuable results [83–85] if ring formation is favored by the geometry of the molecule. It is obvious, to give more favorable consideration to coupling reagents, since these can activate carboxyl groups in the presence of amines. In this respect carbodiimides could be particularly useful because the rate of their reaction with amines is negligible in comparison with the rate of addition of carboxylic acids. On the other hand, while the reactive intermediates are potent acylating agents, if cyclization is impeded by unfavorable geometry, they will produce, through O → N acyl migration, N-acylurea derivatives which contaminate the expected cyclic peptides:

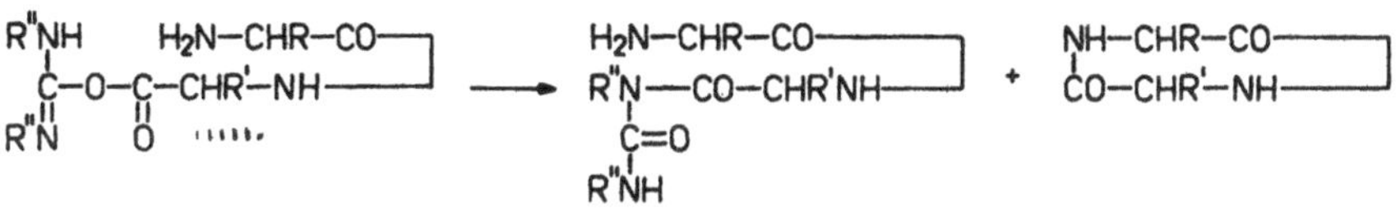

It is not too surprising, therefore, that often low yields were observed in cyclization with carbodiimides even when the reagent was applied in high excess [86]. Once again, somewhat better results can be achieved if the conformation of open chain precursor favors cyclization, as in the case of some hexapeptides [87]. Improvements can be expected if the O → N acyl-migration is suppressed by the addition of auxiliary nucleophiles (cf. Chapter 2) such as N-hydroxysuccinimide [88] or 1-hydroxybenzotriazole. Cyclization with other coupling reagents might suffer from similar disadvantages. Yet, in the synthesis of some simple cyclopeptides fairly good results were obtained [89] when azides were generated from peptide hydrochlorides with the help of diphenylphosphoryl azide [90]. Cyclization of a peptide anchored to an insoluble polymeric support could be carried out [91] with the mixed anhydride producing coupling reagent 1-ethyloxycarbonyl-2-ethyloxy-1,2-dihydroquinoline (EEDQ) [92]. Application of 2-phenylisoxazolium-3′-sulfonate [93] gave fair results in some cyclization experiments [94].

Although moderate cyclization yield was achieved [95] with the help of o-phenylene chlorophosphite [96], at least in principle, this reagent and the analogous derivatives of phosphorous acid should perform well in selective ring closure because the phosphazo intermediates are not useless byproducts but generate additional amounts of the mixed anhydride which is the reactive species in the amide bond forming reaction. Until the advent of the perfect cyclizing reagent, however, it seems to be advisable to seek a clear separation of activation and coupling. Thus, the attempted cyclization of a heptapeptide with the sequence of evolidin gave no tangible product when the reaction was carried out with the aid of dicyclohexylcarbodiimide, but a cyclohexapeptide with the properties of natural evolidin could be secured through the conversion of the N-protected open-chain intermediate (or *seco*-peptide) to the *p*-nitrophenyl ester, deprotection and ring closure [97].

Most cyclizations were carried out in solution, although a case of solid phase synthesis has just been mentioned [91]. An interesting combination of the solid phase technique with the active ester approach was proposed by Fridkin and his associates [98] whose method involves peptidyl derivatives of polymeric nitrophenols. Ring closure and separation from the support occurs concomitantly with separation from the resin:

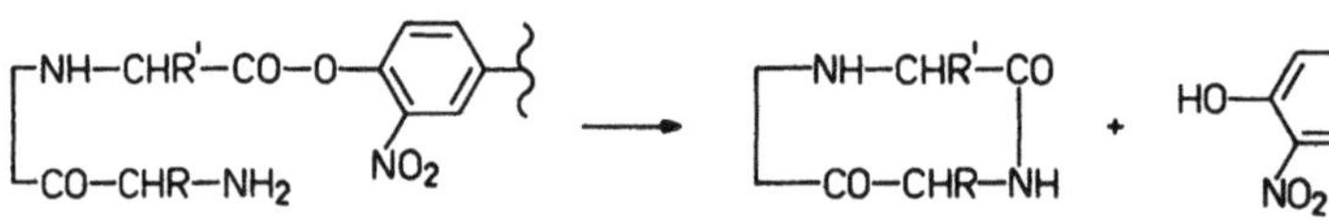

In this brief survey the topic of cyclization cannot be exhausted. To demonstrate the possibility of original approaches to cyclic peptides we include the formation of such compounds through ring expansion of cycloalkanedionedioximes via Beckmann rearrangement [99]:

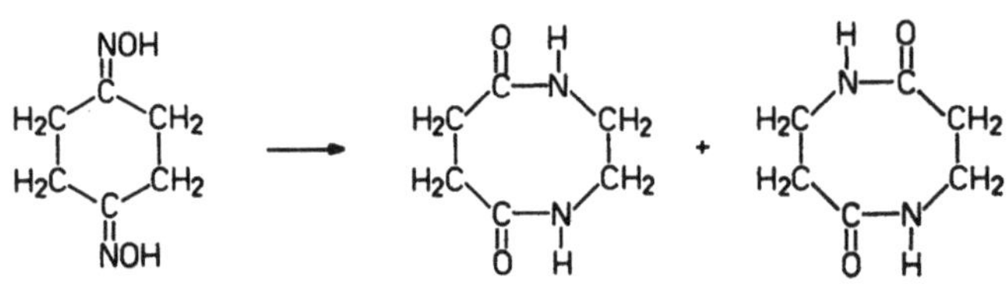

A growing interest in rigid analogs of peptide hormones, which might have stronger or more lasting interaction with specific receptors generates challenging problems in the synthesis of cyclic peptides. An impressive example is the construction of a bicyclic analog [100] of somatostatin in which the disulfide bridge of the parent compound is replaced by a $—CH_2—CH_2$-group:

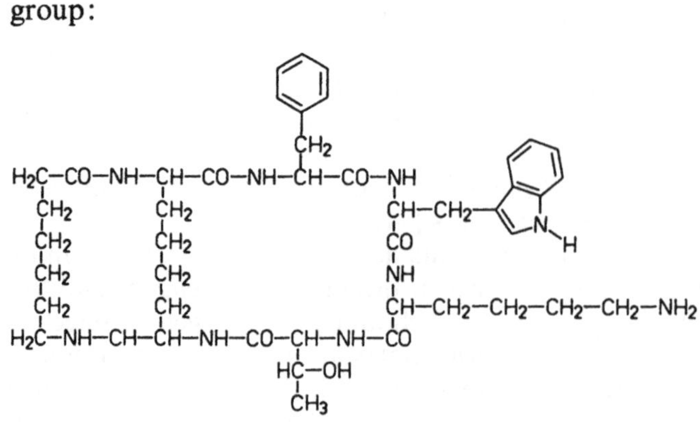

2. Heterodetic Cyclopeptides

In an important group of this class of cyclic peptides a *disulfide* is an integral part of the ring. The hormones oxytocin, vasotocin, vasopressin and somatostatin are characteristic examples, but a cyclic disulfide structure can be found also in the A-chain of insulin. The problems of ring closure through disulfides are the same as those which complicate disulfide bridge formation, discussed earlier in this Chapter. Oxidation of sulfhydryl groups in dilute solution favors cyclization, because it is a unimolecular reaction, while in concentrated solutions dimerization and polymerization are more likely. The oxidizing agents, iodine, diiodoethane, potassium ferricyanide or air have relatively little influence on the distribution of the products, the ring size is more decisive in this respect. The disulfide forms from L-cysteinyl-L-cysteine on oxidation [101] but simple ring closure is impeded in peptides which have one to three amino acid residues between the two cysteines [102]. Cyclization is more facile in hexapeptides with cysteines at both ends but even in these cases some dimerization can take place. Ring formation causes no major difficulties in the synthesis of oxytocin and vasopressins which have this favorable ring size, six amino acid residues and two sulfur atoms forming a twenty atom cycle. Similarly no difficulties were observed in syntheses [103, 104] of the cyclic heptapeptide part of the thyrocalcitonin, in which residues 1 and 7 are connected through a disulfide. Most of the ring-closing reactions were carried out in solution although disulfide formation by oxidation of sulfhydryl groups of peptides attached to insoluble supports has also been reported [105].

Bicyclic peptides, in which one of the two rings is closed through a disulfide, are also known. In the simplest example cyclo-L-cystine [106] the disulfide was prepared by the oxidative removal of the acetamidomethyl group from the sulfhydryl groups. One of the two amide bonds was formed by conventional means, the other through ring closure to the diketopiperazine:

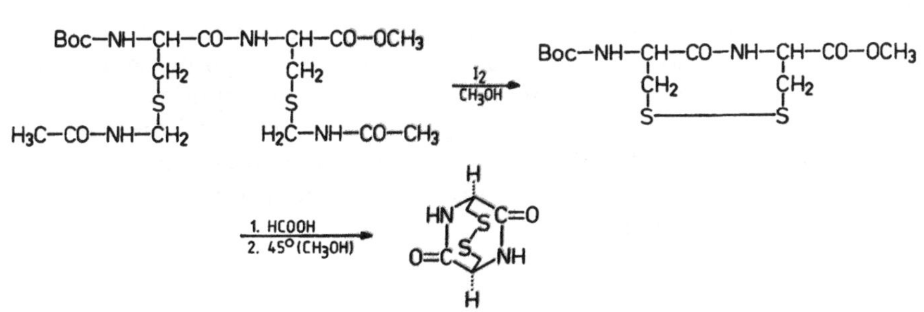

A disulfide between two neighboring cysteine residues was obtained less readily in the microbial cyclopentapeptide malformin [107] and oxidation of the sulfhydryl groups required the use of iodine in dimethylsulfoxide. The final product, malformin A, was obtained in low yield [108]:

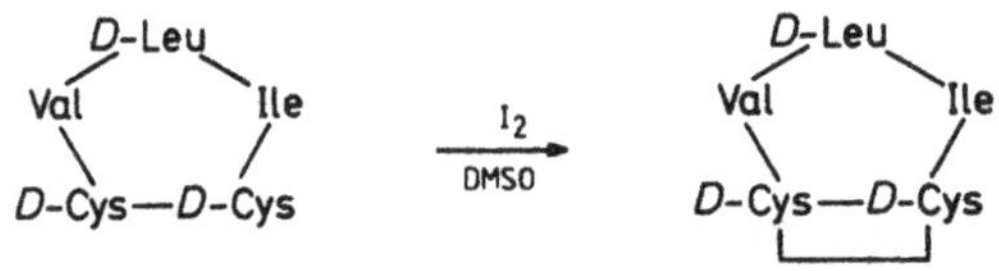

The alternative approach, a simultaneous removal of the two acetamidomethyl groups and oxidation with iodine failed to improve the efficiency of the synthesis [109]. The same method, cleavage of two S-acetamidomethyl groups with concomitant oxidation to the disulfide produced the bicyclodecapeptide 2,7-cystine gramicidin in respectable yield [110]. Thus, the outcome of disulfide formation depends mostly on the conformation of the molecule.

In some heterodetic cyclic peptides a *thioether* is an integral part of the ring. The toxic peptide from the mushroom Amanita phalloides, phalloidin [111], contains a partial structur which originates from the transannular reaction of a cysteine residue with the indole moiety of a tryptophan residue. The thioether thus formed serves as a bridge across a cyclohexapeptide ring and converts the latter to a bicyclic compound:

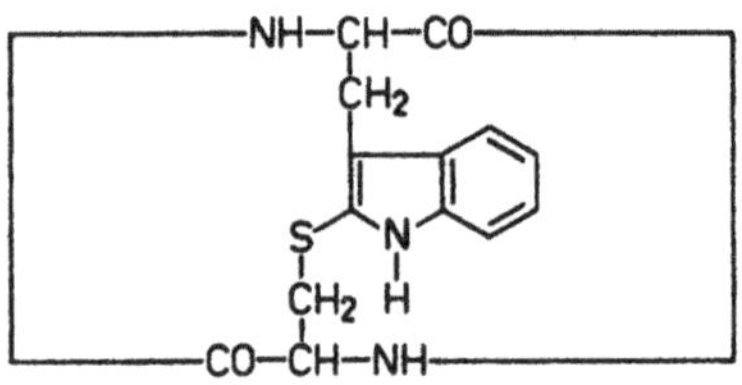

In the related amanitin, from the same mushroom, the thioether appears at a higher oxidation level, as a sulfoxide. In the synthesis of related bicyclic peptides first the thioether containing portion was built and the construction of the target compounds was concluded with the formation of a peptide bond in the homodetic ring [84, 112–115].

In *depsipeptides* the heterodetic nature of the ring follows from the presence of an ester group connecting an amino acid residue with the side chain of a hydroxyamino acid. Such *peptide lactones* can be synthesized by the methods applied in the preparation of homodetic cyclopeptides. Acylation of hydroxyl groups requires, however, potent derivatives of the carboxyl group, for instance mixed anhydrides. Alternatively, moderately reactive intermediates, such as active esters can be used, in the presence of catalysts, e.g. 1-hydroxybenzotriazole [116]:

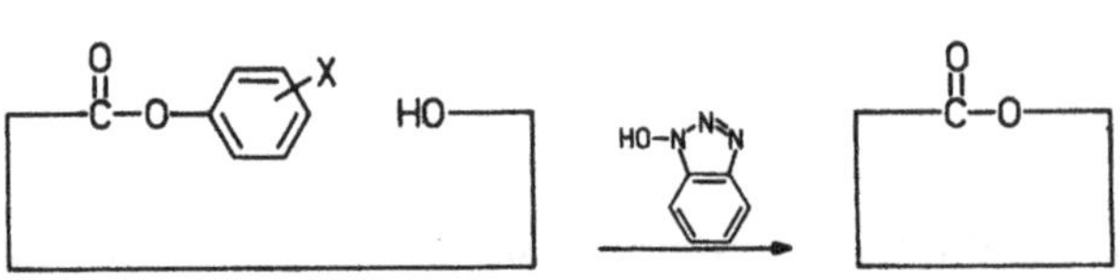

Basic catalysts, imidazole or 4-dimethylaminopyridine, can be even more effective, but they might endanger the chiral integrity of the product.

In certain instances it seems to be advantageous to close the lactone ring through an amide rather than through an ester (lactone) linkage. In such cases first a suitably protected O-aminoacyl derivative of a hydroxyaminoacid is prepared and the chain lengthened. Thereafter, the synthesis is concluded by cyclization via an amide linkage [117]. The same principle can be applied for the synthesis [118] of *peptolides* such as valinomycin:

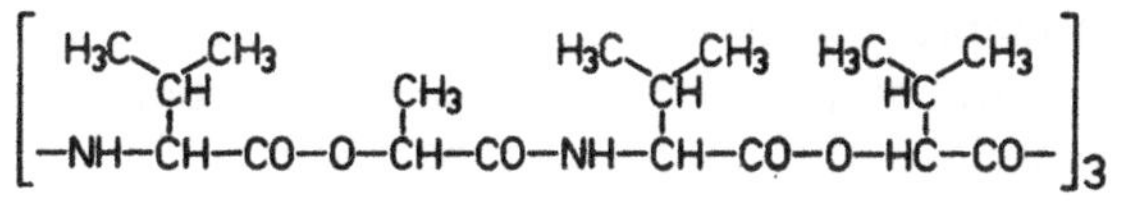

E. Sequential Polypeptides

Preparation of polyamino acids from N-carboxyanhydrides is an extensively practiced process but it transcends the scope of this volume which primarily deals with the synthesis of well defined peptides with specific sequences. Polyamino acids are, at least with respect to degree of polymerization, mixtures which can be heterogeneous in other respects as well. Similar concerns could be valid also for the synthesis of sequential polypeptides, chains in which a few amino acid residues occur in a regularly repeating sequence, such as ABCABCABCABCABCABC ... The conventional approach to such materials is *polycondensation* of activated peptides which have a free N-terminal amino group, e.g. the *p*-nitrophenyl ester of norvalyl-glycyl-proline [119]:

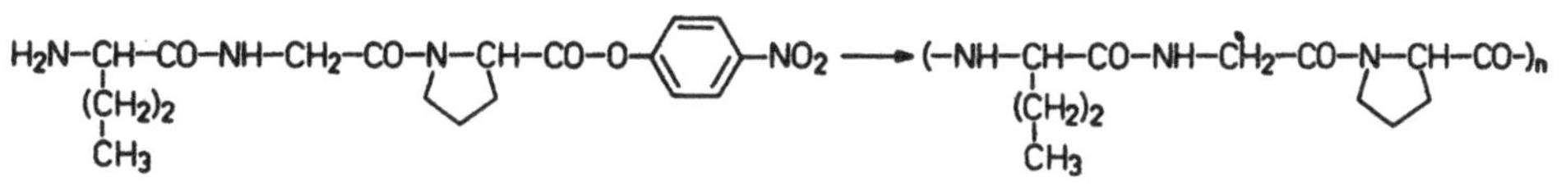

Certain ambiguities are inherent in such schemes. Not only the molecular weight distribution of the polymeric material is mostly problematic, but questions remain also about the C-terminal end-group of the chain. The latter might be an unreacted active ester group or a free carboxyl, if polycondensation was accompanied by some hydrolysis. With initiators, such as primary or secondary amines, added in a small amount, questions about the end group can be solved but the molecular weight distribution is shifted toward lower values. In spite of such ambiguities polycondensation remains a much used avenue, because it yields large molecular weight peptides in a simple manner and within a short time. Activation of the "monomers" in the form of *p*-nitrophenyl esters was quite popular [120–122] and is still practiced [119]. The chiral purity of the products can be trusted only if glycine is the C-terminal residue of the reactive intermediate. Proline, which is not readily racemized is usually accepted in this position without concern. A pronounced tendency of protected peptide pentachlorophenyl esters for crystallization and the possibility of removing amine protecting groups by catalytic reduction without harm to the active ester grouping led to their increasing application in polycondensation [123]. Good results were observed with esters of N-hydroxysuccinimide as well [124]. The use of *o*-hydroxyphenyl esters [125] which can be obtained through unreactive intermediates with a provisionally masked hydroxyl group allows an extension of the polycondensation method to peptides which have a chiral residue at their C-terminus, since racemization is practically absent in this approach. This advantage compensates for the moderate reactivity of *o*-hydroxyphenyl esters. The heterogeneity of the products in the preparation of sequential polypeptides by polycondensation of oligopeptide active esters can be reduced by the addition of initiators and by carefully controlled reaction conditions [126], but better homogeneity can be attained through stepwise chain lengthening. The time consuming nature of such syntheses is justified by the homodisperse character of the products. For instance, a 60-residue chain with the repeating sequence -Pro-Pro-Gly- could be assembled [127] on a solid support through the incorporation of well defined subunits. Sequential peptides secured in this way closely resembled their natural analogs with respect to both physical [128] and biological [129] properties. They are, therefore, clearly superior to the heterogeneous materials obtained by polycondensation.

An interesting and potentially useful strategy for the synthesis of sequential polypeptides is the doubling of a starting sequence by controlled acylation of the suspension of a poorly soluble amino component [130]. The weakly reactive N-ethylsalycilamide ester of a protected tripeptide was allowed to react with the suspension of the corresponding free peptide in dimethylsulfoxide:

Z-Gly-Leu-Gly-X + H-Gly-Leu-Gly-OH ⟶ Z-Gly-Leu-Gly-Gly-Leu-Gly-OH

Z-Gly-Leu-Gly-Gly-Leu-Gly-X + H-Gly-Leu-Gly-Gly-Leu-Gly-OH (susp.)

⟶ Z-(Gly-Leu-Gly-)$_3$-OH

X =

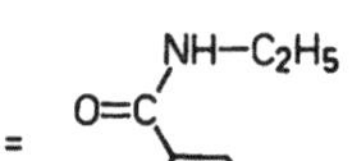

A continuation of this scheme afforded a 48-residue chain in reasonable yield and satisfactory purity. A useful review on the synthesis of sequential polypeptides was written by Jones [131].

F. Partial Synthesis (Semisynthesis)

Many challenging problems of protein chemistry require some modification of complex, biologically active molecules, such as enzymes or protein hormones. Total synthesis of chains of often well above 100 residues is a formidable task. The 51-residue insulin may represent the upper limit of experimentation via synthetic analogs. Therefore, it is an obvious thought to start with a protein itself, to cleave it selectively and to replace one of the fragments with a synthetic peptide [132]. Sometimes it is not necessary to form a covalent bond between a peptide and the remainder of the protein molecule. They can be held together by a series of cooperative interactions between amino acid side chains. A classical example for such recombination without covalent bond(s) is ribonuclease A, which is cleaved by subtilysin to a 104-residue "S-protein" and a 20-residue "S-peptide". Since recombination of the two fragments restores the enzymic activity, the S-peptide can be replaced in the experiment by appropriate analogs and important conclusions can be drawn on the relationships between structure and enzymic acitivity from the hydrolytic potency of the semisynthetic preparations. Similar non-covalent "syntheses" are feasible also with a staphylococcal ribonuclease, cytochrom C or thioredoxin and probably with many other proteins, including hormones, e.g. growth hormones. The methods, however, used in these experiments are not those of peptide synthesis and, therefore, we discuss here only partial syntheses in which covalent bonds are formed. It might be worthwhile to point out in this connection a covalent bond which was produced sometimes unintentionally. When instead of enzymes cyanogen bromide is used for the selective scission of a protein, the homoserine lactone formed in the reaction can be attacked by the newly created amino group to restore the peptide bond [133]:

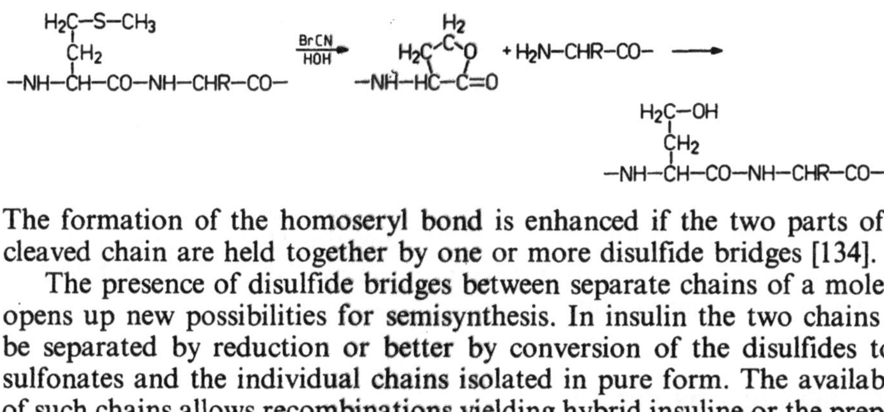

The formation of the homoseryl bond is enhanced if the two parts of the cleaved chain are held together by one or more disulfide bridges [134].

The presence of disulfide bridges between separate chains of a molecule opens up new possibilities for semisynthesis. In insulin the two chains can be separated by reduction or better by conversion of the disulfides to S-sulfonates and the individual chains isolated in pure form. The availability of such chains allows recombinations yielding hybrid insuline or the preparation of insulins in which one chain is a synthetic peptide while the other originates from a natural insulin:

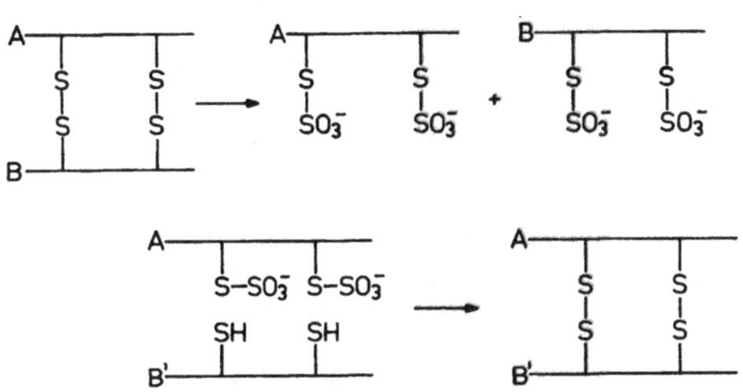

It is more demanding, but also more generally useful to stitch together protein fragments or to attach a peptide to a major portion of a protein molecule through a peptide bond. This route requires, however, special considerations with respect to protection, activation and coupling.

Masking of the functional groups in a protein is not an easy task if it has to be achieved under conditions which do not affect the integrity of the structure or the conformation of the molecule. Hence minimal protection would seem to be attractive. In principle only the protection of the amino groups in lysine side chains is mandatory. This can be done with well established blocking groups such as the benzyloxycarbonyl group or the tert. butyloxycarbonyl group. For reasons of solubility and ready removability some special methods of protection were developed for partial syntheses, e.g. acetimidylation or maleylation [132]. In practice, blocking of carboxyl functions is often necessary. Esterification of insulin with boron trifluoride-methanol was shown to be feasible [135] but a more general approach is based on the use of diazoalkanes [136–138]. Of course, if the two fission products have to be recombined, then protection must precede scission to ensure that the newly liberated amino and carboxyl groups are not blocked.

Most methods of activation and coupling should be applicable in partial syntheses, but peptide bond formation with the help of proteolytic enzymes holds a particular promise. The fascinating idea to use catalysts, which are known for spectacular acceleration of hydrolysis, in the synthesis of peptides tempted investigators for a quarter of a century [139]. As a practical method this approach gained impetus by the recognition that over and above the application of a large excess of one of the starting materials and removal of one of the products from the system (by its insolutility) it is possible to shift the equilibrium of the reaction also by the addition of organic solvents, such as isopropanol, to the aqueous medium [140–143]. Thus, semisynthesis through enzymic cleavage and enzymic condensation [144] became a practical avenue for the preparation of important proteins and of large peptide hormones such as human insulin.

References of Chapter VI

1. Bodanszky, M., Ondetti, M. A.: in "Peptide Synthesis", p. 161, New York: Wiley-Interscience 1966
2. Bodanszky, M., Klausner, Y. S., Ondetti, M. A.: in "Peptide Synthesis (2nd Ed.)", p. 177, New York: Wiley-Interscience 1976
3. Bodanszky, M.: in "Perspectives in Peptide Chemistry" (Eberle, A., Geiger, R., Wieland, T., Eds.), p. 15, Basel: Karger S. 1981
4. Bodanszky, M., du Vigneaud, V.: Nature *183*, 1324 (1959)
5. Barany, G., Merrifield, R. B.: J. Amer. Chem. Soc. *99*, 7363 (1977)
6. Riniker, B., Kamber, B., Sieber, P.: Helv. Chim. Acta *58*, 1086 (1975)
7. Sieber, P., Kamber, B., Hartmann, A., Jöhl, A., Riniker, B., Rittel, W.: Helv. Chim. Acta *57*, 2617 (1974)
8. Schwyzer, R., Kappeler, H.: Helv. Chim. Acta *46*, 1550 (1963)

9. Bodanszky, M., Williams, N. J.: J. Amer. Chem. Soc. *89*, 685 (1967)
10. Bodanszky, M., Ondetti, M. A., Levine, S. D., Williams, N. J.: J. Amer. Chem. Soc. *89*, 6753 (1967)
11. Carpino, L. A., Han, G. Y.: J. Org. Chem. *37*, 3404 (1972)
12. Bodanszky, M., Tolle, J. C., Gardner, J. D., Walker, M. D., Mutt, V.: Int. J. Peptide Protein Res. *16*, 402 (1980)
13. Bodanszky, M., Tolle, J. C., Bednarek, M. A., Schiller, P. W.: Int. J. Peptide Protein Res. *17*, 444 (1981)
14. Bodanszky, M., Bednarek, M. A., Bodanszky, A., Tolle. J. C.: in "Peptides 1980" (Brunfeldt, K., Ed.), p. 93, Copenhagen: Scriptor 1981
15. Carpino, L. A., Tsao, J. H., Ringsdorf, H., Fell, E., Hettisch, G.: J. Chem. Soc. Chem. Commun., 358 (1978)
16. Sieber, P.: Helv. Chim. Acta *60*, 2711 (1977)
17. Kiso, Y., Ukawa, K., Akita, T.: J. Chem. Soc. Chem. Commun., 101 (1980)
18. Kenner, G. W., Seely, J. H.: J. Amer. Chem. Soc. *94*, 3259 (1972)
19. du Vigneaud, V., Gish, D. T., Katsoyannis, P. G., Hess, G. P.: J. Amer. Chem. Soc. *80*, 3355 (1958)
20. Bodanszky, M., Bednarek, M. A.: Int. J. Peptide Protein Res. *20*, 434 (1982)
21. Bednarek, M. A., Bodanszky, M.: Int. J. Peptide Protein Res. *21*, 196 (1983)
22. Royer, G. P., Anantharamaiah, G. M.: J. Amer. Chem. Soc. *101*, 3394 (1979)
23. Pettit, G. R.: in Synthetic Peptides, Vol. 4, p. 22, Amsterdam: Elsevier 1976
24. du Vigneaud, V., Ressler, C., Swan, J. M., Roberts, C. W., Katsoyannis, P. G., Gordon, S.: J. Amer. Chem. Soc. *75*, 4879 (1953)
25. Rittel, W., Iselin, B., Kappeler, H., Riniker, B., Schwyzer, R.: Helv. Chim. Acta *40*, 614 (1957)
26. Schwarz, H., Bumpus, F. M., Page, I. H.: J. Amer. Chem. Soc. *79*, 5697 (1957)
27. Guttmann, S., Boissonnas, R. A.: Helv. Chim. Acta *41*, 1852 (1958); *42*, 1257 (1959)
28. Schwyzer, R., Sieber, P.: Nature *199*, 172 (1963)
29. Hirschmann, R., Mutt, R. F., Veber, D. F., Vitali, R. A., Varga, S. L., Jacob, T. A., Holly, F. W., Denkewalter, R. C.: J. Amer. Chem. Soc. *91*, 507 (1969)
30. Romovacek, H., Dowd, S. R., Kawasaki, K., Nishi, N., Hofmann, K.: J. Amer. Chem. Soc. *101*, 6081 (1979)
31. Yajima, H., Fujii, N.: J. Chem. Soc. Chem. Commun., 115 (1980)
32. Bodanszky, M.: in "Prebiotic and Biochemical Evolution" (Kimball, A. P., Oro, J., Eds.), p. 217, Amsterdam: North Holland Publ. 1971
33. Hardy, P. M., Samworth, D. J.: J. Chem. Soc. Perkin I, 1954 (1977)
34. Ugi, I., Marquarding, D., Urban, R.: in "Chemistry and Biochemistry of Amino Acids, Peptides and Proteins" (Weinstein, B., Ed.), Vol. 6, p. 245, New York: Dekker 1982
35. Urban, R., Eberle, G., Marquarding, D., Rehn, D., Rehn, H., Ugi, I.: Angew. Chem. (Int. Ed.), 627 (1976)
36. Waki, M., Meienhofer, J.: J. Amer. Chem. Soc. *99*, 6075 (1977)
37. Bacher, J. M., Naider, F.: Biopolymers *17*, 2189 (1978)
38. Letsinger, R. L., Kornet, M. J.: J. Amer. Chem. Soc. *85*, 3045 (1963)
39. Felix, A. M., Merrifield, R. B.: J. Amer. Chem. Soc. *92*, 1385 (1970)
40. Inouye, K., Watanabe, K.: J. Chem. Soc., Perkin I, 1911 (1977)
41. Deer, A., Fried, J. H., Halpern, B.: Austral. J. Chem. *20*, 797 (1967)
42. Brenner, M., Zimmermann, J. P., Wehrmüller, J., Quitt, P., Hartmann, A., Schneider, W., Beglinger, U.: Helv. Chim. Acta *40*, 1497 (1957)
43. Brenner, M., Hofer, W.: Helv. Chim. Acta *44*, 1794 (1961)
44. Bodanszky, M., Bodanszky, A.: Chem. Commun., 591 (1967)

45. Sarges, R., Witkop, B.: J. Amer. Chem. Soc. *89*, 2020 (1965)
46. Tilak, M. A.: Tetrahedron Lett., 849 (1970)
47. van Zon, A., Beyerman, H. C.: Helv. Chim. Acta *56*, 1729 (1973); *ibid. 59*, 1112 (1976)
48. Weygand, F., Huber, P., Weiss, K.: Z. Naturforsch. *22b*, 1084 (1967)
49. Mehlis, B., Apelt, H., Bergmann, J., Niedrich, H.: J. Pact. Chem. *314*, 390 (1972)
50. Kisfaludy, L., Schön, I., Szirtes, T., Nyéki, O., Löw, M.: Tetrahedron Lett., 1785 (1974)
51. Hirschmann, R., Schwam, H., Strachan, R. G., Schoenewaldt, E. F., Barkemeyer, H., Miller, S. M., Conn, J. B., Garsky, V., Veber, D. F., Denkewalter, R. B.: J. Amer. Chem. Soc. *93*, 2746 (1971)
52. Dewey, R. S., Schoenewaldt, E. F., Joshua, H., Paleveda, W. J., Jr., Schwam, H., Barkemeyer, H., Arison, B. H., Veber, D. R., Strachan, R. G., Milkowski, J., Denkewalter, R. G., Hirschmann, R.: J. Org. Chem. *36*, 49 (1971)
53. Gutte, B., Merrifield, R. B.: J. Amer. Chem. Soc. *91*, 501 (1969)
54. Bodanszky, M.: Ann. N.Y. Acad. Sci. *88*, 655 (1960)
55. Harington, C. R., Pitt-Rivers, R. V.: Biochem. J. *38*, 417 (1944)
56. Harington, C. R., Mead, T. H.: Biochem. J. *30*, 1598 (1930)
57. Weygand, F., Zumach, G.: Z. Naturforsch. *17b*, 807 (1962)
58. Kamber, B., Rittel, W.: Helv. Chim. Acta *51*, 2061 (1968)
59. Kamber, B.: Helv. Chim. Acta *54*, 398 (1971)
60. Ressler, C., du Vigneaud, V.: J. Amer. Chem. Soc. *76*, 3107 (1954)
61. Wünsch, E.: in Houben-Weyl, Methoden der Organischen Chemie 4th Ed., Vol. 15/1, p. 800, Stuttgart: Thieme 1974
62. Photaki, I.: in "The Chemistry of Polypeptides" (Katsoyannis, P., Ed.), p. 59, New York: Plenum Press 1973
63. König, W., Geiger, R.: in "Perspectives in Peptide Chemistry" (Eberle, A., Geiger, R., Wieland, T., Eds.), p. 31, Basel: Karger 1981
64. Hiskey, R. G., Tucker, W. P.: J. Amer. Chem. Soc. *48*, 4789 (1962)
65. Hiskey, R. G., Mizoguchi, T., Smithwick, E. L.: J. Org. Chem. *32*, 97 (1967)
66. Hiskey, R. G., Sparrow, T. J.: J. Org. Chem. *35*, 15 (1970)
67. Hiskey, R. G., Ward, B. F., Jr.: J. Org. Chem. *35*, 1118 (1970)
68. Brois, S. J., Pilot, J. F., Barnum, H. W.: J. Amer. Chem. Soc. *92*, 7629 (1970)
69. Kamber, B.: Helv. Chim. Acta *56*, 1370 (1973)
70. Hiskey, R. G., Muthukumaraswamy, N., Vunnam, R. R.: J. Org. Chem. *40*, 950 (1975)
71. Abraham, E. P.: Biochemistry of Some Peptide and Steroid Antibiotics, New York: Wiley 1957
72. Schwyzer, R., Iselin, B., Rittel, W., Sieber, P.: Helv. Chim. Acta *39*, 872 (1956)
73. Kopple, K. D.: J. Pharmaceutical Sci. *61*, 1345 (1972)
74. Schwyzer, R., Sieber, P.: Helv. Chim. Acta *41*, 2186 (1958)
75. Schwyzer, R., Sieber, P.: Helv. Chim. Acta *40*, 624 (1957)
76. Kato, T., Izumiya, N.: in "Chemistry and Biochemistry of Amino Acids, Peptides and Proteins", Vol. 2 (Weinstein, B., ed.), p. 1, New York: Dekker 1974
77. Ruttenberg, M. A., Mach, B.: Biochemistry *5*, 2864 (1966)
78. Ziegler, K., Eberle, H., Ohlinger, H.: Liebigs Ann. Chem. *504*, 95 (1933)
79. Waki, M., Izumiya, N.: J. Amer. Chem. Soc. *89*, 1278 (1967)
80. Bodanszky, M., Henes, J. B.: Bioorg. Chem. *4*, 212 (1975)
81. Ramachandran, G. N., Chandrasekaran, R.: in "Progress in Peptide Research" (Lande, S., ed.), Vol. 2, p. 195, New York: Gordon and Breach 1972

82. Gerlach, H., Ovchinnikov, Y. A., Prelog, V.: Helv. Chim. Acta *47*, 2294 (1964).
83. Wieland, T., Faesel, J., Faulstich, H.: Liebigs Ann. Chem. *713*, 201 (1968)
84. Fahrenholz, H., Faulstich, H., Wieland, T.: Liebigs Ann. Chem. *743*, 83 (1971)
85. Wieland, T., Lapatsanis, L., Faesel, J., Konz, W.: Liebigs Ann. Chem. *747*, 194 (1971)
86. Vogler, K., Studer, R. O., Lergier, W., Lanz, P.: Helv. Chim. Acta *43*, 1751 (1960)
87. Wieland, T., Ohly, K. W.: Liebigs Ann. Chem. *605*, 179 (1957)
88. König, W., Geiger, R.: Liebigs Ann. Chem. *727*, 125 (1969)
89. Brady, S. F., Varga, S. L., Freidinger, R. M., Schwenk, D. A., Mendlowski, M., Holly, F. W., Veber, D. F.: J. Org. Chem. *44*, 3101 (1979)
90. Shiori, T., Ninomiya, K., Yamada, S.: J. Amer. Chem. Soc. *96*, 2597 (1972)
91. Isied, S. S., Kuehn, C. G., Lyon, J. M., Merrifield, R. B.: J. Amer. Chem. Soc. *104*, 2632 (1982)
92. Belleau, B., Malek, G.: J. Amer. Chem. Soc. *90*, 1651 (1968)
93. Woodward, R. B., Olofson, R. A.: J. Amer. Chem. Soc. *83*, 1007 (1961)
94. Bláha, K., Rudinger, J.: Collect. Czechoslov. Chem. Commun. *30*, 3325 (1965)
95. Hassal, C. H., Sanger, D. G., Handa, B. K.: J. Chem. Soc. C, 2814 (1971)
96. Anderson, G. W., Blodinger, J., Young, R. W., Welcher, A. D.: J. Amer. Chem. Soc. *74*, 5304 (1952)
97. Studer, R. O., Lergier, W.: Helv. Chem. Acta *48*, 460 (1965)
98. Fridkin, M., Patchornik, A., Katchalski, E.: J. Amer. Chem. Soc. *87*, 4646 (1965)
99. Rothe, M.: Acta Chim. Acad. Sci. Hung. *18*, 449 (1959)
100. Nutt, R. F., Veber, D. F., Saperstein, R.: J. Amer. Chem. Soc. *102*, 6539 (1980)
101. Wade, R., Winitz, M., Greenstein, J. P.: J. Amer. Chem. Soc. *78*, 373 (1956)
102. Jarvis, D., Rydon, H. N., Schofield, J. A.: J. Chem. Soc., 1752 (1961)
103. Guttmann, S., Pless, J., Sandrin, E., Jaquenoud, P. A., Bossert, H., Willems, H.: Helv. Chim. Acta *51*, 1155 (1968)
104. Kamber, B., Rittel, W.: Helv. Chim. Acta *52*, 1074 (1969)
105. Inukai, N., Nakano, K., Murakami, M.: Bull. Chem. Soc. Jpn. *41*, 182 (1968)
106. Kamber, B.: Helv. Chim. Acta *54*, 927 (1971)
107. Curtis, R. W.: Science *128*, 661 (1958)
108. Anzai, K., Curtis, R. W.: Phytochemistry *4*, 263 (1965)
109. Kurath, P.: Helv. Chim. Acta *59*, 1127 (1976)
110. Ludescher, U., Schwyzer, R.: Helv. Chim. Acta *55*, 2052 (1972)
111. Wieland, T.: in "Progress in the Chem. Org. Natural Products" (Zechmeister, L., ed.), Vol. 25, p. 214, Vienna: Springer 1967
112. Wieland, T., Faulstich, H., Fahrenholz, F.: Liebigs Ann. Chem. 743, 77 (1971)
113. Nebelin, E., Faulstich, H., Wieland, T.: Liebigs Ann. Chem., 45 (1973)
114. Faulstich, H., Nebelin, E., Wieland, T.: Liebigs Ann. Chem., 50 (1973)
115. Munekata, E., Faulstich, H., Wieland, T.: J. Amer. Chem. Soc. *99*, 6151 (1977)
116. Klausner, Y. S., Chorev, M.: J. Chem. Soc. Chem. Commun., 973 (1975)
117. Ondetti, M. A., Thomas, P. L.: J. Amer. Chem. Soc. *87*, 4373 (1975)
118. Shemyakin, M. M., Aldanova, N. A., Vinogradova, E. I., Feigina, M. Y.: Tetrahedron Lett., 1921 (1963)
119. Bonora, G. M., Toniolo, C.: Biopolymers *13*, 1055 (1974)
120. Stewart, F. H. C.: Austral. J. Chem. *18*, 887 (1965); ibid. *19*, 489, 1508 (1966)
121. DeTar, D. F., Gouge, M., Honsberg, W., Honsberg, U.: J. Amer. Chem. Soc. *89*, 988 (1967)
122. DeTar, D. F., Vajda, T.: J. Amer. Chem. Soc. *89*, 998 (1967)
123. Kovács, J., Giannotti, R., Kapoor, A.: J. Amer. Chem. Soc. *88*, 2282 (1966)

124. Segal, D. M.: J. Mol. Biol. *43*, 497 (1969)
125. Jones, J. H., Young, G. T.: J. Chem. Soc. (C), 436 (1968)
126. Johnson, B. J.: J. Chem. Soc. (C), 1412 (1969)
127. Sakakibara, S., Kishida, Y., Kikuchi, Y., Sakai, R., Kakiuchi, K.: Bull. Chem. Soc. Jpn. *41*, 1273 (1968)
128. Olsen, B. R., Berg, R. A., Sakakibara, S., Kishida, Y., Prockop, D. J.: J. Mol. Biol. *57*, 589 (1971)
129. Horiuichi, K., Fujimoto, D., Seto, Y., Sakakibara, S.: J. Biochem. *83*, 261 (1978)
130. Kemp, D. S., Bernstein, Z. W., McNeil, G. N.: J. Org. Chem. *39*, 2831 (1974)
131. Jones, J. H.: in "Chemistry and Biochemistry of Amino Acids, Peptides and Proteins", Vol. 4 (Weinstein, B., ed.), p. 29, New York: Dekker 1977
132. Offord, R. E.: in "Semisynthetic Peptides and Proteins" (Offord, R. E., DiBello, C., eds.), p. 3, London: Academic Press 1978
133. Corradin, G., Harbury, H. A.: Biochem. Biophys. Res. Commun. *61*, 1400 (1974)
134. Dyckes, D. F., Creighton, T., Sheppard, R. C.: Nature *247*, 292 (1974)
135. Gattner, H. G., Schmitt, E. W., Naithani, V. K.: Hoppe-Seyler's Z. Physiol. Chem. *356*, 1465 (1975)
136. Offord, R. E.: Nature *221*, 37 (1969)
137. Rees, A. R., Offord, R. E.: Biochem. J. *159*, 467 (1976)
138. Offord, R. E., Storey, H. T., Hayward, C. F., Johnson, W. H., Phasey, M. H., Rees, A. R., Wightman, D. A.: Biochem. J. *159*, 480 (1976)
139. Waldschmidt-Leitz, E., Kühn, K.: Chem. Ber. *84*, 381 (1957)
140. Sealock, R. W., Laskowski, M., Jr.: Biochemistry *8*, 3703 (1969)
141. Kowalski, D., Laskowski, M., Jr.: Biochemistry *15*, 1300, 1309 (1976)
142. Jering, H., Tschesche, H.: Eur. J. Biochem. *61*, 443, 453 (1976)
143. Laskowski, M., Jr.: in "Semisynthetic Peptides and Proteins" (Offord, R. E., DiBello, C., eds.), p. 263, London: Academic Press 1978
144. Fruton, J. S.: in Enzymology, Vol. 53 (Meister, A., ed.), p. 239, New York: Wiley 1982

VII. Techniques for the Facilitation of Peptide Synthesis

Synthesis of a peptide chain comprising a large number of amino acid residues can be a difficult task and proteins with over a hundred and sometimes hundreds of residues are formidable objectives. The steps of protection, coupling, deprotection, isolation and characterization of the intermediates is not only a time consuming but also a discouragingly repetitive endeavor. An additional source of frustration is the scarcity of solvents which are suitable for reactions with high molecular weight intermediates. Such difficulties lead to a search for techniques which can alleviate the burden of peptide chemists, for methods which can facilitate the building of long peptide chains from amino acids. The repetitive character of chain lengthening with active esters prompted speculations [1] about the mechanization of peptide synthesis. The first realization of such ideas came from Merrifield [2] and, simultaneously, from Letsinger and Kornet [3]. In both methods an amino acid is attached to an insoluble polymeric support and the subsequent operations are carried out on the peptidic material linked to the resin. The version proposed by Merrifield [2] developed into a major discipline, solid phase peptide synthesis. One of the most conspicuous characteristics of the new technique is that it stimulated further thoughts toward improvements and analogous solutions to an unprecedented extent. Research grew in this area in an exponential manner and has been reviewed repeatedly. A recent, comprehensive review of solid phase peptide synthesis by Barany and Merrifield [4] required almost 300 pages for a full account. The "solid" phase idea permeated other fields as well and is used now, with success, in the sequencing of proteins and in the synthesis of oligo- and polynucleotides. In this chapter we can give only the outlines of solid phase peptide synthesis particularly because it seems to be necessary to discuss in this connection also alternative techniques of facilitation even if these do not match in importance the solid phase method, at this time.

A. Solid Phase Peptide Synthesis (SPPS)

The insoluble polymeric supports applied in this technique are gels rather than real solids. The reactions carried out on materials attached to the polymers take place not merely on the surface of the "solid" materials but in the inside of the particles, usually beads, as well. This requires, of course, that the polymers swell in the solvents applied and that the reactants enter and leave, by diffusion, these particles. Nonetheless, the term "solid phase peptide synthesis" is a fitting expression, because it gives in a concise way the right impression of the technique and will, therefore, be used, also in abbreviated form (SPPS), in this section.

The essential advantage of SPPS is the ease with which the intermediates are obtained free from starting materials, reagents and most of the by-products. These are separated from the desired intermediate, because the latter, attached to the insoluble polymer, remains undissolved during washing with appropriately selected solvents. Thus, isolation of intermediates by extraction or crystallization becomes superfluous. A serious problem of peptide synthesis, finding solvents for large molecular weight intermediates is similarly circumvented. It is sufficient to use solvents in which the polymer swells. Also, it becomes rather simple to convert a salt of the amino component to the free amine, by treating the peptidyl resin with the solution of a tertiary base and by the removal of the alkylammonium salts formed, by washing. These are significant simplifications of the procedures of peptide synthesis but they are bought at a certain price: purification and characterization of the intermediates is generally omitted. Yet, elimination of the problems of isolation permits mechanization and automation of the process. Numerous instruments have been described for the automatic execution of SPPS. Also, methods were developed for the continuous analytical control of chain building. In this section, however, we do not deal with such technical aspects of SSPS but concentrate on the principles involved.

In the original scheme of SPPS the C-terminal amino acid is anchored to the insoluble support through an ester bond. For this purpose, the resin, a copolymer of styrene and 1–2% divinylbenzene, is converted to a chloromethyl derivative, essentially a benzylchloride moiety connected with the matrix of the polymer. The chloromethyl group is allowed to react with the carboxylate form of a protected amino acid:

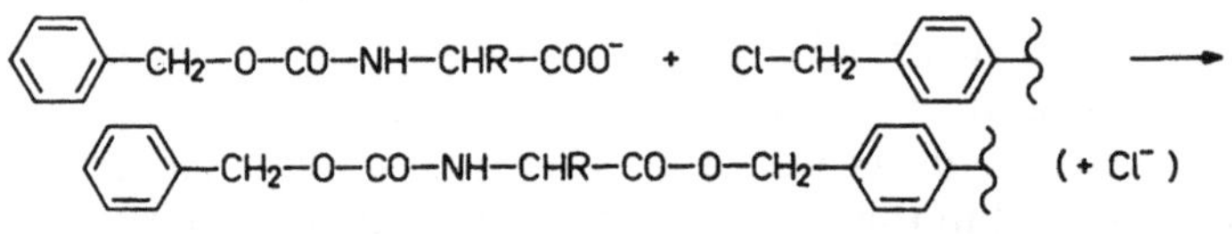

Removal of the benzyloxycarbonyl group by treatment with hydrobromic acid in acetic acid is not entirely selective and some of the newly formed ester bond is also cleaved in the reaction. Initially [2] the desired selectivity was secured by nitration of the polymer. Nitrobenzyl esters are quite resistant to acidolysis. A better solution for this problem was found in the use of the tert.butyloxycarbonyl group for amine protection. The Boc group is removed with a dilute solution of hydrogen chloride in acetic acid or with a mixture of trifluoroacetic acid and dichloromethan. Thus, nitration of the resin became unnecessary:

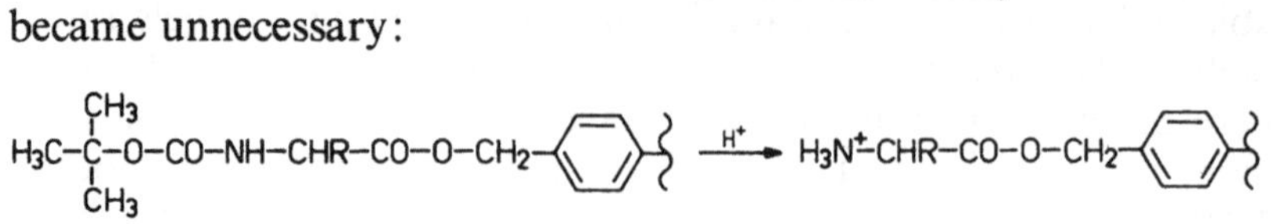

Deprotonation of the amine salt with a solution of a tertiary amine such as triethylamine yields the free amine which, in turn, is acylated with the penultimate amino acid, protected by a Boc group. In most instances dicyclohexylcarbodiimide is used as condensing agent. Excess reagents and the byproduct, N,N′-dicyclohexylurea are removed by dichloromethane washes:

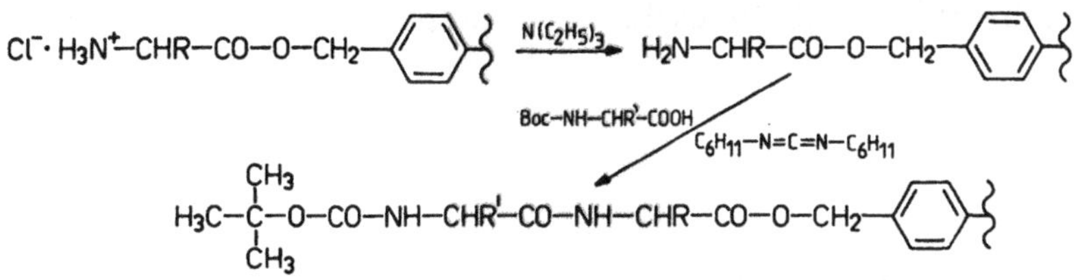

The cycle of deprotection by acidolysis, deprotonation with a tertiary amine and acylation with the next residue to be incorporated, is repeated until the chain of the target peptide is completed. At this point the peptide is cleaved from the insoluble support, usually by acidolysis with strong acids, e.g. hydrogen bromide in trifluoroacetic acid or more recently with liquid hydrogen fluoride. If the side chain functions are masked with acidolytically removable semipermanent blocking groups, then separation of the peptide from the support by cleavage of the benzyl ester bond, is accompanied by complete deprotection and a solution of the unmasked (albeit protonated) peptide is obtained:

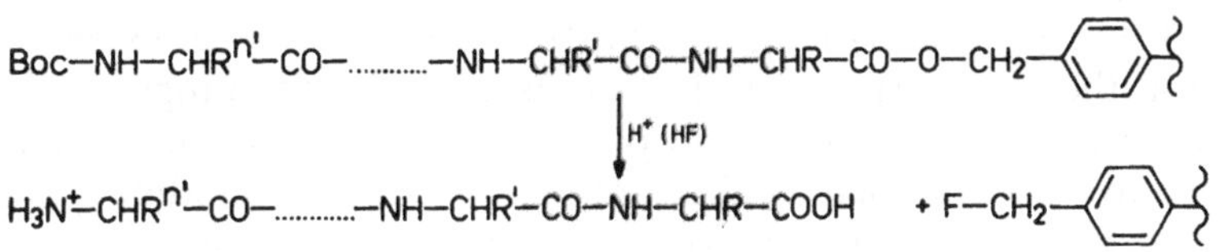

1. The Insoluble Support

From time to time various materials were proposed for the role of insoluble support, but the resin used by Merrifield [2] remains the starting point of most new developments. Copolymerization of styrene with 2% divinylbenzene, added to secure insolubility through crosslinking, yields a mechanically resistant substance, which is chemically sufficiently inert to withstand the operations of peptide synthesis, but also reaetive enough to allow its derivatization. Modification of the copolymer is necessary for the creation of a "handle" which makes possible the anchoring of an amino acid or peptide to the resin through a chemical bond. For better swelling and thus better permeability the amount of divinylbenzene used in copolymerization was lowered to 1% or below. With 1% divinylbenzene the copolymer swells quite well in some organic solvents, e.g. toluene, dimethylformamide or dichloromethane and yet has acceptable mechanical stability. The original chloromethyl derivative [2] is still much in use, although numerous modifications [4] have been introduced. Some of these, the hydroxymethyl-resin, [5–7]

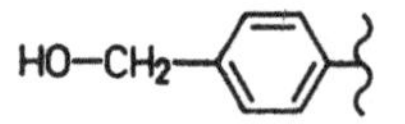

the aminobenzhydryl resin [8, 9], useful in the preparation of peptide amides

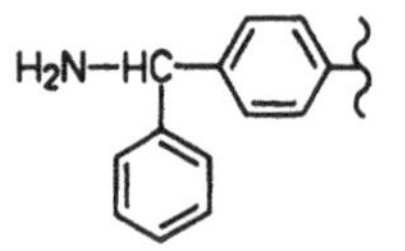

or the resin of Wang [7, 10] which is cleaved by moderately strong acids are

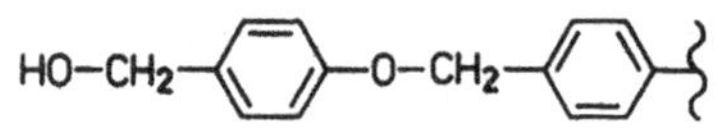

of practical significance.

A more recent thought is to use polymers which are somewhat peptide-like. This should allow a more uniform swelling of the peptidyl resin in selected solvents and could render the sites of subsequent reactions fully accessible both to acylating agents and deblocking reagents. Some of these supports are derivatives of the polystyrene-divinylbenzene copolymer [11–13] while others are based on different macromolecules, such as polydimethyl-acrylamide [14–17] or polyethyleneimine [18]. In the absence of an extensive study needed for the comparison of the merits and shortcomings of the numerous polymeric supports [4] which appeared in the literature it seems to be premature to attempt their evaluation. The futility of predictions is indicated by past experience: quite promising avenues, such as supports which were applied as coats on the *surface* of glass particles of controlled porosity [19] or polystyrene grafted by radiation onto the *surface* of Kel-F [20] remained without followers. In some recent succesful syntheses of complex natural products good results were, at least in part, attributed to the use of improved supports. Since in the same syntheses, also improved methods of activation and new protecting groups were applied it is difficult to discern the influence of the individual modifications.

2. The Bond Between Peptide and Polymer

As described in the introduction of SPPS, the ester bond between the C-terminal residue and the chloromethyl resin was initially [2] obtained by the nucleophilic displacement of the chlorine atom from the polymer-bound benzyl chloride by a carboxylate group:

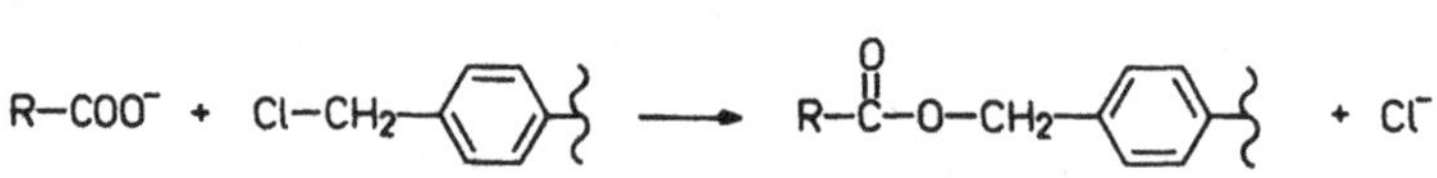

Side reactions can accompany even such a simple formation of a benzyl ester. For instance, in addition to the carboxylate, benzyl chloride can alkylate also certain functions in amino acid side chains, the thioether in methionine [21] or the imidazole nucleus in histidine. When no such complications have to be feared, displacement of the chlorine is quite practical. The efficiency of the exchange, that is the "loading" of the resin, was improved by the use of cesium salts of protected amino acids in dimethylformamide [22]. In a simple alternative, the polymeric benzyl esters are prepared through the acylation of polymer-bound benzyl alcohol [5]

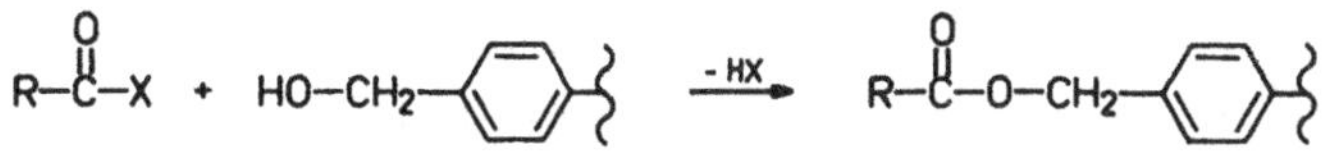

For activation of the protected amino acid carbonyldiimidazole [5], carbodiimides [5], dimethylformamide di-neopentyl acetal [23], symmetrical anhydrides [24] are similarly useful, as is the imidazole catalyzed transesterification of active esters [10] in non-polar solvents [25]. An original approach is the formation of benzyl esters in the reaction between a trialkyl sulfonium salt and a protected amino acid [26]

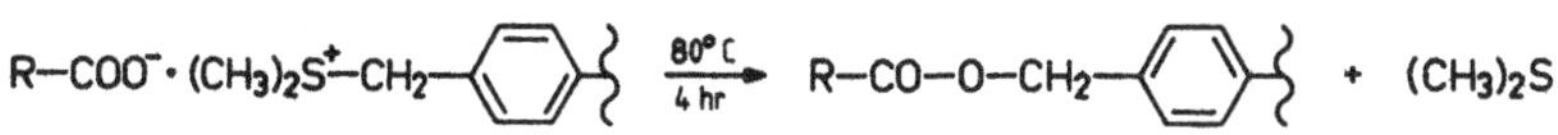

Forming ester bonds by acylation of a hydroxymethyl-polymer rather than by displacement of chloride anions from a chloromethyl resin eliminates the possible alkylation of amino acid side chains by the benzyl chloride moiety. Yet, this approach is also not immune from side reactions. If acylation of the hydroxyl groups is incomplete, the remaining free alcoholic hydroxyls can react with the activated derivative of a protected amino acid introduced at a later stage. This would lead to peptides from which a C-terminal portion of the target compound is missing. To avoid the formation of such "deletion" sequences, hydroxyl groups which did not participate in ester formation during anchoring are acetylated or benzoylated:

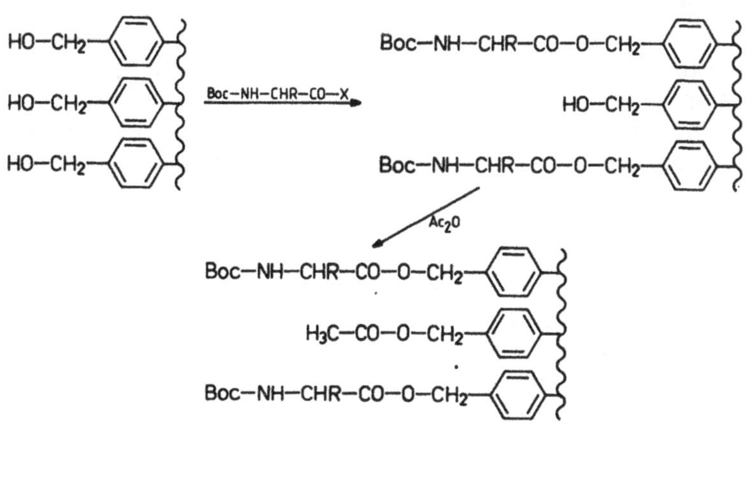

There are also several alternative modes of anchoring the peptide to the polymer. In addition to the, by now classical, ester or amide bond formation between the support and the carboxyl of the C-terminal residue, various side chain functions can also be utilized for this purpose. As an example, we mention the development of a polymer which carries a dinitrofluorobenzene moiety as a handle [27]. This allows the formation of a covalent bond with the sulfhydryl group in the cysteine side chain or with the imidazole in histidine:

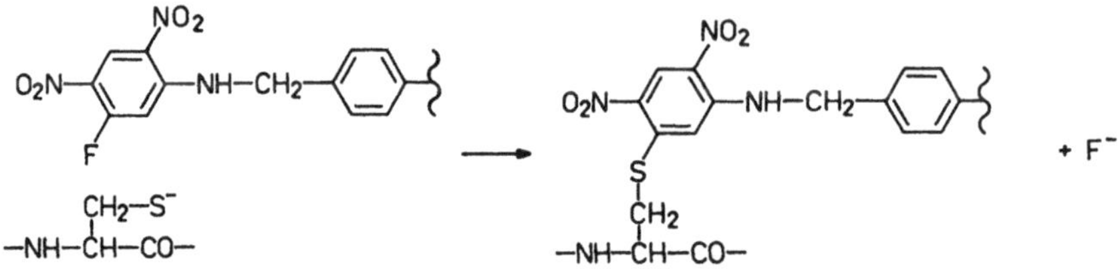

3. Protection and Deprotection

For a considerable period of time the tert. butyloxycarbonyl group was selected for the blocking of α-amino functions by most practitioners of SPPS. A certain tendency for more acid sensitive protection revealed itself in recommendations for the use of the biphenylisopropyloxycarbonyl [28], the phenylisopropyloxycarbonyl [29, 30] or the 3,5-dimethoxyphenylisopropyloxycarbonyl [31] group. The amines thus masked can be deprotected under very mild conditions, e.g. with a dilute solution of trifluoroacetic acid in dichloromethane. It is possible, therefore, to design combinations in which the semipermanent blocking of side chain functions and scission of the peptide from the resin are carried out by acidolysis and yet with not excessively strong acids. This is a highly desirable feature of synthetic schemes, because in the much practiced alternative, the side chain protecting groups had to be modified to render them more acid resistant in order to prevent their partial removal during the unmasking of α-amino groups. The acid resistant protecting groups [32, 33], e.g. the 2-chlorobenzyloxycarbonyl group used for the blocking of amino groups in lysine side chains, require drastic acidic treatment for removal. This, and the cleavage of the peptide from the resin, necessitated the use of liquid hydrogen fluoride as acidic reagent and sometimes conditions under which the integrity of the peptides was endangered (cf. Chapter V). Thus, tactics in which very acid sensitive groups are combined with moderately acid sensitive protection of side chains are probably preferable to schemes which involve drastic cleavage by acidolysis. It seems to us, that a new era was initiated by the discovery of the 9-fluorenylmethyloxycarbonyl group [34] and its introduction in SPPS [24, 35–37]. Incorporation of Fmoc amino acids permits orthogonal combinations, because removal of the Fmoc group after each chain lengthening step requires only a treatment with the solution of a secondary amine, usually piperidine, in dimethylformamide. This reagent and the conditions applied have no effect on acid-labile protecting groups based on the formation of tert. butyl cations and the anchoring of the peptide to the polymeric support also can be so designed that it requires only trifluoroacetic acid for separation. Many of the side reactions experienced in SPPS could be attributed to the action of hydrogen fluoride or similarly strong acids. Hence, the use of a less strong acid, most commonly trifluoroacetic acid, for the scission of the peptide-resin bond and for the simultaneous removal of all protecting groups is a major improvement in SPPS.

4. Methods of Coupling in SPPS

There are probably few methods of activation and coupling which were not tried in SPPS, but dicyclohexylcarbodiimide, adopted by Merrifield [2] for this purpose remains the most widely used reagent so far. The straightforward application of DCC results in efficient incorporation of protected amino acids, yet a certain price has to be paid for this simplicity. The O → N acyl migration in the reactive intermediate (cf. Chapters II and V) leading to the formation of N-acylurea derivatives causes merely a loss on starting materials and is easily compensated by the application of the protected amino acids and the reagent (DCC) in excess. The by-products, ureide derivatives of the protected amino acids are readily removed by washing the peptidyl polymer with organic solvents, an operation needed anyhow for the removal of unreacted starting materials and the by-product, N,N'-dicyclohexylurea. Some additional side reactions associated with DCC-mediated coupling require further measures. Thus, dehydration of the side chains of asparagine and glutamine residues (cf. Chapter V) must be reduced by the addition of auxiliary nucleophiles, such as 1-hydroxybenzotriazole [38]. For complete elimination of this side reaction usually active esters are used instead of DCC-activated intermediates, since the nitriles formed during the preparation of (N-protected) active esters of asparagine and glutamine are removed by recrystallization and active esters, free from nitriles [39], can be incorporated into the peptide chain.

Active esters can be used in the introduction of other amino acid residues as well. Good results were achieved with *p*-nitrophenyl esters [5, 40] but the general application of active esters in SPPS, in spite of their relative selectivity, lags far behind in popularity when compared with the use of DCC as coupling reagent. This is probably due to the longer reaction time required in acylation with active esters, a fairly important consideration in SPPS. Furthermore, the solvent of choice in coupling with active esters is dimethylformamide. Several active esters, e.g. *p*-nitrophenyl esters react quite slowly in the usual medium of SPPS, dichloromethane. Since steric hindrance plays an important role in SPPS, acceleration of the coupling reaction by the use of more reactive esters is a questionable remedy when higher reactivity is counterbalanced by higher bulk as in pentachlorophenyl esters. The introduction in SPPS of *o*-nitrophenyl esters [41] which are less sensitive to solvent effects and which better penetrate crowded environments [42] did not change the preferences of the practitioners of SPPS and DCC remained the reagent of choice. A series of other active esters were tried with no more success in popularity in spite of good results in coupling. Interestingly, even the spectacular acceleration of acylation through the catalysis of active ester reactions with 1-hydroxybenzotriazole [43] or the proposed use of the extremely powerful pentafluorophenyl esters [44, 45] failed to impress the field.

The majority of coupling methods and coupling reagents have been applied also in SPPS, but even sophisticated procedures such as the oxidation-reduction methods [46] gained only occasional application (cf. e.g., ref. 47). A major improvement in the methodology of SPPS was brought about by the modification of the carbodiimide process toward the generation of symmetrical anhydrides rather than O-acylisoureas as the reactive intermediates which achieve the acylation of the polymer-bound amino component. This was readily accomplished by changing the ratio of protected amino acid to carbodiimide from 1:1 to 2:1 and using the reaction mixture as a preformed solution of the symmetrical anhydride:

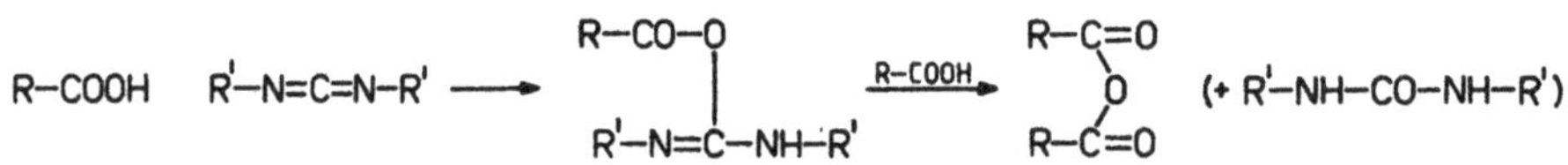

The principle of using symmetrical anhydrides for stepwise chain building [48] could thus readily be extended [49, 50] to SPPS. It is an expensive procedure since in practice usually several times more than the calculated two moles of protected amino acid are needed for the complete incorporation of a single amino acid residue. The better quality of the product is considered by many as sufficient justification for such sacrifice. Further improvements can be expected from the utilization of isolated symmetrical anhydrides [51, 52]. The recently recommended application of mixed anhydrides in SPPS [53] awaits evaluation.

5. Separation of the Completed Peptide Chain from the Polymeric Support

Acidolysis, the originally [2] applied method for breaking the ester bond between peptide and resin is still practiced. The reagent, however, used for the scission of the benzyl ester type linkage, hydrobromic acid in trifluoroacetic acid, has been replaced by the more powerful liquid hydrogen fluoride [54]

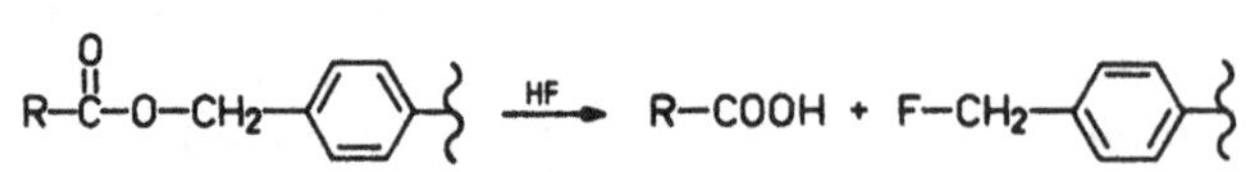

which cleaves not only the bond between peptide and resin but also simultaneously removes the relatively acid resistant semipermanent blocking groups. An alternative approach, the use of highly acid sensitive protecting groups for the masking of the α-amine function allows the application of moderately acid resistant side chain protection and the use of resin types [10] which are separated from the chain by treatment with moderately strong acids, usually trifluoroacetic acid:

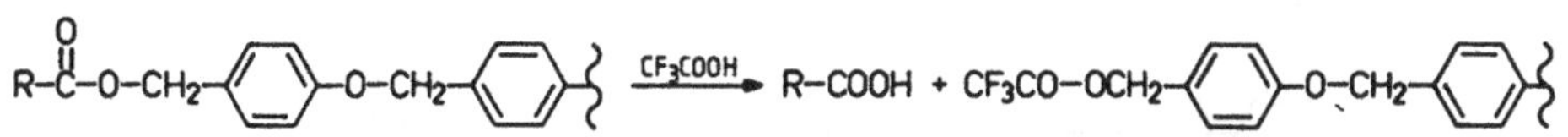

The same resin type is suitable, of course, also for orthogonal schemes in which an acid resistant but base sensitive amine protecting group, such as the 9-fluorenylmethyloxycarbonyl (Fmoc) group [34] is applied [35–37].

In order to increase the versatility of SPPS and to gain freedom from the potent but less than innocuous hydrogen fluoride, a new support, the *p*-alkoxy-benzyloxycarbonylhydrazide resin was proposed [10]. Incorporation of Bpoc-amino acids leads to a chain which on cleavage with a 1:1 mixture of tri-fluoroacetic acid and dichloromethane yields a peptide hydrazide that can be activated to serve as the carboxyl component in a segment condensation:

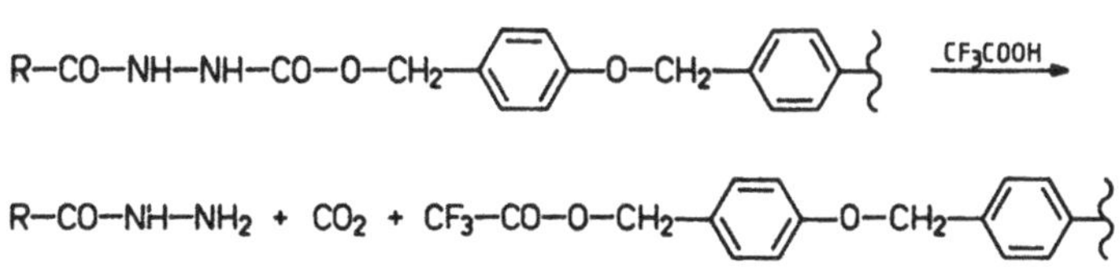

It is equally possible, however, to produce peptide hydrazides via direct hydrazinolysis of peptidyl polymers [40]:

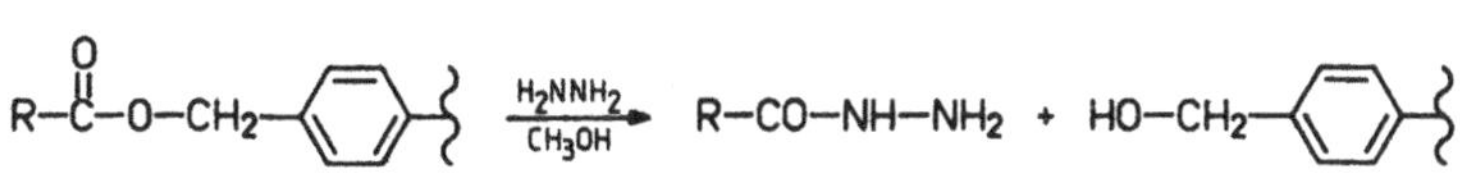

The analogous ammonolysis [40] yields amides[15] but in some cases it is accompanied by alcoholysis and peptide esters are obtained [5]. This, as a method of cleavage, can then be applied for the preparation of methyl [55] and also of benzyl esters [56]. In the latter case KCN was found to be an efficient catalyst of transesterification:

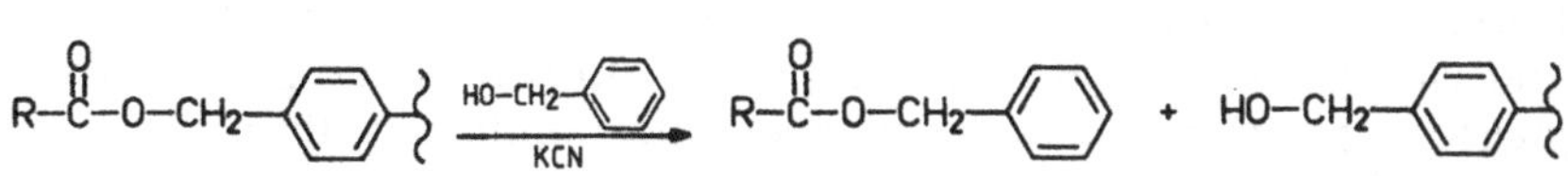

[15] When ammonolysis is complicated by the presence of additional ester groups, or is inhibited by a bulky side chain in the C-terminal residue, it is more practical to use the amino-benzyhydryl resin [8, 9] specially geared for the synthesis of peptide amides.

A particularly interesting process of cleavage [57] involving transesterification with dimethylaminoethanol followed by hydrolysis, is based on intramolecular base catalysis:

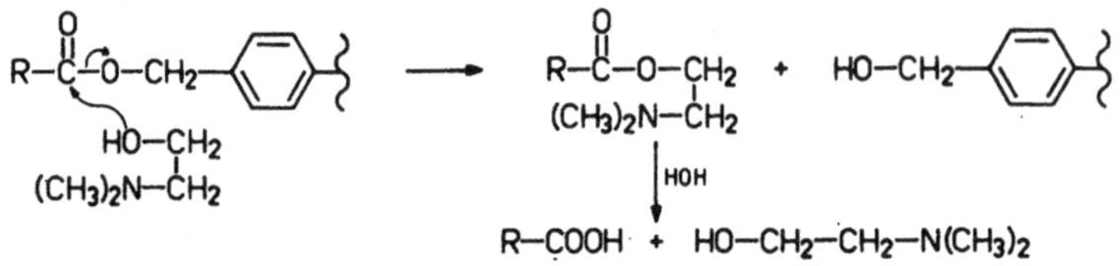

Acidolysis, base catalyzed alcoholysis, ammonolysis and hydrazinolysis are probably the most frequently applied methods for breaking the peptide-resin bond. Several more possibilities were, however, proposed. For instance cleavage of dehydroalanine residues with acids [58] yields the peptide amide:

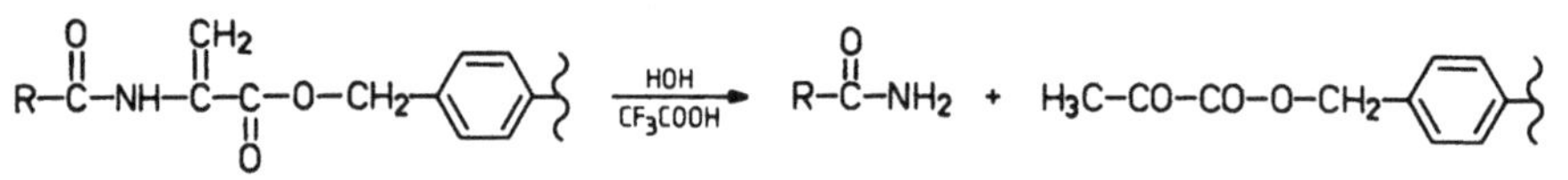

Photolysis of *o*-nitrobenzyl ester type polymers [59] opens a promising avenue:

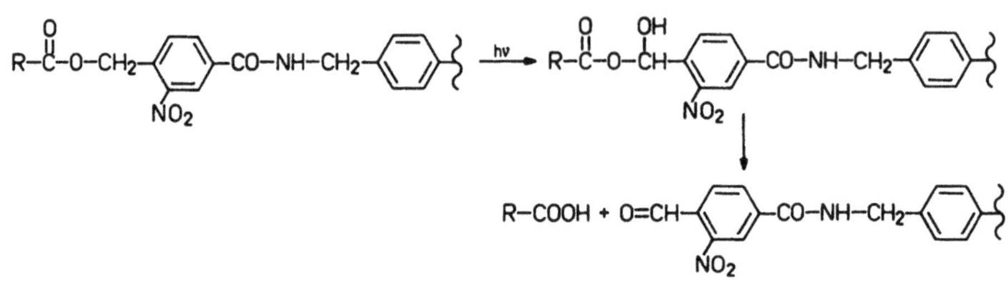

Interestingly, the most classical method of protecting group removal, hydrogenolysis, could be applied only relatively late in the development of SPPS. The difficulties caused by the physical separation of a platinum metal catalyst on a solid support and the polymer-bound peptide could be overcome [60, 61] by the impregnation of the polymer with a solution of palladium acetate in dimethylformamide and generation of the metallic catalyst "*in situ*". Thus, peptide and catalyst share the same support. Removal of the peptide from the polymeric support requires more energetic conditions than the ones generally applied for the hydrogenolysis of benzyl esters:

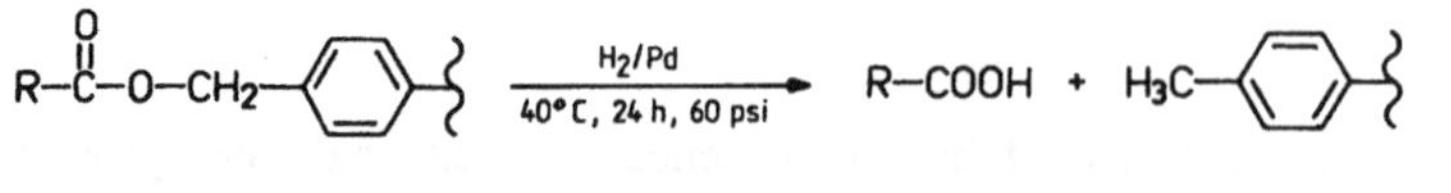

Transfer hydrogenation, with cyclohexene as donor, was found to be similarly useful for this purpose [62].

Anchoring of a peptide to a polymer via a side chain function rather than through the α-carboxyl group of the C-terminal residue abounds in possible variations. Application of side chain carboxyls for this purpose involves chemistry already encountered in connection with α-carboxyls. A certain adaptation of the support is necessary, however, if the imidazole in the histidine side chain [27, 63] or the sulfhydryl group of a cysteine residue [64] has to participate in the link between peptide and polymer. A polymer-bound dinitrofluorobenzene moiety can serve as the reactive site in the formation of the desired bond which, in turn, can be cleaved by thiolysis.

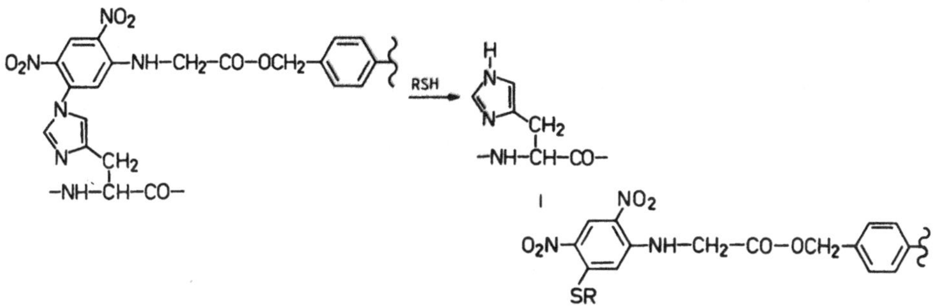

Such side chain attachments, since they are based on methods for the masking of side chain functions, are removed by the processes developed for the removal of the appropriate blocking groups. For instance the dinitrophenyl-imidazole bridge is cleaved by 2-mercaptoethanol.

It is impractical to try to give a full account of the published methods for the scission of the bonds which link peptides to polymers. They are too numerous for brief presentation and are treated better in a specialized article like the one written by Barany and Merrifield [4]. Here we point rather to the trend that can be recognized in the development of such procedures. The initially proposed [65] acidolysis reagent, hydrogen bromide in trifluoroacetic acid, was gradually displaced by the more potent liquid hydrogen fluoride [54]. Such powerful reagents, however, are usually less selective than would be desirable. Thus, HF, in addition to the cleavage of the peptide from the resin, also simultaneously removes practically all semipermanent protecting groups used for side chain protections. While this is a welcome simplification of the process, final deprotection by HF is also accompanied by side reactions. Some of these, like the N → O shift at serine residues, or the transfer of methyl groups from anisole (added as a scavenger) to methionine and tryptophan side chains were discussed in more detail in Chapter V. Additional complications, e.g. Friedel-Crafts acylation of anisole or of the aromatic nuclei of the resin by side chain carboxyls are due to the extreme acidity of the medium as is the formation of succinimide derivatives at aspartyl residues. Other similarly strong acids, such as trifluoromethanesulfonic acid, are equally conducive to undesired reactions. Because of this high price paid for the application of a universal cleaving-deprotecting reagent the use of more acid sensitive peptide-resin bonds seem to gain importance. Accordingly, α-amino functions are blocked by even more acid sensitive masking groups or by groups which are quite resistant to acids and are removed by bases. For instance, a combination of an acid sensitive resin [10] with the incorporation of 9-fluorenylmethyloxycarbonyl (Fmoc) amino acids (particularly in the form of their isolated symmetrical anhydrides [52]) and with side chain protection based on tert. butyl groups represents a scheme which appears to be superior, with respect to homogeneity of the product, to previously practiced approaches.

6. Problems in Solid Phase Peptide Synthesis

Most insoluble polymeric supports used in SPPS swell in selected organic solvents such as dichloromethane or dimethylformamide and allow the diffusion of reactants into the interior of the resin particles. The reactants, acylating agents, catalysts, reagents used for the removal of blocking groups, generally reach the functional groups of the molecules and react with these in the expected way but accessibility of functional groups, e.g. of amino groups to be acylated, is not always perfect. Therefore, in a given time a small portion of the reacting groups, for instance N-terminal amines, might fail to react with the acylating agent and remains unchanged. The same situation can exist during deprotection: a portion of blocked peptide chains remains fully protected after the exposure to the deblocking reagent, e.g. trifluoroacetic acid. The time allowed for the reaction was extended, the reagents were applied in considerable excess, catalysts were added or the acylation and/or deprotection reactions were repeated. The addition of auxiliary solvents, e.g. trifluoroethanol or swelling and shrinking of the resin with suitable organic solvents seem to alleviate the problem. A crucial factor in the prevention of incomplete acylation or deprotection is the polymer itself. Therefore, less crosslinked resins which show more intense swelling and allow a better penetration of the reactants were advocated. Supports were introduced in which the absence of swelling caused the peptides to be attached only to the surface where they are more readily available for the reactants. A correction of incomplete acylation can be achieved by an additional acylation reaction in which the unreacted amino groups are blocked with the help of powerful acylating agents of small molecular weights, such as acetic anhydride or acetylimidazole. Such corrections notwithstanding, the remaining imperfections in acylation or in the removal of blocking groups can lead to the formation of chains from which one of the amino acid residues is absent. Such materials were designated as "failure sequence" or "deletion sequences" but the expression "*deficient peptides*" might be more descriptive. The presence of several deficient peptides in the crude synthetic material creates serious problems in purification, since the properties of such contaminants are generally quite similar to those of the target compound.[16]

[16] Also, the mixture of a series of closely related deficient peptides has an amino acid composition which is close to that of the desired material, a situation which renders a control through quantitative amino acid analysis rather meaningless.

A different type of contaminant is produced by the undesired acylation of α-amino groups, e.g. by trifluoroacetic acid and dicyclohexylcarbodiimide, the former the result of incomplete removal after deprotection, the latter added as the reagent for the incorporation of the next residue:

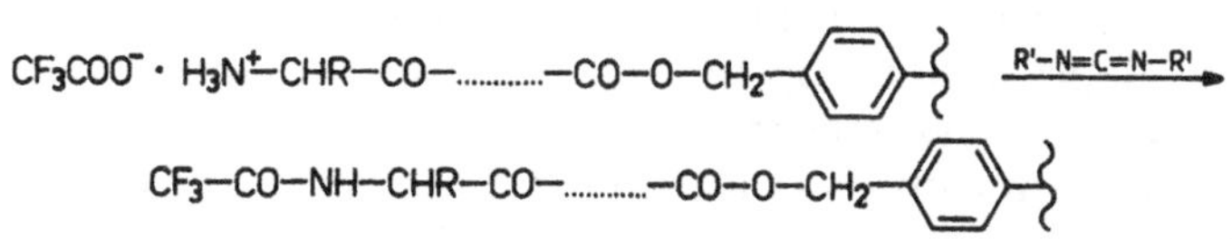

The blocked sequences thus produced remain unchanged during the subsequent steps and will be, therefore, rather different from the target compound with respect both to molecular weight and physical properties. Hence, their separation from the desired product should be less problematic than the elimination of deficient peptides. Yet, these by-products, often called "truncated sequences" but perhaps better designated as "*prematurely terminated chains*", can seriously affect the yield of the total process. It is obvious that analytical methods routinely applied in the synthesis of peptides in solution are not always practical in SPPS. For instance elemental analysis, the study of uv, ir or nmr spectra, etc. are very much impeded by the presence of the support. Yet, to avoid or at least to detect the formation of deficient peptides and prematurely terminated chains, *monitoring* of the process is an important aspect os SPPS.

The literature abounds in methods designed for the monitoring of SPPS and several reviews [2, 66, 67] treat this problem in considerable detail. To establish the completeness of acylation some simple methods, such as staining the beads with ninhydrin, fluorescamine or 2,4,6-trinitrobenzenesulfonic acid are quite useful. In acylation with active esters, the amount of the by-product formed from the leaving group is readily determined, e.g. spectrophotometrically. The results, however, must be evaluated with certain caution, because measurable amounts of some by-products, such as *p*-nitrophenol, can remain adsorbed on polystyrene-divinylbenzene copolymers. It is less obvious to establish the completeness of the reaction in the deblocking of α-amino groups. In the frequently used deblocking by acidolysis, the amount of acid associated with the amine in the form of a salt can be determined after it has been displaced by washing with the solution of a base. The information gained through the analysis of the filtrate is again not necessarily reliable, because strong acids are able to protonate, in addition to free amines, amide nitrogens as well. Therefore the direct titration of the free amines in the deblocked peptide, e.g. with perchloric acid, might provide more trustworthy data. Even more powerful control can be established by the determination of the amino acid sequence through automated Edman degradation or by mass spectrometry. These methods, however, require considerable effort and are quite expensive at this time.

The value of analytical information greatly depends on the point in time when the data were secured. Analysis of the crude peptide after its separation from the polymeric support is worthwhile, because the results can help in the planning of future experiments, but they cannot improve the quality of the synthetic material or eliminate the shortcomings of a synthesis already completed. In order to eliminate the formation of deficient peptides or prematurely terminated chains a continuous monitoring of the procedure is necessary. The fully automated version of SPPS also requires the continuous feed-back of the analytical results and immediate correction of incomplete acylations or imperfect deprotections by the repetition of the appropriate steps. Automation of a process assumes the possibility of generalization. In this respect, however, a certain caution is warranted. Peptides, as shown by their diverse and often highly specific biological activities, are widely different from each other. The individuality of the amino acids is further enhanced by their environment, and the properties of a peptide vary not simply with its amino acid composition but with its sequence as well. Therefore, generalizations in peptide chemistry usually turn out to be unjustified. It might not be too productive to treat amino acids as equals and to calculate [68] the probability of deficient peptide formation by statistical means. The conspicuous differences observed in the rates of acylation between derivatives of valine or isoleucine and of other, less hindered residues should serve as a warning against such endeavors. A biologically active peptide represents something very "unstatistical". Hence, instead of predictions and generalizations, rather a strict control of the synthesis is advisable.

In closing the discussion of the problems of SPPS we point to the *dilution* of the peptide by the polymeric support, particularly at low "loading" of the resin. As a consequence of this dilution large volumes of organic solvents are needed both in the reactions and for the washing of the support after each reaction. The necessity to wash the swollen resin particles by diffusion does not alleviate this situation. Furthermore, to counteract the effect of dilution on the rates of acylation reactions, the acylating agents are used in considerable excess [69]. When symmetrical anhydrides are applied and acylations, to ensure their completeness, are repeated, the excess of a protected amino acid used in the incorporation of a single residue can be many times more than the calculated amount. The ensuing decrease in the economy of the process together with the need for solvent regeneration might limit the adaptation of SPPS in the large scale production of complex peptides.

B. Synthesis in Solution

1. Peptides Attached to Soluble Polymers

The underlying principle of the "liquid phase" method is to adopt the most important feature of solid phase peptide synthesis, that is the facilitation of the separation of intermediates, and to combine it with the not less important asset of syntheses carried out in solution, namely the advantages of reactions performed in a homogeneous phase. In an early version of this approach polystyrene was applied, which, because of the absence of cross-linking, remained soluble in some organic solvents [70]. The peptidyl polymer could be handled as the usual intermediates of peptide synthesis but isolation of the protected intermediates was simplified by rendering them insoluble by dilution of their solutions with solvents in which the peptidyl polymers are insoluble but which remove the various starting materials and the by-products of the reactions. Because of technical difficulties the polystyrene based method found only few applications [71]. Similarly no practical use of polyethyleneimine supports [72] can be found in the literature. A more promising polymer, polyethyleneglycol with molecular weights ranging from 2000 to 20,000 dalton was introduced by Mutter and Bayer [73]:

$$HO{-}CH_2{-}CH_2{-}O{-}CH_2{-}CH_2{-}O{-}CH_2{-}CH_2{-}O{-}CH_2{-}CH_2{-}\ldots\ldots{-}O{-}CH_2{-}CH_2{-}OH$$

Chain building with the help of polyethyleneglycol follows an alternating solution and solid phase technique. Both acylation and deprotection take place in solution but the intermediates are isolated by crystallization-precipitation of the peptidyl polymer, usually by the addition of ether. Up to a certain point the solubility properties of the intermediates are determined by the support rather than by the peptide and this broadens the otherwise often narrow choice of media in which coupling or deblocking can be executed. From a certain chain length on, however, the properties contributed by the peptide portion outweigh those of the supporting polymer and the range of useful solvents again becomes limited.

For anchoring of a terminal residue an acid labile polymer bound protecting group can be applied. For instance an N-terminal amino acid might be linked to polyethyleneglycol via a benzhydryl group [74]:

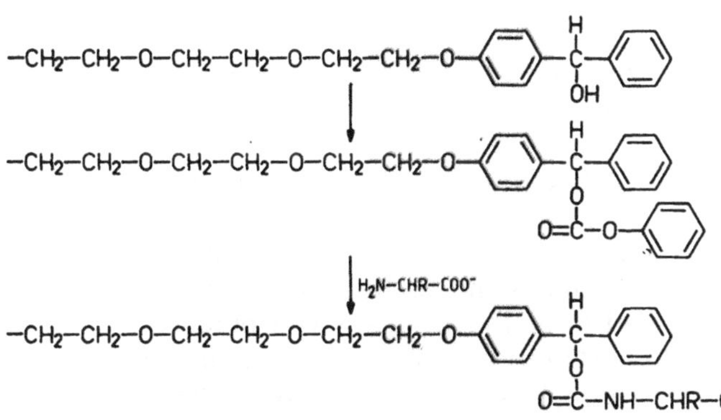

In an analogous manner the carboxyl group of the C-terminal residue can equally be used for the anchoring of the peptide to the polymer. For the many variations envisaged in connection with soluble polymeric supports, anchoring groups, methods of precipitation, etc. and for other technical details of the "liquid phase method"[17] the reader should consult the comprehensive article of Mutter and Bayer [75].

[17] The term "liquid phase" is ambiguous. It is often used to describe syntheses carried out in solution, without the use of polymeric supports. An expression such as "synthesis on soluble polymers" might be more descriptive.

2. The "handle" Method [76]

Incorporation of an ionizable group into a (protected) peptide makes it possible to adsorb the intermediates of a lengthy synthesis on an ion-exchange column. Unreacted starting materials and by-products are removed by washing with appropriate solvents, the peptide intermediates are eluted and exposed, in solution, to the reagents needed for deprotection and acylation. Two such ionizable "handles" were proposed for ion pair formation: 4-dimethylamino-4'-hydroxymethylazobenzene (or *p*-dimethylaminoazobenzyl alcohol) [77]

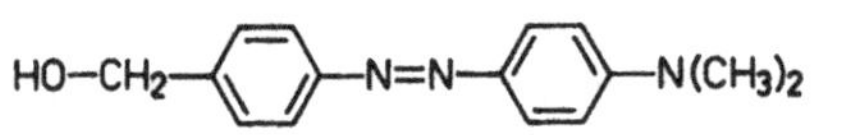

and 4-hydroxymethyl-pyridine (or 4-picolyl alcohol, or 4-pyridylcarbinol) [78]

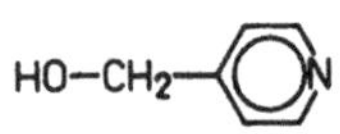

Esterification of the carboxyl group of the C-terminal amino acid with one of these alcohols creates a handle which, on protonation becomes a cationic center[18] readily bound to cation exchange resins such as sulfoethylsephadex or Amberlist-15:

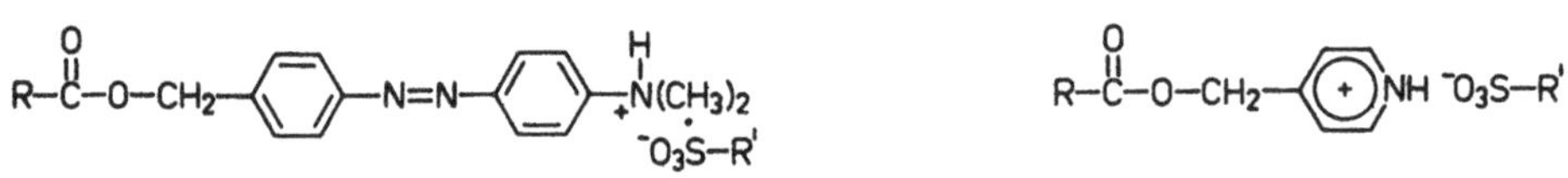

Various refinements of the picolyl ester method, such as selective adsorption on cation exchange resins saturated with 3-bromopyridine or elution with moderately strong organic bases, lead to wider acceptance. After early applications, e.g. for the synthesis of the lutenizing hormone releasing hormone (LH—RH) [80] the method had to be looked upon as a practical approach to complex peptides. A further extension of the picolyl ester method was the introduction of 4-picolyloxycarbonylhydrazides [81]. This handle allows the preparation of peptide hydrazides

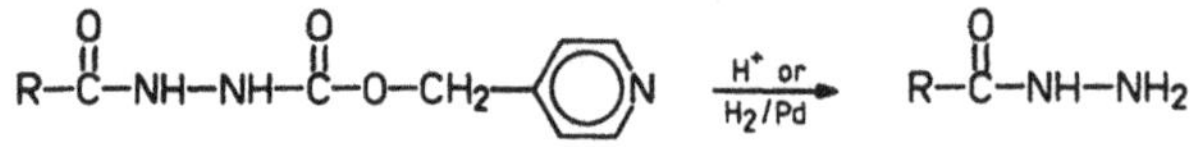

which can be converted to the azides and thus can serve in the synthesis of larger peptides via segment condensation. The same strategy, but with other methods of coupling, is also possible, simply by the hydrogenolytic removal

[18] The Guanidine Group of Arginine Can Similarly Serve as a Cationic "Handle" [79].

of the picolyl ester group and coupling of the peptide through its unmasked carboxyl group to a second segment:

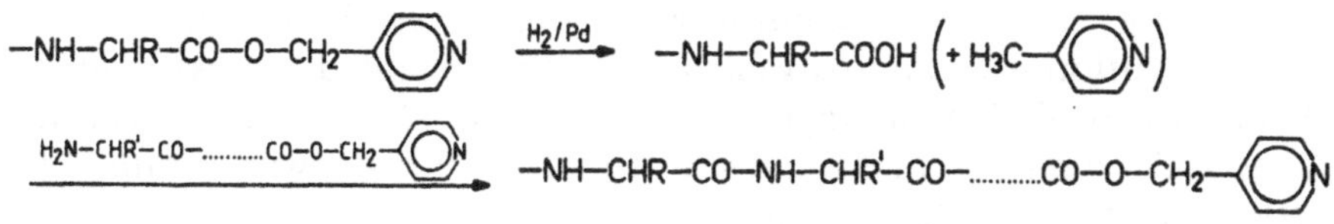

The value of this approach has been demonstrated in the synthesis of a biologically active 22-peptide [82].

3. Synthesis "in situ"

In solid phase peptide synthesis and in syntheses carried out with the help of soluble polymers separation of the protected intermediates from unreacted starting materials and by-products is facilitated by the insolubility of the polymer-bound peptide chains. Yet, in many conventional syntheses, the protected intermediates are also insoluble in most organic solvents even though they are not attached to polymers. In such cases dilution of the mixture with judiciously selected "non-solvents" leads to the precipitation of an intermediate while unreacted starting materials and by-products remain in solution. The intermediates then can be secured, often in pure form, simply by washing with the diluent. This pattern was noted in a synthesis of oxytocin [1, 83] and was followed in the first synthesis of secretin [84]. For a more facile execution of this technique [85] a centrifuge tube provided with a standard taper joint [86] was used. Both coupling and deprotection could be carried out in this simple "reactor" which also permits the removal of solvents by evaporation *in vacuo*. The intermediates were *isolated* and washed by centrifugation. For instance, a tert. butyloxycarbonyl peptide was placed in the reactor, and dissolved in trifluoroacetic acid. On completion of the deprotection step the trifluoroacetic acid was evaporated *in vacuo*, the residue triturated with ether, washed with ether by centrifugation, dried and weighed. A small sample was used for thin layer chromatography and for amino acid analysis, the rest dissolved in dimethylformamide and treated with diisopropylethylamine, hydroxybenzotriazole [43] and with the active ester of the next protected amino acid to be incorporated. When a negative spot test with ninhydrin or fluorescamine indicated complete acylation of the amino component, the solution was concentrated *in vacuo* and diluted with ethyl acetate. The protected intermediate was dried, examined and used in the next cyyle of deprotection, acylation, etc. The characteristic feature of this technique is that the intermediates can be left in the same vessel throughout the chain lengthenining process. It was possible to purify the intermediates, *in situ*, by crystallization or reprecipitation. When benzyloxycarbonylamino acids rather than tert. butyloxycarbonyl derivatives were used, the trialkylammonium bromide by-products were better removed with 95% alcohol than by the usual diluent, ethyl acetate. In syntheses with 9-fluorenylmethyloxycarbonyl amino acids the absence of salts creates less restrictions in the choice of the diluent. The *in situ* technique could be applied for the simultaneous synthesis of four 23-peptides as well [87]. The same principles, deprotection and coupling in solution and isolation of the protected intermediates by dilution with non-solvents can be implemented also without the simplification provided by a single reaction vessel. This was demonstrated in a large scale synthesis of somatostatin [88] by the stepwise strategy [1] with active esters as acylating agents.

4. Synthesis Without Isolation of Intermediates

The tempting possibility to build chains by linking amino acids or peptides to each other and to continue the build-up without isolation of intermediates attracted wide interest. Early schemes for rapid synthesis were based on coupling with water soluble carbodiimides [89, 90], extraction of the solution (e.g. in dichloromethane) with dilute acid, water and bicarbonate, evaporation of the solvent, deprotection by hydrogenation or acidolysis and acylation of the partially deprotected peptide, once again with the aid of carbodiimide. The intermediate protected peptides, even if collected by evaporation, were not purified for the next step. A protected heptapeptide could be secured, in homogeneous form, in this manner [90]. Similar patterns were implemented in two-phase systems and were used for the preparation of antigens [91] and of angiotensin II [92]. In both cases carbodiimide coupling was facilitated by catalysis [38]. Stepwise chain lengthening with N-hydroxyphthalimide esters was also applied for chain building without isolation of intermediates [93] and biologically active peptides were rapidly assembled [94] with the help of the potent pentafluorophenyl esters. An important feature of this latter process is the removal of excess active ester by the addition of N,N-dimethylethylenediamine and extraction of the amide thus formed with dilute aqueous acid:

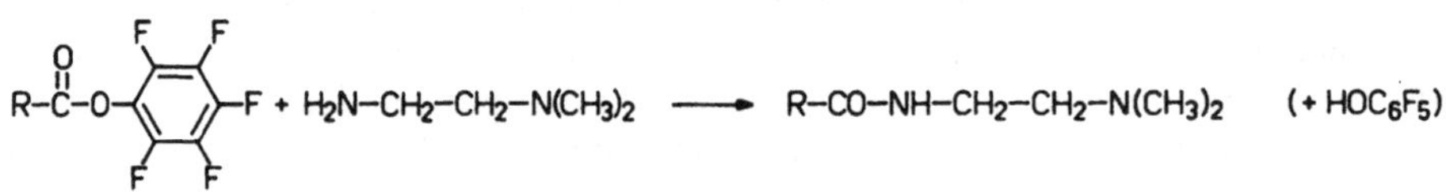

The convenience provided by the powerful acylating agent might be somewhat diminished by the noxious effects of the released pentafluorophenol.

An exceptionally rapid chain building is possible with N-carboxyanhydrides. Under well controlled conditions [95] these highly reactive derivatives of amino acids produce N-carboxypeptides which are "deprotected" by the almost instantaneous decarboxylation:

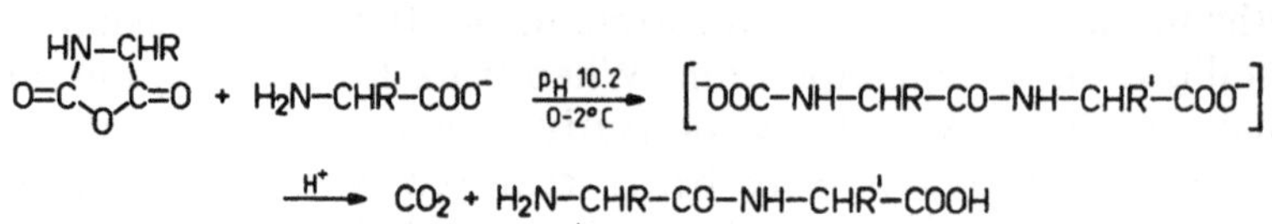

The product is obtained with a free amino group and is ready for acylation with the Leuch's anhydride of the next amino acid residue. It seemed that the well defined conditions (vigorous mixing at 0–2 °C, p_H 10.2) contribute to the homogeneity of the products and that previous concerns [96, 97] about premature decarboxylation and the consequent double incorporation of amino acids were no more justified. Yet, a reexamination of the procedure revealed [98] that chain building with N-carboxyanhydrides is unequivocal with respect to chiral purity of the peptides formed, but it leads to a mixture of materials rather than to a single peptide.

It is a fairly generally held view that isolation of intermediates is a drudgery which, if possible, should be circumvented. Hence the glamour of "single pot" processes. We cannot share this opinion. Intermediates, once isolated, can be examined, analyzed. The results of these studies will unmask potential shortcomings of the methods applied, can point to side reactions not yet detected and thus lead to substantial improvements in the procedures of synthesis. Last, but not least, through the analysis of a series of intermediates a certain assurance is gained about the homogeneity (and sometimes the identity) of the final product.

References of Chapter VII

1. Bodanszky, M.: Ann. N.Y. Acad. Sci. *88*, 655 (1960)
2. Merrifield, R. B.: J. Amer. Chem. *85*, 2149 (1963)
3. Letsinger, R. L., Kornet, M. J.: J. Amer. Chem. Soc. *85*, 3045 (1963)
4. Barany, G., Merrifield, R. B.: in "The Peptides", Vol. 2 (Gross, E., Meienhofer, J., eds.), p. 1, New York: Academic Press 1979
5. Bodanszky, M., Sheehan, J. T.: Chem. Ind., 1597 (1966)
6. Gisin, B. F., Merrifield, R. B.: J. Amer. Chem. Soc. *94*, 6165 (1972)
7. Wang, S. S.: J. Org. Chem. *40*, 1235 (1975)
8. Pietta, P. G., Marshall, G. R.: J. Chem. Soc. (D), 650 (1970)
9. Pietta, P. G., Cavallo, P. F., Takahashi, K., Marshall, G. R.: J. Org. Chem. *41*, 703 (1974)
10. Wang, S. S.: J. Amer. Chem. Soc. *95*, 1328 (1973)
11. Sparrow, J. T.: Tetrahedron Lett., 4637 (1975)
12. Sparrow, J. T.: J. Org. Chem. *41*, 1350 (1976)
13. Mitchell, A. R., Erickson, B. W., Ryabtsev, M. N., Hodges, R. S., Merrifield, R. B.: J. Amer. Chem. Soc. *98*, 7357 (1976)
14. Gait, M. J., Sheppard, R. C.: J. Amer. Chem. Soc. *98*, 8514 (1976)
15. Atherton, E., Clive, D. L. J., Sheppard, R. C.: J. Amer. Chem. Soc. *97*, 6584 (1975)
16. Arshady, R., Atherton, E., Gait, M. J., Lee, K., Sheppard, R. C.: J. Chem. Soc. Chem. Commun., 423 (1979)
17. Arshady, R., Kenner, G. W., Ledwith, A.: Makromol. Chem. *177*, 2911 (1976)
18. Meyers, W. E., Royer, G. P.: J. Amer. Chem. Soc. *99*, 6141 (1977)
19. Parr, W., Grohmann, K., Hägele, K.: Liebigs Ann. Chem., 655 (1974)
20. Tregear, G. W.: in "Chemistry and Biology of Peptides" (Meienhofer, J., ed.), p. 175, Ann Arbor, Michigan: Ann Arbor Sci. Publ. 1972
21. Sieber, P., Iselin, B.: Helv. Chim. Acta *51*, 622 (1968)
22. Gisin, B. F.: Helv. Chim. Acta *56*, 1476 (1973)

23. Schreiber, J.: in "Peptides" (Beyerman, H. C., van de Linde, A., Maassen, van den Brink, eds.), p. 107, Amsterdam: North Holland Publ. 1967
24. Atherton, E., Fox, H., Harkiss, D., Logan, C. J., Sheppard, R. C., Williams, B. J.: J. Chem. Soc. Chem. Commun., 537 (1978)
25. Bodanszky, M., Fagan, D. T.: Int. J. Pept. Protein Res. *10*, 375 (1977)
26. Dorman, L. C., Love, J.: J. Org. Chem. *34*, 158 (1969)
27. Glass, J. D., Schwartz, I. L., Walter, R.: J. Amer. Chem. Soc. *94*, 6209 (1972)
28. Wang, S. S., Merrifield, R. B.: Int. J. Protein Res. *1*, 235 (1969)
29. Ragnarsson, U., Karlsson, S., Sandberg, B. E.: J. Org. Chem. *39*, 3837 (1974)
30. Ragnarsson, U., Karlsson, S., Hamberg, U.: Int. J. Pept. Protein Res. *7*, 307 (1975)
31. Birr, C., Lochinger, W., Stahnke, G., Lang, P.: Liebigs Ann. Chem. *763*, 162 (1972)
32. Yamashiro, D., Li, C. H.: J. Amer. Chem. Soc. *95*, 1310 (1973)
33. Erickson, B. W., Merrifield, R. B.: J. Amer. Chem. Soc. *95*, 3757 (1973)
34. Carpino, L. A., Han, G. Y.: J. Org. Chem. *37*, 3404 (1972)
35. Chang, C. D., Meienhofer, J.: Int. J. Pept. Protein Res. *11*, 246 (1975)
36. Chang, C. D., Felix, A. M., Jimenez, M. H., Meienhofer, J.: Int. J. Pept. Protein Res. *15*, 485 (1980)
37. Atherton, E., Fox, H., Harkiss, D., Sheppard, R. C.: J. Chem. Soc. Commun., 539 (1978)
38. König, W., Geiger. R.: Chem. Ber. *103*, 788 (1970)
39. Bodanszky, M., Denning, G. S., Jr., du Vigneaud, V.: Biochem. Preparations *10*, 122 (1963)
40. Bodanszky, M., Sheehan, J. T.: Chem. Ind., 1423 (1964)
41. Bodanszky, M., Funk, K. W.: J. Org. Chem. *38*, 1296 (1973)
42. Bodanszky, M., Fink, M. L., Funk, K. W., Kondo, M., Lin, C. Y., Bodanszky, A.: J. Amer. Chem. Soc. *96*, 2234 (1974)
43. König, W., Geiger, R.: Chem. Ber. *106*, 3626 (1973)
44. Kisfaludy, L., Roberts, J. E., Johnson, R. H., Mayers, G. L., Kovacs, J.: J. Org. Chem. *35*, 3563 (1970)
45. Kovács, K., Penke, B.: in "Peptides 1972" (Hanson, H., Jakubke, H. D., eds.), p. 187, Amsterdam: North Holland Publ. 1973
46. Mukaiyama, T., Ueki, M., Maruyama, H., Matsueda, R.: J. Amer. Chem. Soc. *90*, 4490 (1968)
47. Matsueda, R., Maruyama, H., Kitazawa, H., Takahagi, H., Mukaiyama, T.: Bull. Chem. Soc. Jpn. *46*, 3240 (1973)
48. Weygand, F., Huber, P., Weiss, K.: Z. Naturforsch. *22b*, 1084 (1967)
49. Wieland, T., Birr, C., Fleckenstein, P.: Angew. Chem. Int. Ed. *10*, 330 (1971)
50. Hagenmeier, H., Frank, H.: Hoppe-Seyler's Z. Physiol. Chem. *356*, 777 (1975)
51. Schüssler, H., Zahn, H.: Chem. Ber. *95*, 1076 (1962)
52. Heimer, E. P., Chang, C. D., Lambros, T., Meienhofer, J.: Int. J. Pept. Protein Res. *18*, 237 (1981)
53. Fuller, W. D., Marr-Leisy, D., Chaturvedi, N. C., Sigler, G. F., Verlander, M. S.: in "Peptides" (Rich, D. H., Gross, E., eds.), p. 201, Pierce Chem. Corp., Rockford, Ill. (1981)
54. Lenard, J., Robinson, A. B.: J. Amer. Chem. Soc. *89*, 181 (1967)
55. Beyerman, H. C., Hindriks, H., Leer, E. W.: Chem. Commun., 1668 (1968)
56. Moore, G., McMaster, D.: Int. J. Peptide Protein Res. *11*, 140 (1978)
57. Barton, M. A., Lemieux, R. U., Savoie, J. Y.: J. Amer. Chem. Soc. *95*, 4501 (1973)
58. Gross, E., Noda, K., Nisula, B.: Angew. Chem. Int. Ed. *12*, 664 (1973)

59. Rich, D. H., Gurwara, S. K.: J. Amer. Chem. Soc. *97*, 1575 (1975)
60. Jones, D. A.: Tetrahedron Lett., 2853 (1977)
61. Schlatter, J. M., Mazur, R. H., Goodmonson, O.: Tetrahedron Lett., 2851 (1977)
62. Khan, S. A., Sivanandaiah, K. M.: Synthesis, 750 (1978)
63. Shaltiel, S., Fridkin, M.: Biochemistry *9*, 5122 (1970)
64. Glass, J. D., Talansky, A., Grzonka, Z., Schwartz, I. L., Walter, R.: J. Amer. Chem. Soc. *96*, 6476 (1974)
65. Merrifield, R. B.: Biochemistry *3*, 1385 (1964)
66. Hirt, J., de Leer, E. W. B., Beyerman, H. C.: in "The Chemistry of Polypeptides" (Katsoyannis, P. G., ed.), p. 363, New York: Plenum Press 1973
67. Birr, C.: Aspects of the Merrifield Peptide Synthesis, Berlin, Heidelberg, New York: Springer 1978
68. Bayer, E., Eckstein, H., Hägele, K., König, W. A., Brüning, W., Hagenmaier, H., Parr, W.: J. Amer. Chem. Soc. *92*, 1735 (1970)
69. Bodanszky, M.: in "Prebiotic and Biochemical Evolution" (Kimball, A. P.,Oro, J., eds.), p. 217, Amsterdam: North Holland-American Elsevier Publ. 1971
70. Shemyakin, M. M., Ovchinnikov, Y. A., Kiryushkin, A. A., Kozhenikova, I. V.: Tetrahedron Lett., 2323 (1965)
71. Andreatta, R. H., Rink, H.: Helv. Chim. Acta *56*, 1205 (1973)
72. Blecher, H., Pfaender, P.: Liebigs Ann. Chem., 1263 (1963)
73. Mutter, M., Bayer, E.: Nature *237*, 512 (1972)
74. Frank, H., Hagenmaier, H.: Experientia *31*, 131 (1975)
p. 285, New York: Academic Press 1980
76. Young, G. T.: in "The Chemistry of Polypeptides" (Katsoyannis, P. G., ed.), p. 43, New York: Plenum Press 1973
77. Wieland, T., Racky, W.: Chimia (Aarau) *22*, 375 (1968)
78. Camble, R., Garner, R., Young, G. T.: Nature *217*, 247 (1968); J. Chem. Soc., 1911 (1969)
79. Schafer, D. J., Black, A. D.: Tetrahedron Lett., 4071 (1973)
80. Burton, J., Fletcher, G. A., Young, G. T.: J. Chem. Soc. (C), 1911 (1969)
81. Macrae, R., Young, G. T.: J. Chem. Soc. Chem. Commun., 446 (1974)
82. Bratby, D. M., Coyle, S., Gregson, R. P., Hardy, G. W., Young, G. T.: J. Chem. Soc. Perkin I, 1901 (1979)
83. Bodanszky, M., du Vigneaud, V.: J. Amer. Chem. Soc. *81*, 5688 (1959)
84. Bodanszky, M., Ondetti, M. A., Levine, S. D., Williams, N. J.: J. Amer. Chem. Soc. *89*, 6753 (1967)
85. Bodanszky, M., Funk, K. W., Fink, M. L.: J. Org. Chem. *38*, 3565 (1973)
86. Bodanszky, M., Kondo, M., Yang Lin, C., Sigler, G. F.: J. Org. Chem. *39*, 444 (1974)
87. Fink, M. L., Bodanszky, M.: J. Amer. Chem. Soc. *98*, 974 (1976)
88. Diaz, J., Guegan, R., Beaumont, M., Benoit, J., Clement, J., Fauchard, C., Galtier, D., Millan, J., Muneaux, C., Muneaux, Y., Vedel, M., Schwyzer, R.: Bioorg. Chem. *8*, 429 (1979)
89. Knorre, D. A., Shubina, T. N.: Dokl. Akad. Nauk SSSR *150*, 559 (1963)
90. Sheehan, J. C., Preston, J., Cruickshank, P. A.: J. Amer. Chem. Soc. *87*, 2492 (1965)
91. Schneider, C. H., Wirz, W.: Helv. Chim. Acta *55*, 1062 (1972)
92. Nozaki, S., Kimura, A., Muramatsu, I.: Chemistry Lett., 1057 (1977)
93. Medzihradszky, K., Radoczy, J.: in "Peptides" (Zervas, L., ed.), p. 49, Oxford: Pergamon Press 1966

94. Kisfaludy, L., Schön, I., Szirtes, T., Nyéki, O., Löw, M.: Tetrahedron Lett., 1785 (1974)
95. Hirschmann, R., Strachan, R. G., Schwam, H., Schoenewaldt, E. F., Joshua, H., Barkemeyer, B., Veber, D., Paleveda, W. J., Jr., Jacob, T. A., Beesley, T. E., Denkewalter, R. G.: J. Org. Chem. *32*, 3415 (1967)
96. Bartlett, P. D., Jones, R. H.: J. Amer. Chem. Soc. *79*, 2153 (1957)
97. Bartlett, P. D., Dittmer, D. C.: J. Amer. Chem. Soc. *79*, 2159 (1957)
98. Pfaender, P., Kuhnle, E., Krahl, B., Backmansson, A., Gnauck, G., Blecher. H.: Hoppe-Seyler's Z. Physiol. Chem. *354*, 267 (1973)

VIII. Recent Developments and Perspectives

During the two years elapsed since writing the first pages of this volume peptide chemistry continued to grow. Only a part of the publications which appeared in this period could be incorporated into chapters already completed and it seemed to be desirable to mention, at least, some more important new contributions to the methodology of peptide synthesis. The author tries also to use this opportunity to include some procedures or aspects which have been overlooked in the discussion of topics of sometimes almost intractable complexity. Also, an attempt will be made in this chapter to assess the present stage of development of peptide synthesis. It would be even better to chart the future course of this discipline, but general failures in futurology warn against such an endeavor and we will confine ourselves to a vague indication of trends, hopes and expectations.

A. Formation of the Peptide Bond

In spite of the almost embarrassing multitude of known coupling methods numerous new procedures were proposed for peptide bond formation. It seems to be particularly tempting to find novel coupling reagents although a survey of the already available tools suggests that it is almost impossible to design a perfect coupling reagent, a compound which can activate the carboxyl group yet is completely inert toward amines. In addition, the ideal reagent should cause no racemization and should not be conducive to side reactions such as the O → N shift in the reactive intermediates formed in coupling with dicyclohexylcarbodiimide.

Several new methods of activation are based on reagents derived from phosphorus. Perhaps stimulated by the diphenylphosphinic acid anhydrides [1] obtained with diphenylphosphinyl chloride

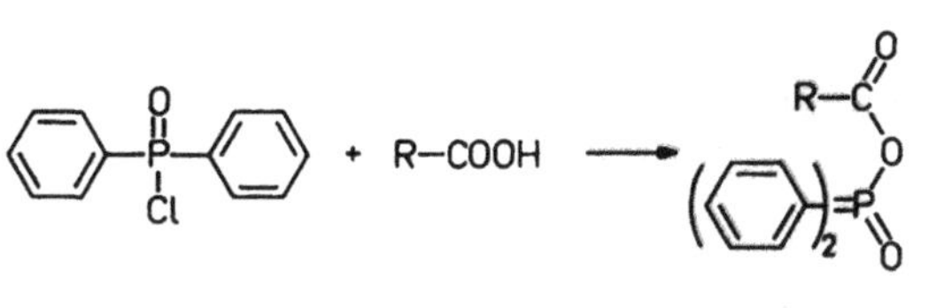

the coupling reagent N,N′-(2-ketooxazolidinyl)-phosphorodiamidic chloride

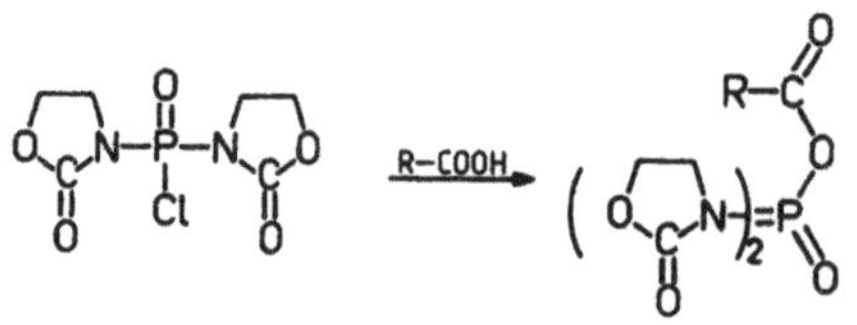

has been proposed [2] for the preparation of amides as were a whole series of related amides or half ester-half amides of phosphoric acid chloride. Several of these are now commercially available. Similarly, derivatives of phosphoric acid were utilized in the formation of the peptide bond via thiol acids. The reaction

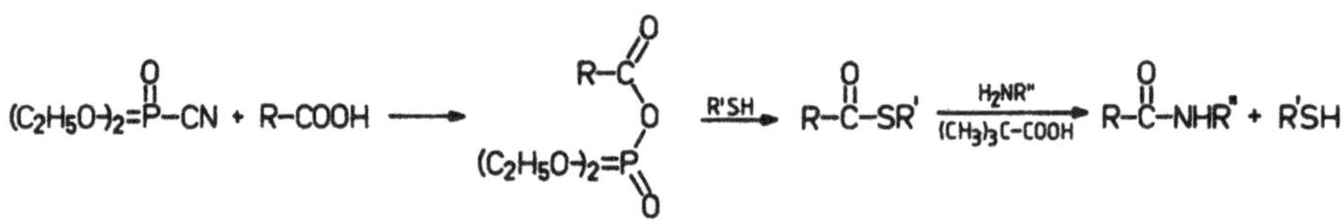

catalyzed by pivalic acid, was claimed [3] to proceed without racemization. Reactive imidoyl phosphites [4]

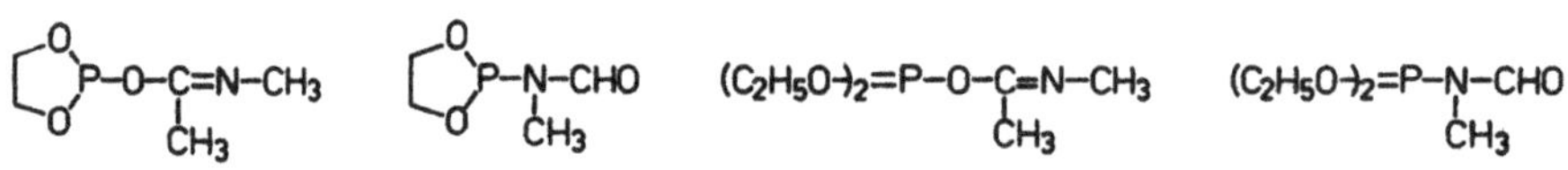

can activate the carboxyl group directly or through a newly formed amide, in a manner which is analogous to the reactions of dialkyl phosphorous acid chlorides discussed in Chapter II. The idea of reactive amides is reminiscent of the suggested use of sulfonamide derivatives [5] of the type

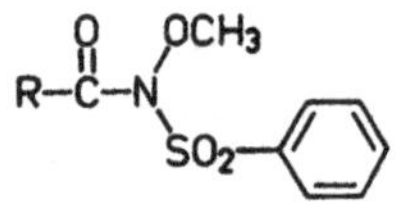

for acylation. Reactive amides are the acylating intermediates also in couplings mediated by the N-acyl thiazolidine-2-thiones [6–8]

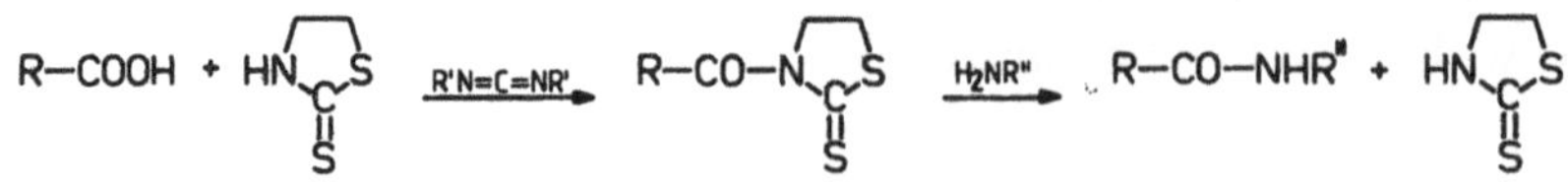

as in the application of the reagents [9]

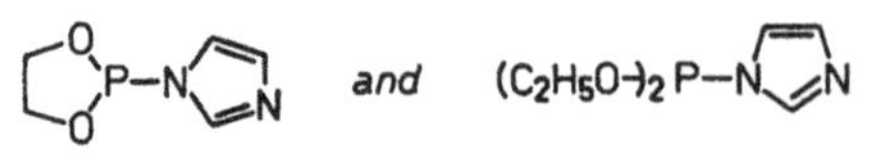

In addition to new variants of already much studied approaches such as reactive N-acyl intermediates novel executions of the most classical methods were attempted several times with various acid chloride forming reagents like the one obtained [10] from dimethylformamide and phosgene:

$$[(CH_3)_2N{=}CHCl]^+ \cdot Cl^-$$

Also, improvements were and are being explored in the preparation of mixed anhydrides in the hope that the fundamental problem of this otherwise excellent method, the formation of a second acylation product can be avoided. Thus, mixed anhydrides can be obtained through acylammonium salts [11]:

$$(CH_3)_2CH{-}CH_2{-}O{-}CO{-}Cl + NR_3 \longrightarrow [(CH_3)_2CH{-}CH_2{-}O{-}\overset{O}{\overset{\|}{C}}{-}N^+R_3]Cl^- \xrightarrow{RCOOH} R{-}\overset{O}{\overset{\|}{C}}{-}O{-}\overset{O}{\overset{\|}{C}}{-}O{-}CH_2{-}CH(CH_3)_2 + R_3NH \cdot Cl$$

Whether or not the mixed anhydrides thus prepared yield a single acylation product remains to be answered by critical experiments in which the order of addition of the reactants is the only variable. Interestingly, in a study [12] on the use of mixed anhydrides in solid phase peptide synthesis no second acylation product could be detected if the mixed anhydrides were prepared in dimethylformamide or tetrahydrofurane and stored at room temperature. An obvious solution for the problem of the second acylation product is the application of symmetrical rather than mixed anhydrides [13]. The formerly expensive symmetrical anhydrides are now readily available. This is true also for the anhydrides of 9-fluorenylmethyloxycarbonyl (Fmoc) amino acids [14] which were secured in crystalline form. Such seemingly minor developments might have major consequences. For unequivocal synthesis well defined, isolated intermediates are desirable and relatively stable acylating agents, such as active esters with moderate reactivity, were not powerful enough to perform well under adverse conditions, e.g. in dilute solution. Only the exceptionally potent pentafluorophenyl esters were sufficiently active toward polymer-bound nuclophiles [15].

Changes in the execution of old methods of activation can bring about improvements in peptide bond formation, but more substantial impact should be expected from procedures which are based on new principles. Thus, the four component condensation method [16], or photochemical coupling [17], e.g.

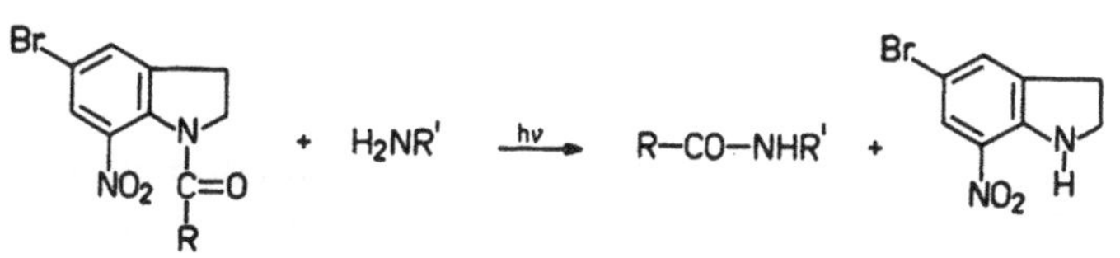

might revolutionize peptide synthesis. A new era is being introduced by the increasing use of proteolytic enzymes [18] and chemical catalysis of the reactions leading to peptide bonds [19] also holds considerable promise.

An unusually original method of peptide bond formation, proposed by Wasserman and Lu [20], is based on photooxygenation of substituted oxazoles. First acyl derivatives of benzoin are prepared

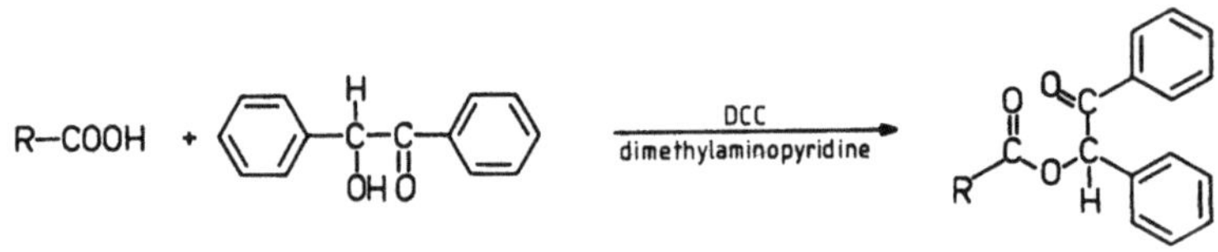

which are then heated with ammonium acetate-acetic acid to yield oxazoles

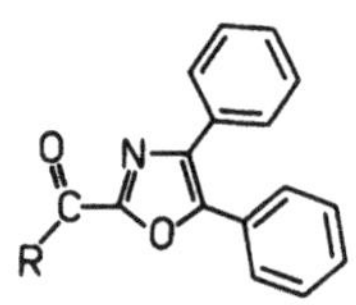

and these, in turn, are irradiated in the presence of methyleneblue and oxygen in dichloromethane to form a reactive triacylamide:

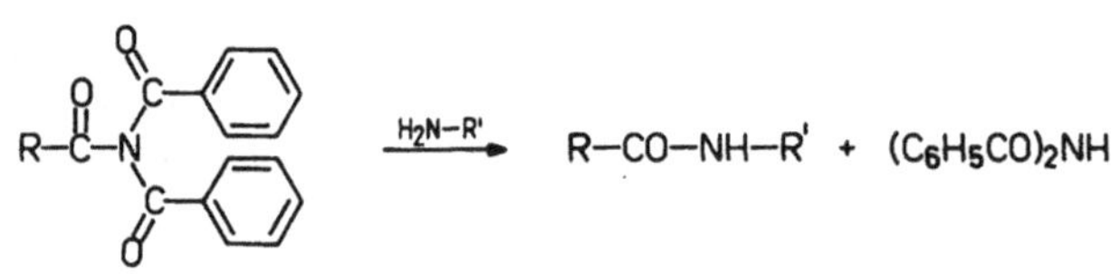

In spite of the fairly drastic conditions prevailing during the formation of the oxazoles, the peptides prepared by the new procedure had good chiral purity. Thus the method certainly deserves further studies.

It is rather difficult to make a general and still valid final statement on the numerous methods of peptide bond formation. The individuality of the amino acids, particularly, when they are parts of a specific sequence is so pronounced that several procedures may be needed for the synthesis of any longer peptide chain. Interestingly, a reactive enol ester of 2-phenyl-1,3-diketoindane

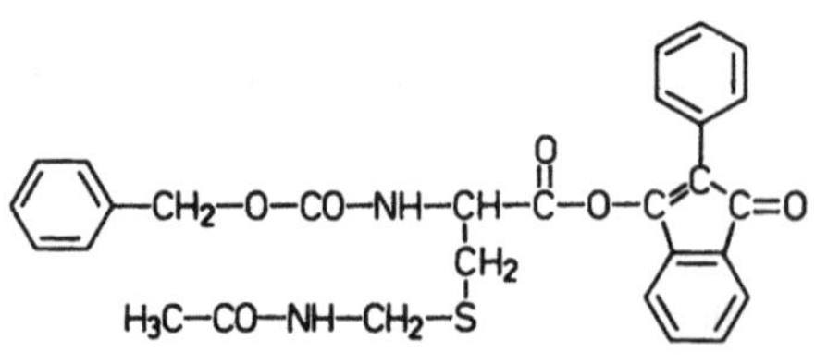

was recommended [21] solely for the incorporation of S-acetamidomethyl-L-cysteine. Yet, if a generalization is desirable, the author has to voice his impression about two approaches to peptide bond formation. It seems to him that methods in which the steps of activation and coupling are well separated give results which are superior to those achieved with coupling reagents, the latter being employed in concurrent activation of the carboxyl group and formation of the amide bond. Thus, the efforts necessary to prepare symmetrical anhydrides or active esters could find their reward in products of better quality.

B. Protection

1. Carboxyl Groups

Protection of the C-terminal α-carboxyl function in the form of 4-sulfobenzyl esters [22, 23] was applied in the synthesis of leucine-enkephalin. The presence of the very polar and strongly acidic sulfo group in the esters enhanced the solubility of the intermediates and facilitated their isolation by ion-exchange chromatography and also by crystallization. Preparation of 4-sulfobenzyl esters is simple: the reaction of the cesium salts of the protected amino acids with substituted benzyl halides afforded the desired intermediates:

This method of esterification, originally introduced [24] for the anchoring of the C-terminal residue to an insoluble support, was extended [25] for the formation of esters in general. New recommendations indicate that esterification of the carboxyl group could be further improved. Thus, the reaction of oxalyl chloride with dimethylformamide affords dimethyl-chloromethylidene ammonium chloride which then activates the carboxyl to form esters under mild conditions [26]:

$$\begin{matrix}CO{-}Cl\\ |\\ CO{-}Cl\end{matrix} + (CH_3)_2N{-}CHO \longrightarrow [(CH_3)_2N{=}CHCl]^{+} \cdot Cl^{-} \xrightarrow{R-COOH} R{-}CO{-}Cl \xrightarrow{R'-OH} R{-}CO{-}OR'$$

Naturally, most methods of activation used for the formation of the peptide bond, are suitable for the preparation of esters as well. (cf. e.g. ref. [27]). Yet, a reactive and hence convenient intermediate might also cause side reactions. For instance the application [28] of ethyl *p*-toluenesulfonate for the preparation of ethyl esters probably leads to alkylation of sensitive centers and the reagent should, at least, alkylate the thioether in methionine. Probably no absolutely safe method is available for esterification. Conversion of protected amino acids to their (base sensitive) 9-fluorenylmethyl esters by imidazole catalyzed transesterification of active esters [29] or by esterification

[30] via activation with carbodiimides and catalysis with *p*-dimethylamino-pyridine [31]

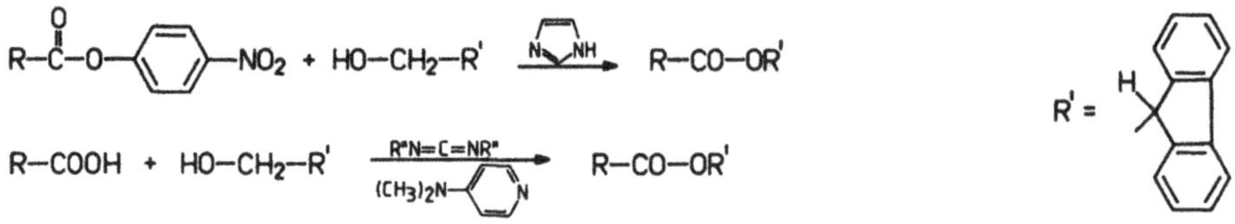

might be accompanied by racemization and in order to secure a product with good chiral purity the conditions of the reactions must be carefully selected, for instance polar solvents should be avoided. Direct esterification of carboxylic acids with alcohols catalyzed by superacids [32] such as Nafion-H (a perfluorinated polymeric sulfonic acid) has to be further scrutinized before it can be adopted for peptide synthesis. The interesting recommendation [33] to aid esterification of protected amino acids with a polymeric support is based on hydrogen bonding produced by potassium fluoride:

deserves further attention because it could be carried out without racemization. A similarly promising avenue is the enhancement of the nucleophilicity of the carboxylate by the addition of crown ethers to the solution of the potassium salt of carboxylic acids in dimethylformamide [34]. Protection of the carboxyl group in the form of metal complexes [35] such as the cobalt derivatives proposed by Isied and Kuehn [36]

$$[\mathrm{R{-}CO{-}O\ \ Co(NH_3)_5}](\mathrm{BF_4})_3$$

might absolve the peptide chemist from problems of esterification.

Some recently introduced carboxyl protecting groups were designed with specific means of removal in mind. Thus, 4,4,4-trichloro-2-butanal [37] reacts in the Ugi-Passerini reaction [38] to give products which can be cleaved by the cobalt(I)phthalocyanine anion [39]. A less complex blocking of carboxyl groups in the form of *β*-chloroethyl esters [40]

$$\mathrm{R{-}CO{-}CH_2{-}CH_2Cl}$$

is removed by the action of sodium sulfide. Since hydrazides can be considered as protected forms of the carboxyl group the simple preparation [41] of hydrazides from the acids by hydroxybenzotriazole catalyzed coupling with carbodiimides to hydrazine is mentioned at this point.

2. Temporary Protection of Amino Groups

The dithiasuccinyl group of Barany and Merrifield [42] discussed in Chapter II

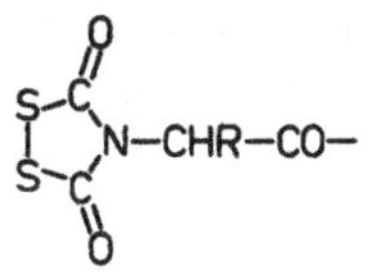

obviously cannot be applied for the protection of the imino group of proline. A logical extension of the concepts leading to the dithiasuccinyl group resulted [43] in a series of alkyl and aryl dithiocarbamoyl derivatives:

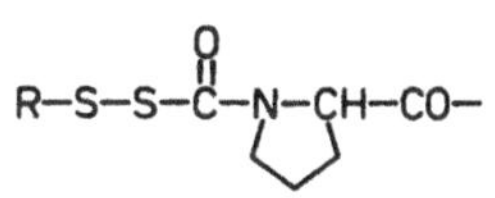

(where R can be methyl, isopropyl, tert. butyl of phenyl). Amino acids with tert-butyl and cyclohexyl dithiocarbamoyl protection were prepared, independently, by a second group of investigators, in Germany [44]. Of course, the new protecting groups can also be used for the blocking of the amino groups of amino acids other than proline. The disulfide is cleaved by thiols, such as thiophenol or 2-mercaptopyridine or by phosphines, e.g. tri-n-butylphosphine.

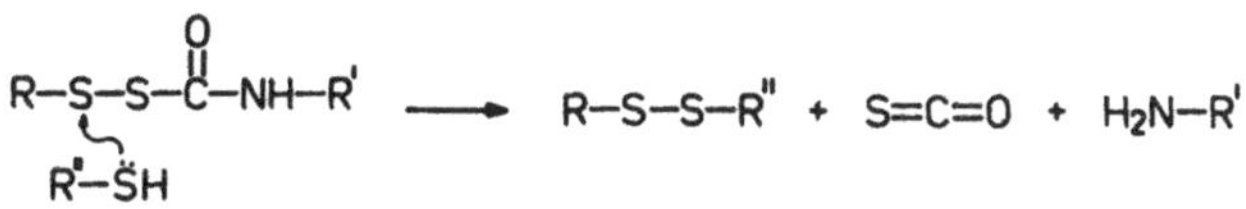

The carbonylsulfide formed in the process is the least attractive feature of the otherwise ingenious method.

3. Semipermanent Blocking of Side Chain Functions

Most protecting groups used for the masking of the ε-amino group of *lysine* have an unfavorable effect on the solubility of the intermediates. Therefore a search for novel approaches to the protection of the side chain amine are certainly warranted. New methods might be needed to facilitate orthogonal protection, that is groups which can be removed independently from the blocking groups applied for the α-amino function. A revival [45] of the earlier recommended [46] 4,5-diphenyl-4-oxazoline-2-one grouping

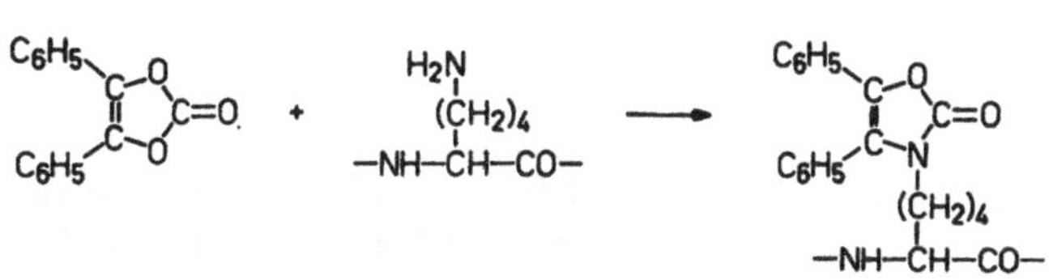

probably does not satisfy these criteria, because it is rather non-polar and is removed by hydrogenation, reduction with sodium in liquid ammonia and also by mild oxidation. Yet, it was applied [45] for the masking of the side chain amine of lysine with good results. Perhaps a more general utilization can be expected from the novel method of blocking the ε-amino group with the phenylacetyl group [47] which can be cleaved by hydrolysis catalyzed with a specific enzyme, penicillin amidohydrolase:

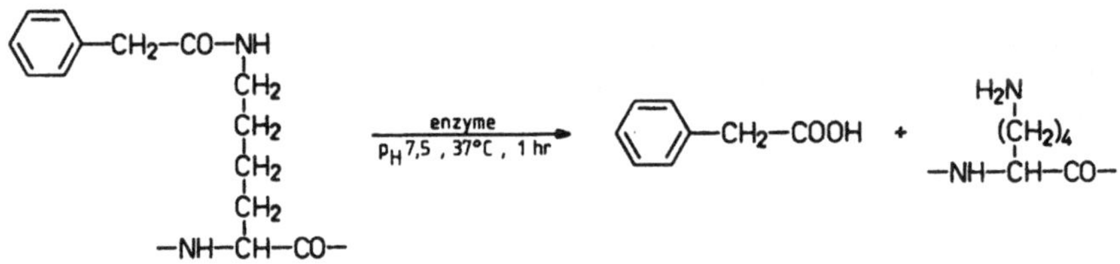

The use of the 3-nitro-2-pyridinesulfenyl group [48] could be extended [49] for the protection of the *thiol function in cysteine*. Conventional masking groups are displaced by the substituted pyridinesulfenyl group which, in turn is removed by thiolysis:

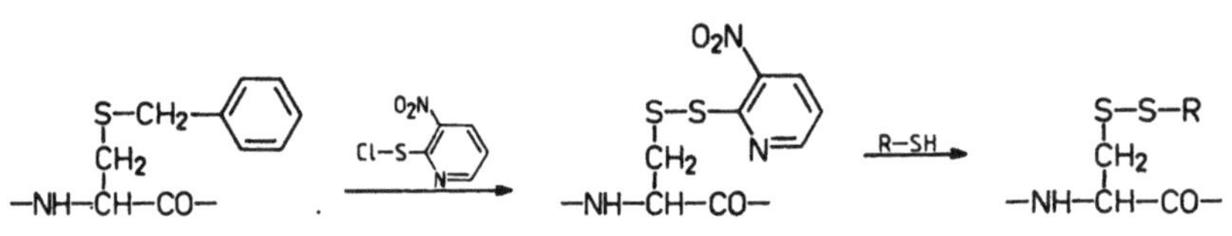

Thus, the 3-nitro-2-sulfenyl group can be quite helpful in the preparation of unsymmetrical disulfides.

A base sensitive blocking of the sulfhydryl function of cysteine could be achieved [50] by the incorporation of S-9-fluorenylmethyl derivatives of the amino acid, through active esters such as

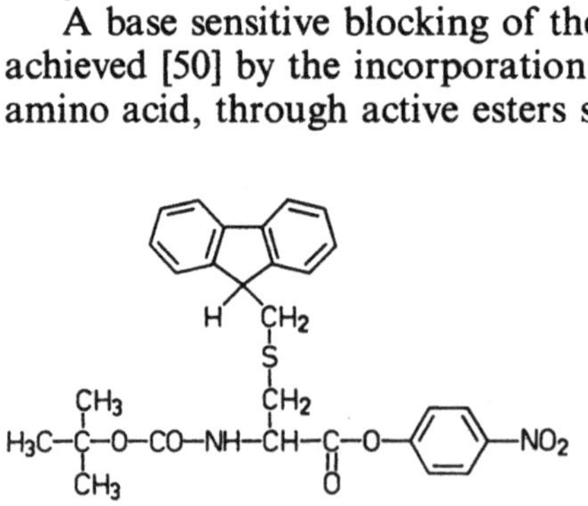

into peptide chains.

The recently recognized ready alkylation of the side chain of *tryptophan* stimulated a new search toward masking of the indole nitrogen. Formation of amides with aromatic sulfonic acids which are cleavable by hydrogen fluoride

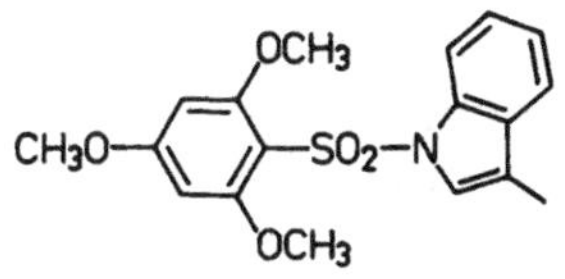

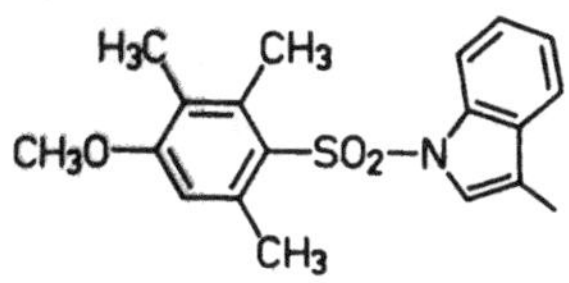

was recommended [51].

The *guanidino group in arginine* presents an undiminished challenge. The 4-methoxy-2,3,6-trimethylbenzenesulfonyl group [52]

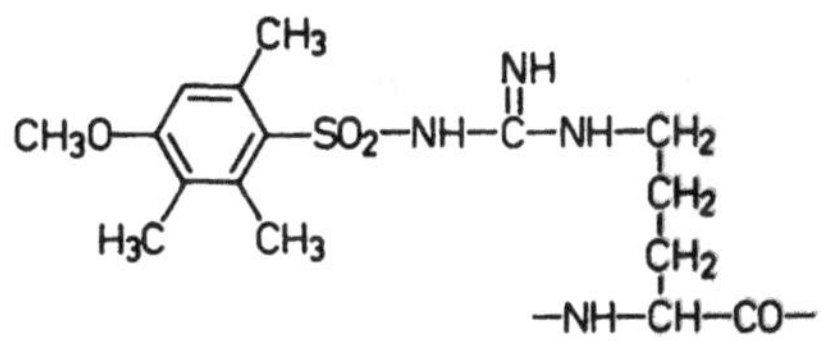

might turn out to be quite useful since it is cleaved by trifluoroacetic acid.

4. Removal of Protecting Groups

For the removal of the semipermanent blocking groups from an assembled peptide chain acidolysis is the most commonly used procedure. Application of strong acids such as hydrogen fluoride or trifluoromethanesulfonic acid is conducive to side reactions, for instance to N → O acyl migration in serine containing peptides or ring closure to aminosuccinyl derivatives in sequences with aspartyl residues. In spite of such difficulties human parathyroid hormone, an 84-peptide, was obtained in an impressive synthesis [53] in which fully protected segments were condensed and HF was used for final deprotection. The need for extensive purification of the synthetic material shows, however, that by-products formed to a not negligible extent. Thus, from one gram of the crude synthetic product only 60 mg purified peptide could be secured by chromatography and after the removal of the analogs in which the methionine residues were oxidized to the sulfoxide 10 mg of homogeneous peptide was obtained by high pressure liquid chromatography. It seems to be likely that these low recoveries in purification are not limited to this actual synthesis but are characteristic for the approach in general. Other strong acids should have similar disadvantages and some procedures, e.g. final deblocking with concentrated (aqueous) hydrochloric acid, practiced in the CIBA-GEIGY laboratories in Basel, require, in all probability, considerable care in their execution. Therefore, it is not surprising that a certain tendency for replacing strong acids with milder reagents can be recognized in the recent literature. The application of very acid labile groups, such as the biphenylisopropyloxycarbonyl group, or base sensitive protection, e.g. by the 9-fluorenylmethyloxycarbonyl group, permits the adoption of moderately acid sensitive groups for the semipermanent blocking of side chain functions and renders the use of fairly acid resistant groups and their removal with very strong acids unnecessary.[19] Of course, more substantial improvements can be expected from a broadening of the range of methods applied for deblocking through the wider application of thiolysis, of specific unmasking reagents such as fluoride ions, metal complexes and particularly of specific enzymes.

C. Side Reactions

Loss of chiral purity of activated histidine residues was repeatedly observed and required further investigations. Recent studies [56] indicate that in N(τ) substituted histidine racemization is catalyzed by the imidazole nitrogen as intramolecular base and involves the equilibrium

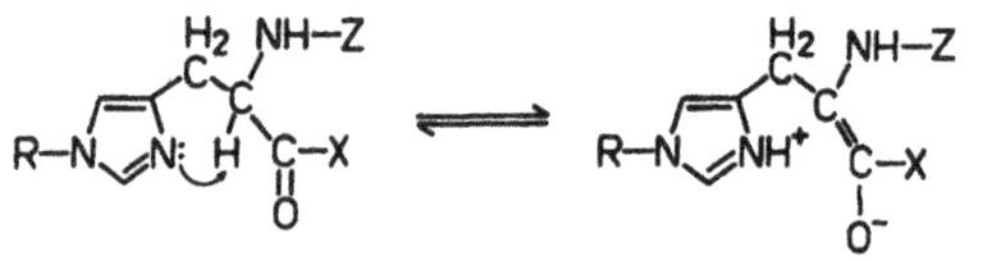

rather than the formation of an optically labile bicyclic N(π)-acyl intermediate.

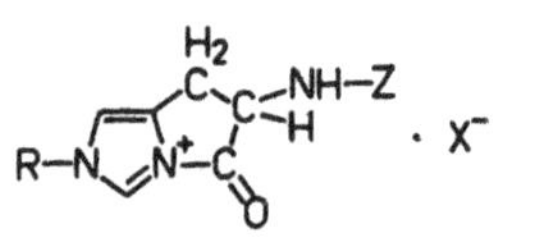

The interesting observation [57] on the prevention of racemization by the addition of fatty acids to the reaction mixture in coupling procedures certainly deserves further attention. Acids with a chain length of 12 to 18 carbon atoms were all effective, but best results were obtained with oleic acid.

Racemization-free coupling of segments, a central problem of peptide synthesis, could be accomplished [58] by carbodiimide activation of thiolacids:

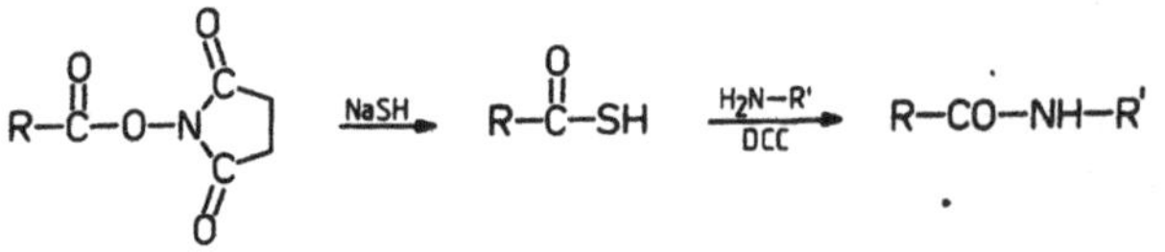

[19] A general problem in the unmasking of peptides by acidolysis is the formation of salts of the liberated amine and hence the need for the addition of tertiary bases in the subsequent acylation step. Attempts to correct this shortcoming were recently reported. The trifluoroacetate anion can be displaced by pentachlorophenolate [54]. Also, the cleavage of highly acid sensitive blocking groups with 1-hydroxybenzotriazole in trifluoroethanol yields amine salts which can be acylated without the addition of base [55].

It is interesting to note that previously undetected side reactions involving side chain functions of amino acids are still being noted in the literature. The ready oxidation of the tioether in methionine is paralleled by the conversion of S-benzyl-cysteine derivatives to the corresponding sulfoxides [59]:

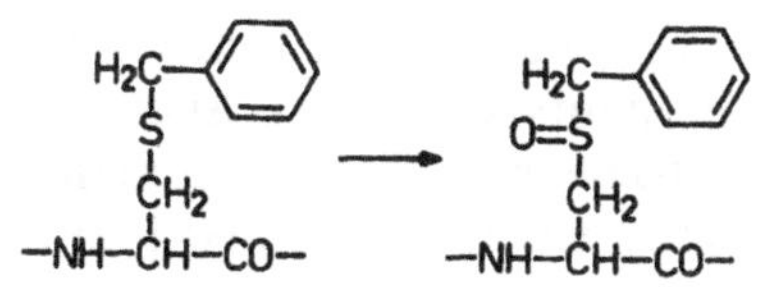

The sulfoxides obtained from S-p-methoxybenzyl or S-acetamidomethyl derivatives of cysteine [60, 61] should be reduced prior to acidolysis of the peptides with strong acids, because in the presence of phenol or anisol as cation scavangers S-p-hydroxyphenyl or S-p-methoxyphenyl derivatives are formed in the process.

D. Solid Phase Peptide Synthesis

The cumulative effect of various improvements became quite conspicuous in recent syntheses by the solid phase method. Some of the more substantial contributions such as the fairly general adoption of symmetrical anhydrides for acylation, the use of acid sensitive anchoring groups which do not require hydrogen fluoride for the cleavage of the peptide from the resin and last, but not least, protection of the α-amino functions with the base sensitive 9-fluorenylmethyloxycarbonyl group and the side chain functions with tert. butyl cation forming blocking groups have already been discussed in Chapter VII. More recently the 3-nitro-2-pyridinesulfenyl group [48] was used [62] for amine protection with good results. It is probably more difficult and perhaps not timely yet to attempt an evaluation of the effect of novel supports [63–65], particularly because direct comparisons of various polymers are only seldom reported. The same can be said about the execution of solid phase peptide synthesis in continuous flow systems [66–68], in some cases under high pressure. Among the interesting new additions to the technique of solid phase synthesis the incorporation of internal standards, such as norleucine, seem to have lasting value, because it allows the monitoring of the progress of a synthesis by simple means. Such new supports with an internal standard are the base sensitive anchoring to a polyacrylamide resin [69]

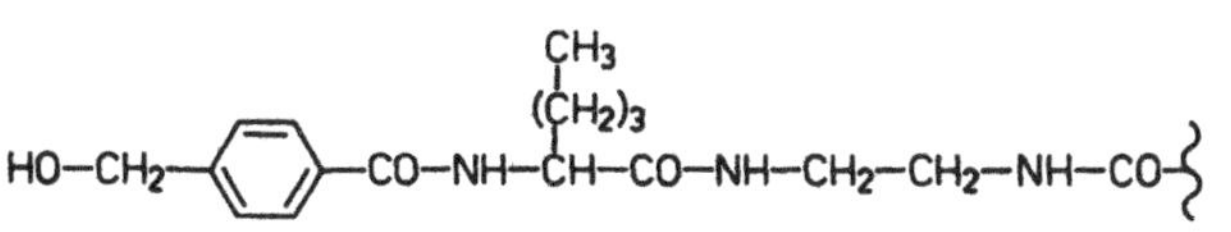

and the corresponding acid sensitive assembly [70]

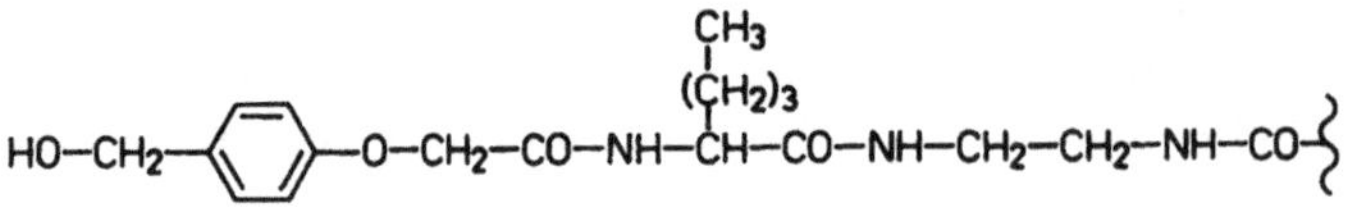

The novel and intriguing thought of mediating acylation reactions between polymeric active esters and amino components which are anchored to a second polymer [71], with imidazole, remains to be tried in actual practice. A new strategy for the synthesis of cyclic peptides was implemented [72] in the preparation of the metal binding cyclopeptide *cyclo*-His-Gly-His-Gly-His-Gly-.

E. Catalysis

It is our impression that catalysis offers the most significant progress in the methodology of peptide synthesis. Acceleration of active ester reactions with 1-hydroxybenzotriazole [19], the rate of acylation with thiol acids by pivalic acid [3] have already been discussed. The enhancement of the rate of acidolysis by the addition of thioanisole to trifluoroacetic acid could be systematically used in major syntheses [73, 74]. In this connection one should not forget that enzymes are highly efficient and very selective catalysts and their sharply increased application both for coupling and deprotection augurs well for the future of peptide synthesis.

References of Chapter VIII

1. Jackson, A. G., Kenner, G. W., Moore, G. A., Ramage, R., Thorpe, W. D.: Tetrahedron Lett., 3627 (1976)
2. Diago-Meseguer, J., Palomo-Coll, A. L., Fenandez-Lizarbe, J. R., Zugaza-Bilbao, A.: Synthesis, 547 (1980); Mestres, R., Palomo-Coll, A. L., *ibid.*, 288 (1982)
3. Yamada, S., Yokoyama, Y., Shiori, T.: Experientia *32*, 907 (1976)
4. Kawanobe, W., Yamaguchi, K., Nakahama, S., Yamazaki, N.: Peptide Chemistry (1981) (Dr. Shiori, T., ed.), p. 1, Osaka: Protein Res. Found. 1982
5. Nakonieczna, L., Makowski, Z., Berezowska, I., Taschner, E.: (Loffet, A., ed.), Editions de L'Université de Brusells (1976), p. 117
6. Li, C. H., Yieh, Y. H., Lin, Y., Lu, Y. J., Chi, A. H., Hsing, C. Y.: Acta Biochim. Biophys. Sinica *12*, 398 (1980); Tetrahedron Lett., 3467 (1981)
7. Yajima, H., Akajii, K., Hirotâ, Y., Fujii, N.: Chem. Pharm. Bull. Jpn. *28*, 3140 (1980)
8. Yajima, H., Akajii, K., Fujii, N., Moriga, M., Murakami, M., Kizuta, K., Takagi, A.: Chem. Pharm. Bull. *29*, 3080 (1981)
9. Anderson, G. W.: Ann. N.Y. Acad. Sci. *88*, 676 (1960)
10. Zaoral, M., Arnold, Z.: Tetrahedron Lett., 9 (1960)
11. Naithani, V. K., Naithani, A. K.: Abstr. 17th Eur. Pept. Symp., Prague, p. 18 (1982)

12. Fuller, W. D., Marr-Leisy, D., Chaturvedi, N. C., Sigler, G. F., Verlander, M. S.: in Peptides, Synthesis-Structure-Function (Rich, D. H., Gross, E., eds.), p. 201, Pierce Chem. Comp., Rockford, Ill. (1981)
13. Benoiton, N. L., Chen, F. M. F.: Febs. Letters *125*, 104 (1981); cf. also Wieland, T., Flor, F., Birr, C.: Liebigs Ann. Chem., 1595 (1973)
14. Heimer, E. P., Chang, C. D., Lambros, T., Meienhofer, J.: Int. J. Peptide Protein Res. *18*, 237 (1981)
15. Kovács, K., Penke, B.: in Peptides 1972 (Hanson, H., Jakubke, H. D., eds.), p. 187, Amsterdam: North Holland Publ. 1973
16. Ugi, I., Marquarding, D., Urban, R.: in Chemistry and Biochemistry of Amino Acids, Peptides and Proteins, Vol. 6 (Weinstein, B., ed.), p. 245, New York: Dekker 1982
17. Pass, Sh., Amit, B., Patchornik, A.: J. Amer. Chem. Soc. *103*, 7674 (1981)
18. Fruton, J. S.: in "Advances in Enzymology", Vol. 53 (Meister, A., ed.), p. 239, New York: Wiley 1982
19. König, W., Geiger, R.: Chem. Ber. *106*, 3626 (1973)
20. Wasserman, H. H., Lu, T. J.: in press
21. Minchev, S., Sofroniev, N.: Abstract, 17th Eur. Peptide Symp., Prague, p. 21 (1982)
22. Hubbuch, A., Bindewald, R., Föhles, J., Naithani, V. K., Zahn, H.: Angew. Chem. *92*, 394 (1980)
23. Bindewald, R., Danho, W., Büllesbach, E. E., Zahn, H.: Abstracts, 17th Eur. Pept. Symp., Prague, p. A4 (1982)
24. Gisin, B. F.: Helv. Chim. Acta *56*, 1476 (1973)
25. Wang, S. S., Gisin, B. F., Winter, D. P., Makofske, R., Kulesha, I. D., Tzougraki, C., Meienhofer, J.: J. Org. Chem. *42*, 1286 (1977)
26. Stadler, P. A.: Helv. Chim. Acta *61*, 1675 (1978)
27. Itoh, M., Hagiwara, D., Notani, J.: Synthesis, 456 (1975)
28. Ueda, K.: Bull. Chem. Soc. Jpn. *52*, 1879 (1979)
29. Bednarek, M. A., Bodanszky, M.: Int. J. Peptide Protein Res. *21*, 196 (1983)
30. Kessler, H., Siegmeier, R.: Tetrahedron Lett. 281 (1983)
31. Steglich, W.: Angew. Chem. *90*, 556 (1978)
32. Olah, G. A., Keumi, T., Meidar, D.: Synthesis, 929 (1978)
33. Horiki, K., Igano, K., Inouye, K.: Chemistry Lett., 165 (1978)
34. Roeske, R. W., Gesellchen, P. D.: Tetrahedron Lett., 3369 (1976)
35. Ugi, I., Aigner, H., Beier, B., Ben-Efraim, D., Burghard, H., Bukall, P., Eberle, G., Eckert, H., Marquarding, D., Rehn, D., Urban, R., Wackerle, L., von Zichlinsky, H.: in Peptides 1976, (Loffet, A., ed.), p. 159, Editions de l'Université de Bruxelles, Brussels (1976)
36. Isied, S. S., Kuehn, C. G.: J. Amer. Chem. Soc. *100*, 6752 (1978)
37. Zahr, S., Ugi, I.: Synthesis, 266 (1979)
38. Marquarding, D., Gokel, G., Hoffmann, P., Ugi, I.: in Ugi, I. "Isonitrile Chemistry", Chapter 7, New York: Academic Press 1971
39. Eckert, H., Ugi, I.: Angew. Chem. *14*, 825 (1975)
40. Barbedo, M. I. M. R. E., Amaral-Trigo, M. J. S. A.: Abstract, 17th Eur. Pept. Symp., Prague, p. 11 (1982)
41. Wang, S. S., Kulesha, I. D., Winter, D. P., Makofske, R., Kutny, R., Meienhofer, J.: Int. J. Peptide Protein Res. *11*, 297 (1978)
42. Barany, G., Merrifield, R. B.: J. Amer. Chem. Soc. *102*, 3084 (1980)
43. Barany, G.: Int. J. Peptide Protein Res. *19*, 321 (1982)
44. Wünsch, E., Moroder, L., Nyfeler, R., Jaeger, E.: Hoppe-Seyler's Z. Physiol. Chem. *363*, 197 (1982)

45. Benoiton, N. L., Mathiaparanam, P.: in Peptides 1980 (Brunfeldt, K., ed.), p. 221, Copenhagen: Scriptor 1981
46. Sheehan, J. C., Guziec, F. S.: J. Org. Chem. *39*, 3034 (1973)
47. Brtnik, F., Barth, T., Jost, K.: Collection Czechoslovak Chem. Commun. *46*, 1983 (1981)
48. Matsueda, R., Walter, R.: Int. J. Peptide Protein Res. *16*, 392 (1980)
49. Matsueda, R., Higashida, S., Ridge, R. J., Matsueda, G. R.: Peptide Chemistry 1981 (Shiori, T., ed.), p. 31, Protein Res. Found., Osaka (1981)
50. Bodanszky, M., Bednarek, M. A.: Int. J. Peptide Protein Res. *20*, 437 (1982)
51. Fukuda, T., Wakimasu, M., Kobayashi, S., Fujino, M.: in Peptide Chemistry 1981 (Shiori, T., ed.), p. 47, Protein Res. Res. Found., Osaka (1981)
52. Fujino, M., Wakimasu, M., Kitada, C.: Chem. Pharm. Bull. *29*, 2825 (1981)
53. Kimura, T., Morikawa, T., Takai, M., Sakakibara, S.: J. Chem. Soc. Chem. Commun., 340 (1982)
54. Izdebski, J., Drabarek, S.: Abstr. 17th Eur. Pept. Symp., Prague, p. 13 (1982)
55. Bodanszky, M., Bednarek, M. A., Bodanszky, A.: Int. J. Peptide Protein Res. *20*, 387 (1982)
56. Jones, J. H., Ramage, W. I., Witty, M. J.: Int. J. Peptide Protein Res. *15*, 301 (1980)
57. Przybylski, J., Miecznikowska, H., Strube, M., Jeschkeit, H., Kupryszewski, G.: Abstr. 17th Eur. Peptide Symp., Prague, p. 20 (1982)
58. Yamashiro, D., Blake, J.: Int. J. Peptide Protein Res. *18*, 383 (1981)
59. Schultz, R. M., Huff, J. P., Anagnastaras, P., Olsher, U., Blout, E. R.: Int. J. Peptide Protein Res. *19*, 454 (1982)
60. Funakoshi, S., Fujii, N., Akaji, K., Irie, H., Yajima, H.: Chem. Pharm. Bull. *27*, 2151 (1979)
61. Yajima, H., Akaji, K., Funakoshi, S., Fujii, N., Irie, H.: Chem. Pharm. Bull. *28*, 1942 (1980)
62. Wang, S. S., Matsueda, R., Matsueda, G. R.: Peptide Chemistry 1981 (Shiori, T., ed.), p. 37, Osaka: Protein Research Found. 1982
63. Arshady, R., Atherton, E., Clive, D. L. J., Sheppard, R. C.: J. Chem. Soc. Perkin I, 529 (1981)
64. Williams, B. J.: Abstr. 17th Eur. Pept. Symp., Prague, p. 30 (1982)
65. Atherton, E., Brown, E., Sheppard, R. C., Rosevear, A.: J. Chem. Soc. Chem. Commun., 1151 (1981)
66. Sheppard, R. C.: Abstr. 17th Eur. Pept. Symp., Prague, p. 9 (1982)
67. Chaturvedi, N., Sigler, G., Fuller, W., Verlander, M., Goodman, M.: in Chemical Synthesis and Sequencing of Peptides and Proteins (Liu, P., Schechter, A., Heinrickson, R., Condliffe, P., eds.), p. 169, New York: Elsevier 1981
68. Lukas, T. J., Prystowsky, M. B., Erickson, B. W.: Proc. Natl. Acad. Sci., U.S.A. *78*, 2791 (1981)
69. Atherton, E., Logan, C. J., Sheppard, R. C.: J. Chem. Soc. Perkin I, 538 (1981)
70. Sheppard, R. C., Williams, B. J.: J. Chem. Soc. Chem. Commun., 587 (1982)
71. Patchornik, A.: presentation at the 17th Eur. Pept. Symp., Prague (1982); cf. also Cohen, B. J., Kraus, M. A., Patchornik, A.: J. Amer. Chem. Soc. *103*, 7620 (1981)
72. Isied, S. S., Kuehn, C. G., Lyon, J. M., Merrifield, R. B.: J. Amer. Chem. Soc. *104*, 2632 (1982)
73. Yajima, H., Kanaki, J., Kitajima, M., Funakoshi, S.: Chem. Pharm. Bull. *28*, 1214 (1980)
74. Takeyama, M., Koyama, K., Inoue, K., Kawano, T., Adachi, H., Tobe, T., Yajima, H.: Chem. Pharm. Bull. *28*, 1873 (1980)

Author Index

Subject Index

Bei Fragen zur Produktsicherheit wenden Sie sich bitte an:
If you have any questions regarding product safety,
please contact:

Walter de Gruyter GmbH
Genthiner Straße 13
10785 Berlin
productsafety@degruyterbrill.com